THREE-DIMENSIONAL VELOCITY AND VORTICITY MEASURING AND IMAGE ANALYSIS TECHNIQUES

ERCOFTAC SERIES

VOLUME 4

Series Editors

Aims and Scope of the Series

ERCOFTAC (European Research Community on Flow, Turbulence and Combustion) was founded as an international association with scientific objectives in 1988. ERCOFTAC strongly promotes joint efforts of European research institutes and industries that are active in the field of flow, turbulence and combustion, in order to enhance the exchange of technical and scientific information on fundamental and applied research and design. Each year, ERCOFTAC organizes several meetings in the form of workshops, conferences and summerschools, where ERCOFTAC members and other researchers meet and exchange information.

The ERCOFTAC Series will publish the proceedings of ERCOFTAC meetings, which cover all aspects of fluid mechanics. The series will comprise proceedings of conferences and workshops, and of textbooks presenting the material taught at summerschools.

The series covers the entire domain of fluid mechanics, which includes physical modelling, computational fluid dynamics including grid generation and turbulence modelling, measuring-techniques, flow visualization as applied to industrial flows, aerodynamics, combustion, geophysical and environmental flows, hydraulics, multi-phase flows, non-Newtonian flows, astrophysical flows, laminar, turbulent and transitional flows.

The titles published in this series are listed at the end of this volume.

Three-Dimensional Velocity and Vorticity Measuring and Image Analysis Techniques

Lecture Notes from the Short Course held in Zürich, Switzerland, 3-6 September 1996

Edited by

TH. DRACOS

Institute of Hydromechanics and Water Resources Management,
Swiss Federal Institute of Technology,
Zürich, Switzerland

Springer-Science+Business Media, B.V.

A C.I.P. Catalogue record for this book is available from the Library of Congress.

Printed on acid-free paper

ISBN 978-90-481-4757-1 ISBN 978-94-015-8727-3 (eBook)
DOI 10.1007/978-94-015-8727-3

Originally published by Kluwer Academic Publishers in 1996.
Softcover reprint of the hardcover 1st edition 1996

Table of Contents

Chapter III
Particle Tracking Velocimetry

PREFACE

The contents of this volume reflect to a large extent the efforts made by a group of Institutes at the ETH Zürich to develop new techniques for measurements of flows in fluids in the last decade. The motivation came from the study of transport and mixing processes in natural and industrial systems. One of the characteristic properties of turbulence is its high mixing efficiency. The techniques developed are therefore suitable, although not exclusively, for turbulence measurements. They can be subdivided into point-measurements and field-measurements.

The aim of the point-measurements developed is to determine the three components of the velocity and all their first derivatives with good temporal resolution and accuracy in turbulent flows. The old and well established method of hot-wire anemometry was used for this purpose. One of the main achievements in this context is the construction of miniature multi-wire probes. This technique was introduced to the Institute of Hydromechanics and Water Resources Management of ETH Zürich by Profs. A. Tsinober and E. Kit from Tel-Aviv University. This was made possible by the generous financial support by ETH, for which I would like to express my gratitude on this occasion. In addition, Dr. F.E. Joergensen from DANTEC contributed an example of recent developments in the hardware of Constant Temperature Anemometry (CTA), for which I am very thankful.

Field-measurements of flow velocities are mainly based on non-intrusive optical techniques. Particle Image Velocimetry is one of them. It is already well developed and widely used. It was ERCOFTAC's wish to include a chapter on this technique in the present volume. I would like to thank Dr. W.T. Lai from TSI, Drs. D.R. McCluskey and E. Hayes from DANTEC as well as my colleagues, Dr. J. Kompenhans from DLR Göttingen and Dr. K.D. Hinsch from the University of Oldenburg and their co-workers, who made this possible by their substantial contributions.

With the financial support by the Swiss National Science Foundation a three-dimensional "Particle Tracking" technique was developed at ETH Zürich. This task was accomplished in close cooperation between the Institute of Geodesy and Photogrammetry, Prof. A. Grün, and the Institute of Hydromechanics and Water Resources Management, Prof. Th. Dracos. The method is based on photogrammetric principles for the determination of the spatial coordinates of particles at a given time and on hydrodynamic principles allowing to track the movement of particles in time. The main contributions to this work came from Drs. H.-G. Maas, N. Malik and D. Papantoniou.

The most recent technique is based on three-dimensional Laser Induced Fluorescence Imaging. The method was developed in a common effort of the Department of Industrial Chemistry and Chemical Engineering, Prof. P. Rys, the Institute of Geodesy and

Photogrammetry, Prof. A. Grün, and the Institute of Hydromechanics and Water Resources Management, Prof. Th. Dracos. The project was supported by ETH Zürich. All participants in this project would like to thank Prof. R. Hütter, Vice-President Research of ETH Zürich, for his interest in this project and for the generous support granted. The main contributions were made by S. Deusch and Drs. H.-G. Maas, G. Merkel and A. Stefanidis.

I would like to thank all those who were prepared to offer their precious time to write a contribution to this volume. When contributions of many authors to the same subject are integrated in a volume, it is the task of the editor to eliminate extensive overlapping between the individual presentations. In the present case this could be achieved only to some extent. I hope that the reader will not find occasional overlapping too disturbing.

This volume will appear in the ERCOFTAC Series of Kluwer Academic Publishers. I would like to thank Prof. P. Monkevitz, President of the Leonard Euler ERCOFTAC Center in Switzerland, for his support and the editors of the Series, P. Hutchinson and W. Rodi, for their consent to publish this contribution in the ERCOFTAC Series.

The publication of this volume would not have been possible without the competent help of Mrs. A. Vyskocil, who did all the secretarial work and a great part of the editorial work.

Last but not least, I would like to thank the publishers for the good cooperation and their patience with the editor.

Zürich, June 1996 Prof. Th. Dracos

INTRODUCTION

TH. DRACOS
Prof. em. ETHZ Zürich
Untergasse 14
CH - 8126 Zumikon

The title of this book refers to three-dimensional measurements in turbulent flows. The question is, is turbulence so important and if so, are three-dimensional measurements needed?

The first of these questions is equivalent to the question, what is turbulence? Unfortunately, a definition of turbulence does not exist. One can only describe its characteristic properties. Some characteristic properties of turbulent flows are given below as a synthesis of statements of different well known authors.

A turbulent flow is a randomly fluctuating vorticity containing (viscous) flow

A.S. Monin

With fluctuations (continuously) spreading in all wavelengths between a minimum determined by viscous forces and a maximum determined by the boundary conditions of the flow

P. Bradshaw

It (thus) must involve a wide range of spatial wavelengths

M. Lesieur

It possesses far greater ability (than laminar flow) to transfer momentum, heat, solutes, suspended particles, and to propagate chemical reactions

A.S. Monin, A.H. Yaglom

Turbulence proper is tied to three-dimensionality

J. v. Neuman

Turbulence generation and dissipation are associated with vortex stretching. The vorticity balance equation for incompressible Newtonian fluids

$$\frac{\partial \underline{\Omega}}{\partial t} + (\underline{U} \cdot \nabla)\underline{\Omega} - \underline{\Omega} S_{ij} - \nu \nabla^2 \underline{\Omega} = 0$$

shows that the local change of vorticity $\underline{\Omega} = \nabla \times \underline{U}$ is due to vorticity advection, vorticity change caused by the strain rate and viscous vorticity diffusion.

The rate of strain turns the vortex lines until they coincide with one of the principle directions of the rate of strain tensor resulting in stretching or compression of vortex filaments leading to amplification or damping of vorticity. This process can only occur in three-dimensional turbulence. In two-dimensional flow $\underline{\Omega}$ is perpendicular to the plane of the rate of strain and stretching or compression cannot occur. This is the reason why turbulence is tied to three-dimensionality. Three-dimensional measurements are therefore necessary for the study of turbulence in general and for the study of vorticity in particular.

The lower bound of the spatial wavelengths depends on the Reynolds number of the flow. Super computers made possible the simulation of simple turbulent flows by direct numerical integration of the Navier-Stokes equations. The range of wavelengths which can be resolved depends on the resolution of the discretisation grid. The computation time is nearly proportional to the third power of the number of discretisation steps in one direction. This number determines the number of collocation points in pseudo spectral schemes and thus the spectral resolution. It also imposes an upper bound for the Reynolds number of the flow studied.

Even with the high computational performance of today's super computers computations of this kind are limited to rather low Reynolds numbers. However, they yield whole-field information for the three components of the velocity and for the pressure with high spatial resolution and even higher temporal resolution.

On the one hand, three-dimensional measurements are needed to check the numerical result. On the other hand, numerical results can be used for transferring computational results to flows with higher Reynolds numbers or to more complicated flows.

Topology of geometrical features of turbulent flows is another topic of interest in turbulence research. Topology studies based on two-dimensional information can often be misleading.

One of the most important properties of turbulence is its ability to transfer very efficiently vector and scalar properties as well as suspended and dissolved matter. For example, the dispersion of suspended particles is strongly influenced by the turbulence of the flow. Particle dispersion is a typical three-dimensional process and can only be studied by following trajectories of particles in three-dimensional space.

The propagation of chemical reactions between reactants dissolved in a fluid is strongly enhanced by the turbulence in the fluid. This is particularly true when the Schmidt number is large. In these case material surfaces which delimit the parts of the fluid containing one of the reactants from the part of the fluid containing the other one are stretched and folded by turbulence. The extent of these surfaces is significantly increased in a short time, and molecular diffusion can then efficiently act to initiate the chemical reaction. This is also a three-dimensional process which is not only important for chemical reactions but also in combustion.

The list of processes influenced by turbulence can be continued for long. It involves natural as well as industrial processes. The few examples presented here demonstrate the need for the development of three-dimensional measurements, and what is even more important, the need for whole-field measurements.

The development of different sensors of fast response and high resolution on the one hand and the development of digital acquisition and handling systems on the other made such measurements possible.

One of the oldest measuring techniques is hot-wire anemometry. This technique is intrusive and was partially replaced and complemented by Laser Doppler Anemometry (LDA), a non-intrusive optical technique. Both techniques offer the possibility to measure all three components of the velocity vector at one location.

Improvements in probe construction technology, in the electronic equipment and in the data acquisition and handling systems made the measurement of all velocity components in neighbouring points by hot-wires possible. Derivatives of the velocity components can thus be computed and derivative depending quantities like vorticity determined. The progress achieved in this field is presented in the first chapter of this book.

Floating objects have been used for a very long time for velocity measurements. Flow visualisation by suspended or floating particles has produced many nice images and helped to understand at least qualitatively many flow phenomena. But only in the last ten to fifteen years it became possible to develop visualisation techniques for quantitative field measurements of the velocity. Two of them will be treated to some extent in chapter two and three.

The first method is designed to determine the mean movement of a small group of particles embedded in a fluid element during a short time. The particles are illuminated by a thin sheet of laser light. The method is called Particle Image Velocimetry (PIV). It is a well established technique. It allows to determine with high accuracy and high spatial resolution the projection of the velocity vectors on the illuminated plane. It can be used in air and water or other liquids and for a large range of velocities. Its extension to three-dimensional measurements in a plane is in progress. Holographic applications allow to measure velocity vectors in three-dimensional space.

The other method is a particle tracking technique. Here the paths of individual particles are followed inside an illuminated volume. Velocity vectors, i.e. all their three components, are determined in three-dimensional space. The accuracy of the method is nearly as good as the one of PIV but the spatial resolution is smaller. So far it has been applied only in liquids and at small to moderate flow velocities. It has, however, the capability to determine long particle trajectories if the movement of the particles is observed with a camera arrangement following their movement.

The study of mixing processes is important when mixing sensitive chemical reactions are investigated. In recent years mixing has been studied using fluorescent dyes and laser sheet illumination (LIF). It allows to determine the concentration distributions and their local derivatives. Therefrom images of the surfaces can be deduced on which maximum chemical activity is expected to occur. In the method presented in the last chapter Laser Induced Fluorescence Tomography (LIFT) is used to construct three-dimensional images of concentration. The grey level variation occurring in small fluid elements due to the concentration distribution in these elements is then used to determine velocity, vorticity and the rate of strain associated with these elements. This novel method is based on tracking in time the fluid elements selected and allows to measure all three quantities directly and simultaneously without performing differentiation of the velocity components. Its spatial resolution is high and its temporal resolution sufficient. With the equipment available at the time it is restricted to low velocity flows with moderate Reynolds numbers.

CHAPTER I

MULTI HOT-WIRE ANEMOMETRY

DETERMINATION OF 3-D VELOCITY AND VORTICITY VECTORS IN TURBULENT FLOWS BY MULTI-HOTWIRE ANEMOMETRY

G. LEMONIS[1], T. DRACOS[2]
1 *Laboratory of Applied Thermodynamics, Department of Mechanical Engineering, University of Patras, Greece*
2 *Prof. em. ETH Zurich, Switzerland*

1. Introduction

The flow of viscous fluids is described by the Navier-Stokes equations. These equations are non-linear. Analytical solutions exist only for some simple flows. Physical and numerical experiments are therefore indispensable tools in fluid mechanics. It is not the intention to give any kind of introduction into fluid mechanics or the theory of turbulence. Excellent treatments of these topics are found in Batchelor (1960, 1967), Monin & Yaglom (1971), Hinze (1975), Tennekes & Lumley (1992) and others. The presentation of some well known and widely used balance equations for momentum, energy and enstrophy should merely illustrate what kind of expressions one wishes to determine by velocity measurements in turbulent flows.

For Newtonian incompressible fluids the Navier-Stokes equations can be written in the following form :

continuity equation

$$\frac{\partial \widetilde{U}_i}{\partial x_i} = 0 \,, \tag{1.1}$$

momentum balance equation

$$\frac{D\widetilde{U}_i}{Dt} = -\frac{1}{\rho} \cdot \frac{\partial \widetilde{P}}{\partial x_i} + \nu \cdot \frac{\partial^2 \widetilde{U}_i}{\partial x_j \cdot \partial x_j} + \widetilde{F}_i \,, \tag{1.2}$$

where $\widetilde{U}_i$, i=1,2,3 is the velocity, $\widetilde{P}$ the pressure, $\widetilde{F}_i$ an external body force, ρ the density and ν the kinematic viscosity of the fluid, both assumed to be constant. A tilde denotes instantaneous value of a variable on which no Reynolds decomposition into a mean value and fluctuations (see discussion to eqs. (1.11)) has been performed. Carte-

Th. Dracos (ed.), Three-Dimensional Velocity and Vorticity Measuring and Image Analysis Techniques, 1–42.

sian tensor notations are used and the summation rule applies. The term on the left hand side of eq. (1.2) is the material derivative of the velocity

$$\frac{D\tilde{U}_i}{Dt} = \frac{\partial \tilde{U}_i}{\partial t} + \tilde{U}_j \cdot \frac{\partial \tilde{U}_i}{\partial x_j} \quad . \tag{1.3}$$

The spacial derivatives $\partial \tilde{U}_i / \partial x_j$ form the second order velocity gradient tensor, which can be decomposed into a symmetrical and an antisymmetrical part

$$\frac{\partial \tilde{U}_i}{\partial x_j} = \frac{1}{2} \cdot \left(\frac{\partial \tilde{U}_i}{\partial x_j} + \frac{\partial \tilde{U}_j}{\partial x_i} \right) + \frac{1}{2} \cdot \left(\frac{\partial \tilde{U}_i}{\partial x_j} - \frac{\partial \tilde{U}_j}{\partial x_i} \right) \quad . \tag{1.4}$$

The symmetrical part is the rate of strain tensor

$$\tilde{S}_{ij} = \frac{1}{2} \cdot \left(\frac{\partial \tilde{U}_i}{\partial x_j} + \frac{\partial \tilde{U}_j}{\partial x_i} \right) \quad . \tag{1.5}$$

The antisymmetrical part defines the vorticity of the flow

$$\tilde{\Omega}_k = \varepsilon_{ijk} \cdot \frac{\partial \tilde{U}_i}{\partial x_j} \quad , \tag{1.6}$$

where ε_{ijk} is the alternating unit tensor.

In view of eqs. (1.3), (1.5) and (1.6) eq. (1.2) becomes

$$\frac{\partial \tilde{U}_i}{\partial t} + \frac{1}{2} \cdot \frac{\partial}{\partial x_j} \left(\tilde{U}_j \tilde{U}_j \right) - \left(\underline{\tilde{U}} \times \underline{\tilde{\Omega}} \right)_i = -\frac{1}{\rho} \cdot \frac{\partial \tilde{P}}{\partial x_i} + 2 \cdot \nu \cdot \frac{\partial \tilde{S}_{ij}}{\partial x_j} + \tilde{F}_i \; . \tag{1.7}$$

Taking the curl of this equation, assuming that the external body force $\tilde{F}_i$ is conservative and noting that the second term on the left hand side and the first term on the right hand side are gradients of scalars, one obtains the vorticity balance equation :

$$\frac{\partial \tilde{\Omega}_i}{\partial t} - \left[\nabla \times \left(\underline{\tilde{U}} \times \underline{\tilde{\Omega}} \right) \right]_i = \nu \cdot \frac{\partial^2 \tilde{\Omega}_i}{\partial x_j \partial x_j} \quad , \tag{1.8}$$

or with

$$\left[\nabla\times\left(\underline{\tilde{U}}\times\underline{\tilde{\Omega}}\right)\right]=-\underline{\tilde{U}}\cdot\nabla\underline{\tilde{\Omega}}+\underline{\tilde{\Omega}}\cdot\nabla\underline{\tilde{U}}\ , \tag{1.9}$$

$$\frac{\partial\tilde{\Omega}_i}{\partial t}+\tilde{U}_j\cdot\frac{\partial\tilde{\Omega}_i}{\partial x_j}=2\cdot\tilde{\Omega}_j\cdot\tilde{S}_{ij}+\nu\cdot\frac{\partial^2\tilde{\Omega}_i}{\partial x_j\partial x_j}\ . \tag{1.10}$$

In a turbulent flow all dependent variables change with time, even if the mean flow is steady. Following Reynolds, these variables are split into a time independent averaged part and a time dependent fluctuating part with zero average.

$$\tilde{U}_i=U_i+u_i,\quad \tilde{P}=P+p,\quad \tilde{S}_{ij}=S_{ij}+s_{ij},\quad \tilde{\Omega}_i=\Omega_i+\omega_i\ . \tag{1.11}$$

Introduction of these variables in the previous equations and averaging the individual terms yield equations for the mean and the fluctuating part of the flow. Some of these equations will be given and shortly discussed in the following.

The momentum balance for the mean flow reads :

$$\frac{\partial U_i}{\partial t}+U_j\frac{\partial U_i}{\partial x_j}=-\frac{1}{\rho}\cdot\frac{\partial P}{\partial x_i}+\frac{\partial}{\partial x_j}\left(2\cdot\nu\cdot S_{ij}-\left\langle u_i\cdot u_j\right\rangle\right)+F_i\ . \tag{1.12}$$

The first term on the left hand side vanishes when the flow is steady. Due to the fluctuating part of the velocity an additional term appears on the right hand side of this equation. It contains double correlations of the components of the fluctuating part of the velocity. When compared with the other term in the parenthesis these correlations can be interpreted as additional stresses. They form a second order tensor called Reynolds stress tensor.

Multiplication of (1.2) or (1.7) with U_i gives the balance equation for the kinetic energy $1/2\cdot U_i\cdot U_i$ of the flow. Neglecting the body force, application of Reynolds decomposition and subsequent averaging yields :

(i) for the kinetic energy of the mean flow per unit mass and time :

$$\underset{(I)}{\frac{\partial}{\partial t}\left(\frac{1}{2}\cdot U_i\cdot U_i\right)}+\underset{(II)}{U_j\cdot\frac{\partial}{\partial x_j}\left(\frac{1}{2}\cdot U_i\cdot U_i\right)}=\underset{(III)}{-\frac{\partial}{\partial x_j}\left(\frac{P}{\rho}\cdot U_j\right)}+\underset{(IV)}{\left\langle u_i\cdot u_j\right\rangle\cdot S_{ij}}$$

$$\underset{(V)}{-\frac{\partial}{\partial x_j}\left(U_i\cdot\left\langle u_i\cdot u_j\right\rangle\right)}+\underset{(VI)}{2\cdot\nu\cdot\frac{\partial}{\partial x_j}\left(U_i\cdot S_{ij}\right)}\underset{(VII)}{-2\cdot\nu\cdot S_{ij}\cdot S_{ij}}\ . \tag{1.13}$$

The meaning of the various terms in this equation is : (I) local change of kinetic energy (in steady flows this term is zero); (II) advective transport of kinetic energy; (III) work done by pressure; (IV) this term represents transfer of kinetic energy from the mean flow to the fluctuating part of the flow giving a negative contribution; (V) work done by the Reynolds stresses; (VI) work done by viscous stresses and (VII) viscous dissipation in the mean flow.

(ii) for the fluctuating part of the flow per unit mass and time :

$$\underset{(I)}{\frac{\partial}{\partial t}\left(\frac{1}{2}\cdot\langle u_i\cdot u_i\rangle\right)}+\underset{(II)}{U_j\cdot\frac{\partial}{\partial x_j}\left(\frac{1}{2}\cdot\langle u_i\cdot u_i\rangle\right)}=\underset{(III)}{-\frac{\partial}{\partial x_j}\left(\frac{\langle p\cdot u_j\rangle}{\rho}+\frac{\langle u_i\cdot u_i\cdot u_j\rangle}{2}\right)}$$

$$+\underset{(IV)}{\nu\cdot\frac{\partial^2}{\partial x_j\partial x_j}\left(\frac{1}{2}\cdot\langle u_j\cdot u_j\rangle\right)}\underset{(V)}{-\langle u_i\cdot u_j\rangle\cdot S_{ij}}\underset{(VI)}{-2\cdot\nu\cdot\langle s_{ij}\cdot s_{ij}\rangle}, \qquad (1.14)$$

and the meaning of the terms is : (I) local change; (II) advective transport by the mean motion; (III) convective turbulent diffusion of the total mechanical energy of turbulence; (IV) work done by the turbulent shear stresses; (V) turbulence production term (see term (IV) in eq. (1.13)); VI viscous dissipation of turbulence.

Now, instead of applying the same procedure to the vorticity equation (1.10), one can apply it to the equation for the enstrophy $1/2\cdot\Omega_i\cdot\Omega_i$ which originates from (1.10) by multiplication with Ω_i. The balance of enstrophy $1/2\cdot\Omega_i\cdot\Omega_i$ per unit mass and time for the mean flow is given by

$$\underset{(I)}{\frac{\partial}{\partial t}\left(\frac{1}{2}\cdot\Omega_i\cdot\Omega_i\right)}+\underset{(II)}{U_j\cdot\frac{\partial}{\partial x_j}\left(\frac{1}{2}\cdot\Omega_i\cdot\Omega_i\right)}=\underset{(III)}{-\frac{\partial}{\partial x_j}\left(\Omega_i\cdot\langle\omega_i\cdot u_j\rangle\right)}+\underset{(IV)}{\langle\omega_i\cdot u_j\rangle\cdot\frac{\partial\Omega_i}{\partial x_j}}+\underset{(V)}{\Omega_i\cdot\Omega_i\cdot S_{ij}}$$

$$\underset{(VI)}{+\Omega_i\cdot\langle\omega_j\cdot s_{ij}\rangle}+\underset{(VII)}{\nu\cdot\frac{\partial^2}{\partial x_j\partial x_j}\left(\frac{1}{2}\cdot\Omega_i\cdot\Omega_i\right)}\underset{(VIII)}{-\nu\cdot\frac{\partial\Omega_i}{\partial x_j}\cdot\frac{\partial\Omega_i}{\partial x_j}}. \qquad (1.15)$$

Terms (I) and (II) give the local and advective change of enstrophy respectively. The terms on the right hand side mean : (III) transport of mean vorticity by turbulent vorticity - velocity interaction; (IV) gradient production of turbulent vorticity; (V) and (VI) stretching (or compression) of mean vorticity by the mean and turbulent rate of strain respectively; (VII) viscous diffusion of mean vorticity and (VIII) viscous dissipation of mean vorticity.

The balance of the fluctuating part of enstrophy $1/2 \cdot \omega_i \cdot \omega_i$ is :

$$\frac{\partial}{\partial t}\left(\frac{1}{2}\cdot\omega_i\cdot\omega_i\right)+U_j\cdot\frac{\partial}{\partial x_j}\left(\frac{1}{2}\cdot\omega_i\cdot\omega_i\right)=-\langle\omega_i\cdot u_j\rangle\cdot\frac{\partial\Omega_i}{\partial x_j}-\frac{1}{2}\cdot\frac{\partial}{\partial x_j}\langle\omega_i\cdot\omega_i\cdot u_j\rangle$$

(I) (II) (III) (IV)

$$+\langle\omega_i\cdot\omega_j\cdot s_{ij}\rangle+\langle\omega_i\cdot\omega_j\rangle\cdot S_{ij}+\Omega_i\cdot\langle\omega_i\cdot s_{ij}\rangle$$

(V) (VI) (VII)

$$+\nu\cdot\frac{\partial^2}{\partial x_j\partial x_j}\left\langle\frac{1}{2}\cdot\omega_i\cdot\omega_i\right\rangle-\nu\cdot\left\langle\frac{\partial\omega_i}{\partial x_j}\cdot\frac{\partial\omega_i}{\partial x_j}\right\rangle. \qquad (1.16)$$

(VIII) (IX)

The terms (I) and (II) on the left hand side give again the local change and the advective transport of fluctuating enstrophy by the mean flow respectively. The meaning of the terms on the right hand side is : (III) the same as term (IV) in (1.15) but with opposite sign; it represents the vorticity exchange between the mean and the fluctuating part of the flow; (IV) transport of turbulent vorticity by turbulent velocity fluctuations; (V) and (VI) intensification (or damping) of turbulent vorticity by mean and turbulent stretching (or compression); term (VII) appears in equation (1.15) with the same sign and shows that the turbulent rate of strain produces mean and turbulent vorticity at the same rate (VIII) viscous diffusion of turbulent vorticity; (IX) viscous dissipation of turbulent vorticity.

In all these equations double and even triple correlations of fluctuating quantities appear. They are either velocity components or derivatives of velocity components. These correlations can be interpreted as additional unknowns of the problem. This is particularly true for the Reynolds stresses in the momentum equation. The description of the flow based on the Reynolds approach leads to a non-closed system of equations.

With the exception of direct numerical simulations of turbulent flows all other numerical techniques require modelling of turbulent terms, e.g. the Reynolds stresses in the momentum equations or the kinetic energy of turbulence and of its dissipation in the energy balance equation of turbulence. Parameter estimation for these models often rely on experimental results. Experimental validation of numerical computations based on turbulence models also requires 3-D measurements of high accuracy. One of the techniques capable of measuring all three components of the velocity and all their first derivatives at one position is hot-wire anemometry.

2. Measuring Principle, Probe Design and Operation

2.1. BASIC PRINCIPLES OF HOT-WIRE ANEMOMETRY

The principle of the hot-wire anemometer was introduced at the beginning of the current century by King (1914). Modern systems, their capabilities and limitations are described by Hinze (1975), Perry (1982) and Lomas (1986). In this section some explanatory concepts will be presented. The discussion will be restricted to constant temperature anemometry methods (CTA).

The relationship between the anemometer bridge voltage and the fluid velocity was first derived by King (King's law)

$$E^2 = A + B \cdot U^n. \tag{2.1}$$

The exponent n can be determined only experimentally and is typically about 0.5.

To take into account the direction of the incident flow, Jørgensen (1971) introduced the effective cooling velocity U_{eff}, defined as

$$U_{eff}^2 = U_n^2 + k_b \cdot U_b^2 + k_t \cdot U_t^2. \tag{2.2}$$

U_n, U_b and U_t are the components of the velocity vector $\underline{U}$ with reference to a wire-supported cartesian coordinate system (x_w, y_w, z_w) with the x_w-axis in the plane formed by the wire and its supporting prongs. The y_w-axis is coincident with the wire (*Fig. 2.1*).

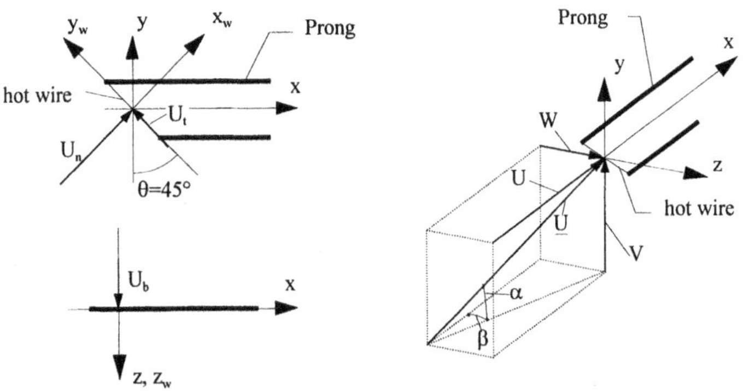

Figure 2.1. Velocity components on probe-supported (x,y,z) and wire-supported (x_w,y_w,z_w) cartesian coordinates for a 45°-slanted hot-wire

U_n, U_b and U_t are called normal, binormal and tangential velocity components, respectively and factors k_b and k_t are called binormal and tangential directional sensitivity coefficients, respectively.

The distinction between stream velocity $\underline{U}$ and effective cooling velocity U_{eff} is important because of the different contributions of the components U_n, U_b und U_t to the heat transfer rate from the wire to the incident flow. Coefficients k_t und k_b can be determined experimentally by an angular probe calibration in uniform flow. Typical values for the directional sensitivity factors are $k_b = 1.0 - 2.0$ and $k_t = 0.02 - 0.2$.

Expressed in terms of U_{eff}, King´s law (2.1) reads

$$E^2 = A + B \cdot U_{eff}^{\,n}. \tag{2.3}$$

From eq. (2.2) and (2.3) it is possible to derive a mathematical model for the wire response of the form

$$E = f(|\underline{U}|, \alpha, \beta). \tag{2.4}$$

The angles α and β are the pitch and yaw angles which determine the direction of the velocity vector with respect to the probe supported coordinate system (*Fig. 2.1*). Theoretically, using probes consisting of three wires with known response functions, one could construct a mathematical model in three equations to evaluate directly the three-dimensional velocity vector from a measured data set of three anemometer voltages E_1, E_2, E_3.

King´s law (2.3) and Jørgensen's equation (2.2) are valid for ideal probe geometries and heat transfer properties. The application of these relationships to real probes can lead to significant errors. In addition, as shown in recent works (Bruun & Tropea, 1985, Müller, 1987, Chew & Ha, 1988, Wagner & Kent, 1988), the directional sensitivity coefficients k_t and k_b, which in Jørgensen's equation are assumed to be constants, depend in many cases on the velocity magnitude as well as on the angle of attack. These coefficients can deviate considerably from their mean values.

2.2. COMPENSATION OF AMBIENT TEMPERATURE VARIATIONS

An important error source of the hot-wire anemometer is its sensitivity to temperature variations of the onset flow. Flow temperature fluctuations cause changes of the electrical wire resistance, which may be interpreted erroneously as velocity variations. Many experimentators suggest therefore to calibrate hot-wire probes at different flow temperatures. This should be useful at large changes in the ambient temperature. In the case of small temperature variations of the order of +/- 0.5 °C analytical correction formulas for low temperature changes may also yield sufficient accuracy.

Different temperature compensation methods for hot-wire anemometers have been developed by Bearmann (1971), Kavence & Oka (1973), and others. Abdel-Rahman et al. (1986) have tested experimentally a number of temperature compensation methods over a large range of ambient temperatures (22-98 °C) and for a low range of velocities (0.35-9 m/sec). They found that, for this range, the most consistent compensation is achieved using the relationship proposed by Collis & Williams (1959). Here, the fluid properties are evaluated at the „film temperature" T_f, the arithmetic mean value of the wire temperature T_w and the ambient temperature T

$$T_f = \frac{T_w + T}{2}. \tag{2.5}$$

The wire temperature T_w is given by

$$R_w = R_0 \cdot [1 + \alpha_R \cdot (T_w - T_0)], \tag{2.6}$$

where R_w is the electrical resistance of the heated wire, R_0 is the electrical resistance of the "cool" wire under a reference ambient temperature T_0 and α_R the thermal coefficient of the electrical resistance for the wire material, which for tungsten is about $4.82 \cdot 10^{-3}$ grd^{-1}.

A derivation of a corresponding correction formula is given by Rosemann (1989). To describe the problem of forced convection around a heated wire, the empirical formula found by Kramers (1946)

$$Nu = 0.42 \cdot Pr^{0.20} + 0.57 \cdot Pr^{0.33} \cdot Re_d^{0.5} \tag{2.7}$$

is used, which expresses the relationship between the Nußelt number

$$Nu = \frac{E^2}{R_w \cdot \pi \cdot l_w \cdot \lambda \cdot (T_w - T)}, \tag{2.8}$$

and the Reynolds number

$$Re_d = \frac{U_{eff} \cdot d_w}{\nu}, \tag{2.9}$$

with Pr the Prandtl number, d_w the wire diameter, l_w the wire length and λ the thermal conductivity of the fluid. Substituting (2.7) and (2.9) into (2.8) yields

$$\frac{E^2}{\pi \cdot l_w \cdot \lambda_f \cdot (T_w - T) \cdot R_w} = 0.42 \cdot Pr^{0.20} + 0.57 \cdot Pr^{0.33} \cdot \left(\frac{U_{eff} \cdot d_w}{\nu_f} \right)^{0.5}. \tag{2.10}$$

Eq. (2.10) is an approximation for the anemometer bridge voltage E at velocity U_{eff} and ambient temperature T. Index „f" characterizes fluid properties taken at the film temperature T_f. The relationship for the „correct“ voltage E_0 at the same velocity but at a reference temperature T_0 reads

$$\frac{E_0^2}{\pi \cdot l_w \cdot \lambda_{f,0} \cdot (T_w - T_0) \cdot R_w} = 0.42 \cdot Pr^{0.20} + 0.57 \cdot Pr^{0.33} \cdot \left(\frac{U_{eff} \cdot d_w}{\nu_{f,0}} \right)^{0.5} . \tag{2.11}$$

In this equation, index „f,0“ stands for properties taken at the film temperature $T_{f,0}$ at reference ambient temperature of the onset flow T_0

$$T_{f,0} = \frac{T_w + T_0}{2} . \tag{2.12}$$

The Prandtl number Pr of air is in the temperature range of interest practically constant. Substituting (2.10) into (2.11), rearranging terms and considering that

$$\left(\frac{\rho_f}{\rho_{f,0}} \right) = \frac{p}{p_0} \cdot \left(\frac{T_f}{T_{f,0}} \right)^{-1} , \quad \left(\frac{\mu_f}{\mu_{f,0}} \right) = \left(\frac{T_f}{T_{f,0}} \right)^{0.76} , \quad \left(\frac{\lambda_f}{\lambda_{f,0}} \right) = \left(\frac{T_f}{T_{f,0}} \right)^{0.76} , \tag{2.13}$$

and that for low temperature variations

$$\left(\frac{T_f}{T_{f,0}} \right)^{-0.38} \cong 1, \tag{2.14}$$

we obtain a correction formula, valid for small ambient temperature variations

$$\frac{E_0}{E} = \sqrt{\frac{T_w - T_0}{T_w - T}} , \tag{2.15}$$

where E is the measured anemometer bridge voltage and E_0 the corrected one.

2.3. OPTIMUM OVERHEAT RATIO

An important parameter which will affect significantly the sensitivity and the accuracy of hot-wire anemometers is the overheat ratio OHR, defined as

$$OHR = \frac{R_w}{R_0} , \tag{2.16}$$

where R_w is the electrical resistance of the heated wire and R_0 the resistance of the "cool" wire at reference atmospheric conditions „0". The range suggested for OHR is 1.2 - 4.0, as given by most anemometer manufacturers for measurements in air flows.

In order to evaluate the influence of the overheat ratio on the sensor sensitivity to velocity variations and its dependence on ambient temperature fluctuations we use the mathematical model for the wire response as given by eq. (2.10). Substituting eq. (2.6) and eq. (2.16) into eq. (2.10) we can express the anemometer output voltage in terms of OHR. Replacing T with $T_0+\tau$, where τ is the departure of the flow temperature from the reference temperature T_0, we get

$$\frac{E^2}{\pi \cdot l_w \cdot R_0 \cdot \lambda_f \cdot OHR \cdot \left(\frac{OHR-1}{\alpha_R} + \tau \right)} = 0.42 \cdot Pr^{0.20} + 0.57 \cdot Pr^{0.33} \cdot \left(\frac{d_w}{\nu_f} \right)^n \cdot U^n_{eff}, \quad (2.17)$$

where n = 0.5. Substituting the constants

$$C_1 = \pi \cdot l_w \cdot R_0,\; C_2 = 0.42 \cdot Pr^{0.20},\; C_3 = 0.57 \cdot Pr^{0.33} \cdot d_w{}^n, \quad (2.18)$$

into (2.17) we obtain

$$E^2 = C_1 \cdot \lambda_f \cdot OHR \cdot \left(\frac{OHR-1}{\alpha_R} + \tau \right) \cdot \left(C_2 + C_3 \cdot \frac{U^n_{eff}}{\nu^n_f} \right). \quad (2.19)$$

The quantities λ_f and ν_f depend on the film temperature T_f, which varies with OHR. For the temperature range of interest for investigations in most subsonic wind tunnels of $T_0 = 10$ °C to $T_0 = 25$ °C, T_f can reach values ranging from a minimum of $T_f \approx 30$ °C at OHR = 1.2 to a maximum of $T_f \approx 130$ °C at OHR = 2. For this range for T_f and for ambient pressure of 1 atm, the kinematic viscosity of air may be approximated by

$$\nu_f = \nu_1 \cdot e^{\nu_2 \cdot T_f}, \quad (2.20)$$

with $\nu_1 = 13.656$ m^2/s and $\nu_2 = 0.0053$ grd^{-1}, and the thermal conductivity of air by

$$\lambda_f = \lambda_1 \cdot T_f + \lambda_2, \quad (2.21)$$

with $\lambda_1 = 6 \cdot 10^{-5}$ W·m^{-1}·grd^{-2} and $\lambda_2 = 0.0247$ W·m^{-1}·grd^{-1}. Formulas (2.20) and (2.21) approximate the tabulated values of ν and λ given in literature (Idelchik, 1986) for the range of T_f given above with a deviation of max. 1.5 %.

In order to express the wire sensitivity with respect to the effective cooling velocity, $\partial E/\partial U_{eff}$, as function of the overheat ratio OHR, eqs (2.20), (2.21) and

$$T_f = \frac{T_w + T_0 + \tau}{2} = \frac{1}{2} \cdot \left(\frac{OHR - 1}{\alpha_R} + 2 \cdot T_0 + \tau \right) \tag{2.22}$$

are substituted into (2.19) :

$$E^2 = C_1 \cdot \left[\lambda_1 \cdot \frac{1}{2} \cdot \left(\frac{OHR - 1}{\alpha_R} + 2 \cdot T_0 + \tau \right) + \lambda_2 \right] \cdot OHR \cdot \left(\frac{OHR - 1}{\alpha_R} + \tau \right) \cdot \left(C_2 + C_3 \cdot \frac{U_{eff}^n}{v_1^n \cdot e^{\frac{n \cdot v_2}{2} \cdot \left(\frac{OHR-1}{\alpha_R} + 2 \cdot T_0 + \tau \right)}} \right) . \tag{2.23}$$

Differentiation of E^2 with regard to U_{eff} yields

$$\frac{\partial(E^2)}{\partial U_{eff}} = C_1 \cdot \left[\lambda_1 \cdot \frac{1}{2} \cdot \left(\frac{OHR - 1}{\alpha_R} + 2 \cdot T_0 + \tau \right) + \lambda_2 \right] \cdot OHR \cdot \left(\frac{OHR - 1}{\alpha_R} + \tau \right) \cdot \left(C_3 \cdot \frac{n \cdot U_{eff}^{n-1}}{v_1^n \cdot e^{\frac{n \cdot v_2}{2} \cdot \left(\frac{OHR-1}{\alpha_R} + 2 \cdot T_0 + \tau \right)}} \right) . \tag{2.24}$$

Finally, $\partial E/\partial U_{eff}$ is found using the relation

$$\frac{\partial(E^2)}{\partial U_{eff}} = 2 \cdot E \cdot \frac{\partial E}{\partial U_{eff}} \Leftrightarrow$$

$$\frac{\partial E}{\partial U_{eff}} = \frac{1}{2 \cdot E} \cdot \frac{\partial(E^2)}{\partial U_{eff}} . \tag{2.25}$$

To obtain a relationship between the wire sensitivity with respect to ambient temperature variations, $\partial E/\partial \tau$, and the overheat ratio OHR eq. (2.23) is differentiated with regard to τ :

$$\frac{\partial(E^2)}{\partial\tau}=C_1\cdot\frac{\lambda_1}{2}\cdot OHR\cdot\left(\frac{OHR-1}{\alpha_R}+\tau\right)\cdot\left(C_2+C_3\cdot\frac{U_{eff}^n}{v_1^n\cdot e^{\frac{n\cdot v_2}{2}\cdot\left(\frac{OHR-1}{\alpha_R}+2\cdot T_0+\tau\right)}}\right)+$$

$$C_1\cdot\left[\frac{\lambda_1}{2}\cdot\left(\frac{OHR-1}{\alpha_R}+2\cdot T_0+\tau\right)+\lambda_2\right]\cdot OHR\cdot\left(C_2+C_3\cdot\frac{U_{eff}^n}{v_1^n\cdot e^{\frac{n\cdot v_2}{2}\cdot\left(\frac{OHR-1}{\alpha_R}+2\cdot T_0+\tau\right)}}\right)+$$

$$C_1\cdot\left[\frac{\lambda_1}{2}\cdot\left(\frac{OHR-1}{\alpha_R}+2\cdot T_0+\tau\right)+\lambda_2\right]\cdot OHR\cdot\left(\frac{OHR-1}{\alpha_R}+\tau\right)\cdot\left(C_3\cdot\frac{-\frac{1}{2}\cdot n\cdot v_2\cdot U_{eff}^n}{v_1^n\cdot e^{\frac{n\cdot v_2}{2}\cdot\left(\frac{OHR-1}{\alpha_R}+2\cdot T_0+\tau\right)}}\right). \tag{2.26}$$

The wire sensitivity to ambient temperature variations is then found from

$$\frac{\partial E}{\partial\tau}=\frac{1}{2\cdot E}\cdot\frac{\partial(E^2)}{\partial\tau}. \tag{2.27}$$

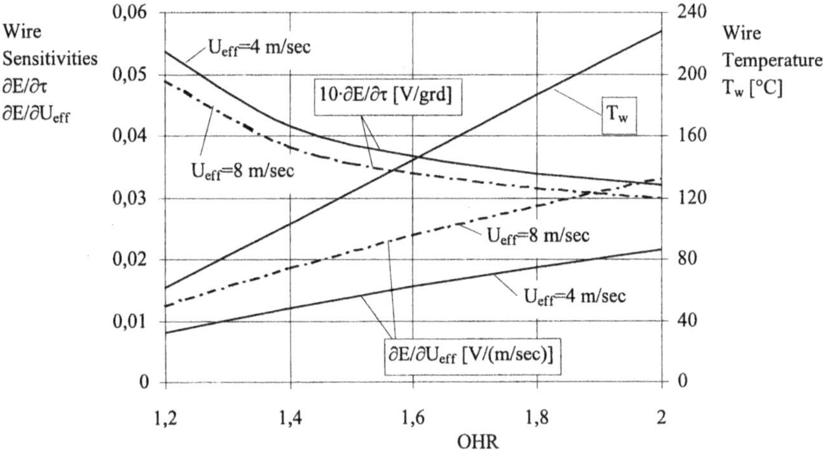

Figure 2.2. Wire sensitivities $\partial E/\partial\tau$ and $\partial E/\partial U_{eff}$, and wire temperature T_w as functions of the overheat ratio OHR

Equations (2.25) and (2.27) are plotted in *Fig. 2.2* for $U_{eff} = 4$ m/s and 8 m/sec, $T_0 = 20$ °C, $d_w = 2.5$ μm, $l_w = 0.5$ mm, $n = 0.5$, $\alpha_R = 4.82\cdot10\text{-}3$ grd^{-1}, $R_0 = 6\ \Omega$, $Pr = 0.72$, and constants v_1, v_2, $\lambda_1$and λ_2 as given for eq. (2.20) and eq. (2.21). It is seen in *Fig. 2.2* that high values of OHR favour the wire sensitivity with respect to the stream velocity, and reduce its sensitivity to ambient temperature variations. However, notice must be taken of the fact that the wire temperature rises with increasing OHR. In the case of a multi-wire probe, where many wires are operated simultaneously in a very narrow area, high wire temperatures can cause thermal interferences between the wires. It is therefore recommended to choose OHR not too high, in order to avoid thermal interference which can reduce the system accuracy significantly. Values of OHR between 1.4 and 1.5 are a good compromise. From *Fig. 2.2* follows that, for this OHR range, $\partial E/\partial \tau$ is significantly lower than for OHR = 1.2-1.3. The important sensitivity to velocity variations, $\partial E/\partial U_{eff}$, at OHR = 1.4 is about 50% higher than at OHR = 1.2 and the wire temperature is relatively low, close to 100 °C. Values of OHR above 1.6 do not significantly reduce the wire sensitivity to temperature fluctuations and cause high wire operating temperatures.

2.4. DIRECTIONAL SENSITIVITY OF HOT-WIRES

To improve the measurement accuracy it is important to operate the wires in the direction of their maximum angular sensitivity with respect to the incident flow. In order to investigate this relationship we consider a wire inclined at 45° (*Fig. 2.1*) to derive expressions describing the wire response E as function of the pitch angle α and yaw angle β. From eq. (2.10) follows that, for a single wire, the variation of E with regard to α and β is determined only by the variation of U_{eff} with respect to these variables, since the remaining quantities involved in eq. (2.10) are independend of the wire orientation in the flow.

To express U_{eff}, eq. (2.2), in terms of α and β the velocity vector $\underline{U} = (U,V,W)$, whose Euklidian form is $Q = |\underline{U}| = (U^2 + V^2 + W^2)^{0.5}$, is decomposed in components normal, binormal and tangential (parallel) to the wire

$$\begin{aligned} U_n &= -U \cdot \cos 45° + V \cdot \sin 45° \ , \\ U_b &= W \ , \\ U_t &= -U \cdot \sin 45° + V \cdot \cos 45° . \end{aligned} \tag{2.28}$$

The velocity components with respect to the probe-supported coordinate system are :

$$\begin{aligned} U &= Q \cdot \cos\beta \cdot \cos\alpha \ , \\ V &= Q \cdot \sin\alpha \ , \\ W &= Q \cdot \cos\alpha \cdot \sin\beta \ . \end{aligned} \tag{2.29}$$

Substituting (2.29) into (2.28) we obtain :

$$
\begin{aligned}
U_n &= \frac{Q}{\sqrt{2}} \cdot (\cos\beta \cdot \cos\alpha + \sin\alpha)\,, \\
U_b &= Q \cdot \cos\alpha \cdot \sin\beta\,, \\
U_t &= \frac{Q}{\sqrt{2}} \cdot (-\cos\alpha \cdot \cos\beta + \sin\alpha).
\end{aligned}
\tag{2.30}
$$

Thus, for a wire slanted at 45° Jørgensen's equation (2.2) can be written as

$$
\begin{aligned}
U_{eff}^2 = \frac{Q^2}{2} \cdot \big[(\cos\beta \cdot \cos\alpha + \sin\alpha)^2 &+ 2 \cdot k_b \cdot \cos^2\alpha \cdot \sin^2\beta \\
&+ k_t \cdot (-\cos\alpha \cdot \cos\beta + \sin\alpha)^2\big]\,.
\end{aligned}
\tag{2.31}
$$

Differentiation of eq. (2.10) with respect to the velocity direction yields

$$
d(E^2) = B \cdot \left\{ \frac{\partial\left(U_{eff}^n\right)}{\partial\alpha} d\alpha + \frac{\partial\left(U_{eff}^n\right)}{\partial\beta} d\beta \right\}\,, \tag{2.32}
$$

with n = 0.5 and

$$
B = 0.57 \cdot \pi \cdot l_w \cdot \lambda_f \cdot (T_w - T) \cdot R_w \cdot Pr^{0.33} \cdot \left(\frac{d_w}{\nu_f} \right)^{0.5}. \tag{2.33}
$$

Substituting eq. (2.2) into eq. (2.32), recalling that for f = f(x)

$$
\frac{\partial(f^2)}{\partial x} = 2 \cdot f \cdot \frac{\partial f}{\partial x},
$$

assuming that k_t and k_b are constants, and considering each angular direction separately we obtain

$$
\frac{\partial E}{\partial\alpha} = \frac{1}{2} \cdot \frac{n \cdot B \cdot U_{eff}^{n-1}}{E} \cdot \left(\frac{\partial U_n}{\partial\alpha} + k_b \cdot U_b \cdot \frac{\partial U_b}{\partial\alpha} + k_t \cdot U_t \cdot \frac{\partial U_t}{\partial\alpha} \right)\,, \tag{2.34}
$$

for the sensitivity of the wire in direction of the pitch angle. This corresponds to the tangential sensitivity of the wire in *Fig. 2.1*. For the sensitivity of the wire in direction of the yaw angle one gets

$$\frac{\partial E}{\partial \beta}=\frac{1}{2}\cdot\frac{n\cdot B\cdot U_{eff}^{n-1}}{E}\cdot\left(\frac{\partial U_n}{\partial \beta}+k_b\cdot U_b\cdot\frac{\partial U_b}{\partial \beta}+k_t\cdot U_t\cdot\frac{\partial U_t}{\partial \beta}\right)\ , \tag{2.35}$$

which corresponds to the binormal sensitivity of the wire. The differentials of U_n, U_b and U_t with regard to α and β involved in eqs. (2.34) and (2.35) read:

$$\begin{aligned}\frac{\partial U_n}{\partial \alpha}&=\frac{Q}{\sqrt{2}}\cdot(-\sin\alpha\cdot\cos\beta+\cos\alpha)\ ,\\ \frac{\partial U_b}{\partial \alpha}&=-Q\cdot\sin\alpha\cdot\sin\beta\ ,\\ \frac{\partial U_t}{\partial \alpha}&=\frac{Q}{\sqrt{2}}\cdot(\sin\alpha\cdot\cos\beta+\cos\alpha)\ ,\end{aligned} \tag{2.36}$$

and

$$\begin{aligned}\frac{\partial U_n}{\partial \beta}&=-\frac{Q}{\sqrt{2}}\cdot\cos\alpha\cdot\sin\beta\ ,\\ \frac{\partial U_b}{\partial \beta}&=Q\cdot\cos\alpha\cdot\cos\beta\ ,\\ \frac{\partial U_t}{\partial \beta}&=\frac{Q}{\sqrt{2}}\cdot\cos\alpha\cdot\sin\beta\ .\end{aligned} \tag{2.37}$$

The anemometer voltage E as function of α and β and its differentials with respect to α and β was computed under consideration of eqs. (2.10), (2.31), and (2.33) to (2.37) and is plotted in *Fig. 2.3.* For the remaining quantities appearing in these equations the following values were chosen: $Q = 6$ m/s, $T = 20$ °C, $l_w = 0.5$ mm, $d_w = 2.5$ μm, $R_0 = 6\ \Omega$, $\alpha_R = 4.82\cdot10^{-3}$ grd^{-1}, OHR = 1.4, $n = 0.5$, $k_b = 1.2$, and $k_t = 0.04$. In a large part of the

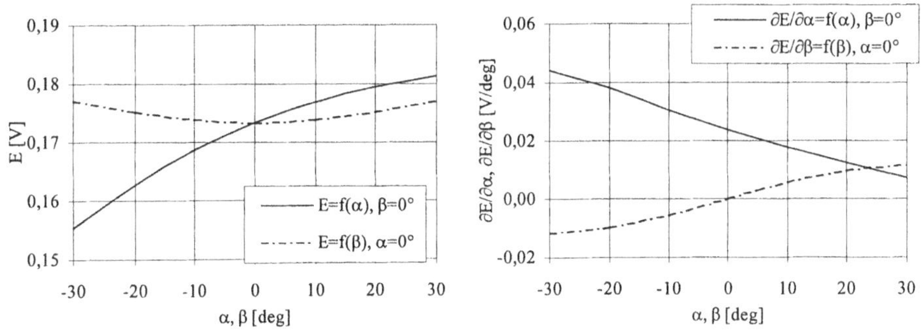

Figure 2.3. Computed response of a wire slanted at 45°, left : anemometer voltage as function of pitch and yaw; right : angular wire sensitivites as function of pitch and yaw

variation range of angles α and β the pitch (or tangential) sensitivity of the wire, $\partial E/\partial\alpha$, is considerably higher than the yaw (or binormal) sensitivity, $\partial E/\partial\beta$. In the range $\alpha \in [-15°,15°]$, where about 90-95% of the data are expected in most investigations, $\partial E/\partial\beta$ is very small. This means that in this region the resolution accuracy of the wire is insufficient.

Fig. 2.4 shows measured values of E, $\partial E/\partial\alpha$ and $\partial E/\partial\beta$ at different pitch and yaw orientations of a 45°-slanted wire. The measurements were performed at $Q = 13.3$ m/sec and an ambient temperature of $T_0 = 20$ °C. The wire had a length of $l_w = 0.5$ mm, a diameter of $d_w = 2.5$ μm and a cool resistance of $R_0 = 6\ \Omega$. The overheat ratio used was OHR = 1.4. In agreement with the above analysis the measured wire response with respect to the pitch angle is satisfactory. On the contrast, the wire is practically insensitive to yaw changes for yaw angles between -5° and 5°. The voltage distribution with regard to the yaw angle is symmetrical in respect to zero, so that the anemometer output is the same for positive and negative yaw angles of the same magnitude. This can cause ambiguities.

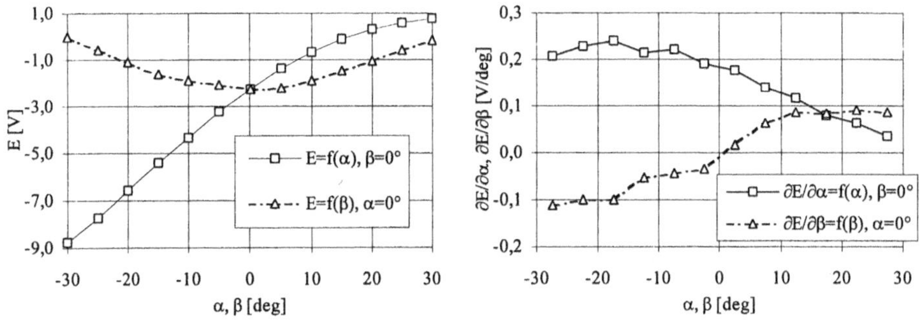

Figure 2.4. Measured response of a wire slanted at 45°; left : anemometer voltage as function of pitch and yaw; right : angular wire sensitivites as functions of pitch and yaw

The conclusion of this analysis is that sufficient accuracy in the measurement of a velocity component is only then guaranteed, when the wire is oriented with respect to this component so, that the component can be decomposed only in directions normal and tangential to the wire.

2.5. LAYOUT OF HOT-WIRE PROBES FOR 3-D VELOCITY AND VORTICITY MEASUREMENTS

First attempts to measure vorticity using hot-wire anemometry were undertaken by Kovasznay (1954) and later by Wyngaard (1969), Foss (1976, 1981), Kastrinakis & Eckelmann (1983, 1984), Antonia et al. (1987, 1988) and others. The probes developed in these works were capable of detecting one or two of the three components of the vorticity vector. In the last decade Wallace et al. (1986, 1994) and later Dracos et al. (1989)

and Tsinober et al. (1992, 1993) developed sensors with which all three components of the velocity and vorticity vectors and of the rate of strain tensor could be determined simultaneously.

For an accurate measurement of all three velocity components one needs a hot-wire array with at least three wires not parallel to the same plane. Once the velocity components are measured, their derivatives in flow direction (x or x_1) can be computed using Taylor's hypothesis. This hypothesis is valid with good accuracy in homogeneous, isotropic turbulence. Piomelli et al. (1989) have, however, shown that it also yields results with good accuracy in boundary layer flows even relatively close to the wall.

To determine derivatives in cross-stream directions (y or x_2 and z or x_3) the velocity vector has to be measured simultaneously at at least three neighbouring points not on a straight line in the plane. A hot-wire probe capable of measuring all derivatives of the velocity vector must consequently consist of at least three arrays of three hot-wires each, a total of nine hot-wires. To get an optimal directional sensitivity the hot-wires of an array should be arranged orthogonally. Two of them should in this case be located in one of the planes x_1/x_2 or x_1/x_3 and the third in the other one.

Correlations like the ones found in the equations presented in the introduction make sense only if the correlated quantities are measured at the same point. A possible arrangement of a 3-D velocity-vorticity probe with 9 slanted wires bundled in three arrays is shown in *Fig. 2.5(a)*.

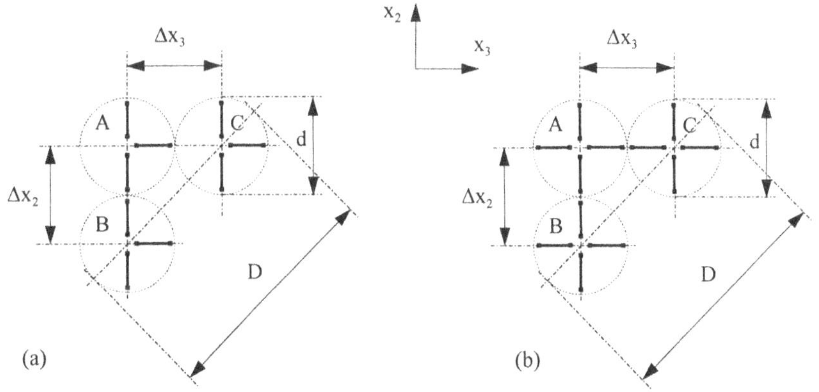

Figure 2.5. Front view of a nine- and a twelve-wire probe with slanted wires.

Each array has a sensing diameter d. The velocity components measured by such an array are averaged over an area of approximately $\pi \cdot d^2/4$. The discretization lengths Δx_2, Δx_3 depend on the size of the individual arrays and are slightly larger than d. Accurate measurements of the velocity components and their derivatives demand a miniaturization of the arrays. When wires with 2.5 μm diameter and 500 μm length are used, the sensing diameter of a 3-wire array is $d \approx 0.8$ mm, the overall sensing diameter of the

The advantages of this probe are that it is relatively easy to manufacture and that the discretization lengths are rather small. However, its deficiencies were soon recognized :

i) As will be shown later, velocity components in planes in which two hot-wires are present are measured with higher accuracy than components in planes with only one wire. This affects to a much higher degree the derivatives of these components.
ii) To get velocity components and their derivatives at the same location the derivatives must be determined at the center of array A. The same array must be used for the computation of the derivatives in x_1 direction. The derivatives are in this case computed using first order forward or backward finite differentials.

Adding a fourth wire to each array (*Fig. 2.5(b)*) eliminates the first deficiency (Dracos et al., 1989). The dimensions of this probe are about the same as those of the previous one, and its construction is slightly more complicated.

To attain a better approximation of the derivatives, keeping in mind that velocity components and all their derivatives must be determined at the same point, two more arrays must be added. This increases the number of wires to 20. Two arrangements of the five 4-wire arrays are shown in *Fig. 2.6(a,b)*. Both have been used. They differ in that the arrays are turned by 45° in the second (b) relative to the first (a).

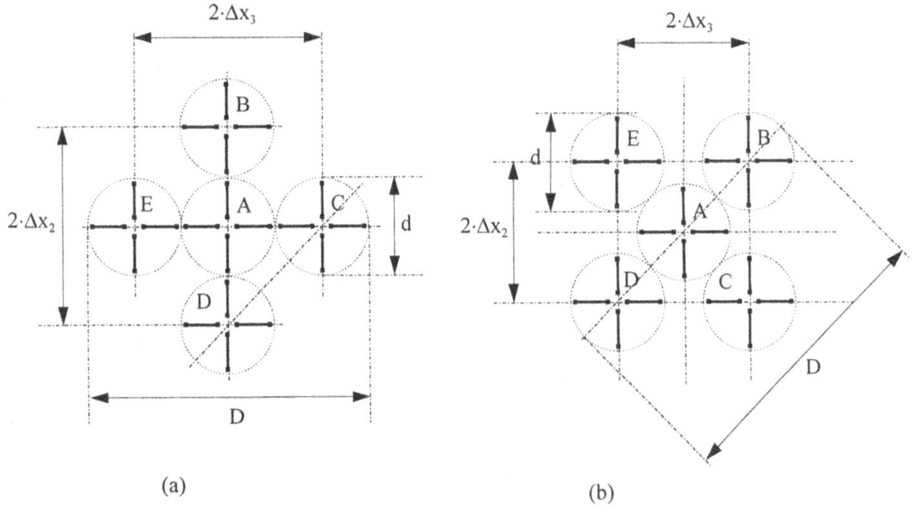

Figure 2.6. Front view of twenty-wire probes.

The sensing diameter of an array did not change. The sensing diameter of the probe increases only slightly to 3.0 mm. The velocity components and their derivatives are now determined at the center of array A, which is also used for the computation of the derivatives in x_1-direction.

In the arrangement depicted in *Fig. 2.6(a)* the discretization lengths $2 \cdot \Delta x_2 = 2 \cdot \Delta x_3$ are twice as large as in the three-array probes of *Fig. 2.5*. However, the accuracy in the determination of velocity derivatives is increased, as second order central differences can be used for their computation. Furthermore, this arrangement makes also possible the computation of second order velocity derivatives.

Except for the considerable increase in the discretization length, this arrangement shares a further inperfection with the nine- and twelve-wire probes shown in *Fig. 2.5*. Some of the prongs of the individual three- or four-wire arrays are very close to prongs of adjacent arrays. Interference between them cannot be excluded.

The design presented in *Fig. 2.6(b)* brings some improvements in this respect. The discretization lengths $2 \cdot \Delta x_2$, $2 \cdot \Delta x_3$ are reduced by a factor $2^{-1/2}$ compared to probe in *Fig. 2.6(a)*, and interference between prongs is eliminated. For the computation of derivatives in the x_2/x_3 plane a second order five point central difference scheme is used. In x_1-direction the derivatives are computed as in the previous case from array A using Taylor's hypothesis and a second order central difference scheme. This arrangement does not allow to compute second order derivatives but gives the possibility to compute mixed derivatives.

In both cases the price for the higher degree of approximation is a considerable increase of the discretization lengths in x_2 and x_3 directions and an increase of the number of wires from 12 to 20. The construction of the probe becomes much more cumbersome. The increase in the accuracy with which derivatives can be determined is, however, considerable and makes this effort acceptable.

2.6. CONSTRUCTION OF MULTI-HOTWIRE PROBES

To resolve velocity fluctuations induced by the smallest eddies the size of the sensor used must be extremely small. When derivatives of the velocity have to be determined, subminiature arrays must be bundled to miniature probes. Miniature multi-hotwire probes with 9 and more wires are not available commercially. Manufacturing techniques for subminiature arrays are described by Perry (1982) and Lomas (1986). Here, a probe-construction method will be presented which is an improved version of the method first described by Vukoslavĉević et al. (1991) and further developed by Tsinober et al. (1992).

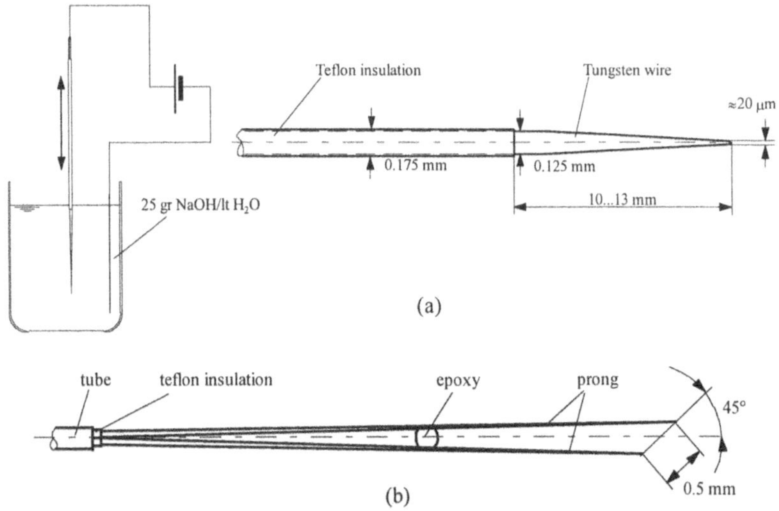

Figure 2.7. Supporting prongs and 45°-slanted hot-wire

For the wire supporting prongs, teflon-insulated tungsten wire with a diameter of 175 μm is used. The insulation is removed from both ends of the wire. One end is electrochemically etched in a solution of 25 gr NaOH/lt H_2O at 5.0 Amp. to a conical shape with a diameter of ca. 20 μm at the tip (*Fig. 2.7(a)*). Eight prongs are then glued together to a bundle which is inserted and glued into a stainless steel tube of 1.0 mm outer and 0.8 mm inner diameter. The conical ends of the prongs are bent to the desired position and fixed at this position with a small drop of epoxy (*Fig. 2.7(b)*). A sketch of the tip of an orthogonal four-wire array with 45°-slanted wires is shown in *Fig. 2.8*.

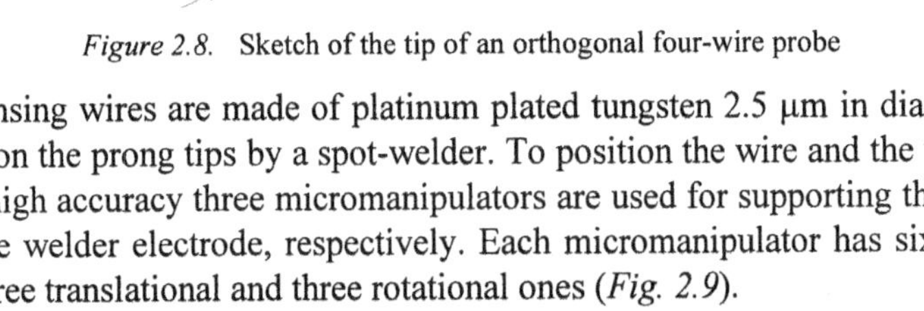

Figure 2.8. Sketch of the tip of an orthogonal four-wire probe

The sensing wires are made of platinum plated tungsten 2.5 μm in diameter. They are welded on the prong tips by a spot-welder. To position the wire and the welder electrode with high accuracy three micromanipulators are used for supporting the probe, the wire and the welder electrode, respectively. Each micromanipulator has six degrees of freedom, three translational and three rotational ones (*Fig. 2.9*).

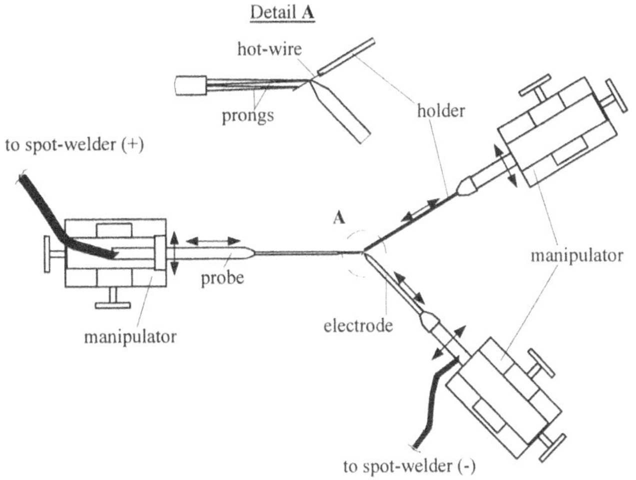

Figure 2.9. Wire welding facility

Five four-hot-wire arrays are fitted together to a twenty-wire probe. The disadvantage of this design is that the probe becomes useless even if a single prong is damaged. However, designs with exchangeable arrays are very costly. Furthermore, because of the additional mechanical parts needed to fix the single arrays the sting diameter of a dismantleable probe would become much larger. This would increase sting induced aero-

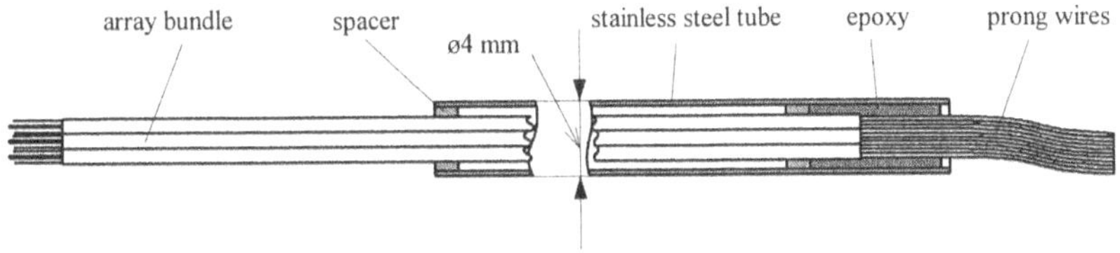

Figure 2.10. Cross section of the 20-wire vorticity probe

dynamic interference in the area of the probe head. In order to fix the five arrays in the proper position with high accuracy special miniature presses were manufactured. After the five arrays were glued to a single bundle, the bundle was inserted and glued into a stainless steel tube of 4.0 mm outer diameter. The bundle was centred in the tube by two spacers. To avoid damage of the teflon insulation on the outcoming prong wires at the rear part of the array bundle, the prong wires are coated over a length of about 1 cm in epoxy (*Fig. 2.10*). The entire work is performed under a microscope. At the uninsulated rear end of the prongs, sticking out of the stainless steel tubes thin, flexible cables leading to an 8-pin plug are soldered.

3. Calibration of Hot-Wire Probes and Determination of 3-D Velocity

3.1. CALIBRATION PROCEDURE AND DATA REDUCTION

The relation (2.3) between the output voltage E of an anemometer connected to a hot-wire and the effective cooling velocity U_{eff} is non-linear. As stated earlier, theoretically, eqs. (2.2) and (2.3) can be used to construct a mathematical model for the wire response to $\underline{U}$ of the form given in eq. (2.4). Subsequently, the analyst could derive inverse functions of the form $\underline{U} = \underline{f}(E_1, \ldots, E_4)$, where the indices denote wire numbers in a four-hotwire probe, to evaluate 3-D velocity vectors from anemometer voltages measured in turbulent flows. This approach, however, demands perfect probe geometry, which can never be achieved for miniature multi-hotwire probes in practice. Therefore, an experimental, assumption-free probe calibration procedure was developed, based on the numerical approximation of the probe response functions to known velocity vectors (Lemonis & Dracos, 1995). The procedure involves the simultaneous measurement of the output voltages $E_1, \ldots, E_4$ of the anemometers connected to the probe, for a regular grid of pitch and yaw angles at different values of the velocity magnitude Q. The calibration procedure does not require any knowledge about the probe geometry and its physical properties.

The 3-D calibration of an orthogonal four-wire probe, which is the building block of the twenty-wire probes shown in *Fig. 2.6*, yields trivariate, discrete functions $\{F_w\}$ of the anemometer output voltages $\{E_w\}$ of the form $\{F_w\}=\{f_w(Q, \alpha, \beta)\}$, with w=1-4 being the wire number. Using expressions approximating functions $\{F_w\}$ it is possible to calculate voltages $\{E_w\}$ at any triple of values Q, α and β. Thereafter, numerical methods are applied to inversely evaluate three-dimensional velocity vectors corresponding to anemometer voltage data sets $\{E_w\}$ sampled in a turbulent flow field.

The calibration functions $\{F_w\}=\{f_w(Q, \alpha, \beta)\}$, w=1-4, are approximated by trivariate Bernstein-Bézier tensor-product polynomials in a least-squares sense (Lasser,1987). The approximating polynomial P_w of each calibration function F_w is defined as the tensor-product of three Bernstein polynomials $B_i^l(x) = \binom{l}{i} \cdot x^i \cdot (1-x)^{l-i}$:

$$P_w(\underline{x}) = \sum_{i=0}^{l}\sum_{j=0}^{m}\sum_{k=0}^{n} b_{w_{ijk}} \cdot B_i^l(x_1) \cdot B_j^m(x_2) \cdot B_k^n(x_3), \text{ w=1-4,} \tag{3.1}$$

with (l+1)·(m+1)·(n+1) unknown polynomial factors $b_{w_{ijk}}$ and coordinate vector $\underline{x} = (x_1, x_2, x_3)$ on the calibration axes, x_1 corresponding to Q, x_2 to α and x_3 to β.

Least-square approximation demands the minimization of the square of error D

$$D^2 = \sum_{\nu=1}^{N} \left\{ \sum_{i=0}^{l} \sum_{j=0}^{m} \sum_{k=0}^{m} b_{w_{ijk}} \cdot B_i^l(x_{1_\nu}) \cdot B_j^m(x_{2_\nu}) \cdot B_k^n(x_{3_\nu}) - F_{w_\nu} \right\}^2, \text{ w=1-4,} \tag{3.2}$$

with $N = n_1 \cdot n_2 \cdot n_3$; n_1, n_2 and n_3 number of discretization points on the calibration axes. Differentiation of (3.2) yields

$$\sum_{\nu=1}^{N} \sum_{i=0}^{l} \sum_{j=0}^{m} \sum_{k=0}^{n} b_{w_{ijk}} \cdot B_{ijk}^{lmn}(\underline{x}_\nu) \cdot B_{pqr}^{lmn}(\underline{x}_\nu) = \sum_{\nu=1}^{N} F_{w_\nu} \cdot B_{pqr}^{lmn}(\underline{x}_\nu), \text{ w=1-4,} \tag{3.3}$$

where

$$B_{ijk}^{lmn}(\underline{x}) \equiv B_i^l(x_1) \cdot B_j^m(x_2) \cdot B_k^n(x_3).$$

Eq. (3.3) reads in matrix notation

$$\begin{vmatrix} \sum_{\nu=1}^{N} B_{000}^{lmn}(\underline{x}_\nu) \cdot B_{000}^{lmn}(\underline{x}_\nu) & \cdots & \sum_{\nu=1}^{N} B_{lmn}^{lmn}(\underline{x}_\nu) \cdot B_{000}^{lmn}(\underline{x}_\nu) \\ \vdots & \ddots & \vdots \\ \sum_{\nu=1}^{N} B_{000}^{lmn}(\underline{x}_\nu) \cdot B_{lmn}^{lmn}(\underline{x}_\nu) & \cdots & \sum_{\nu=1}^{N} B_{lmn}^{lmn}(\underline{x}_\nu) \cdot B_{lmn}^{lmn}(\underline{x}_\nu) \end{vmatrix} \cdot \begin{pmatrix} b_{w_{000}} \\ \vdots \\ b_{w_{lmn}} \end{pmatrix} = \begin{pmatrix} \sum_{\nu=1}^{N} F_{w_\nu} \cdot B_{000}^{lmn}(\underline{x}_\nu) \\ \vdots \\ \sum_{\nu=1}^{N} F_{w_\nu} \cdot B_{lmn}^{lmn}(\underline{x}_\nu) \end{pmatrix}, \text{ w=1-4.} \tag{3.4}$$

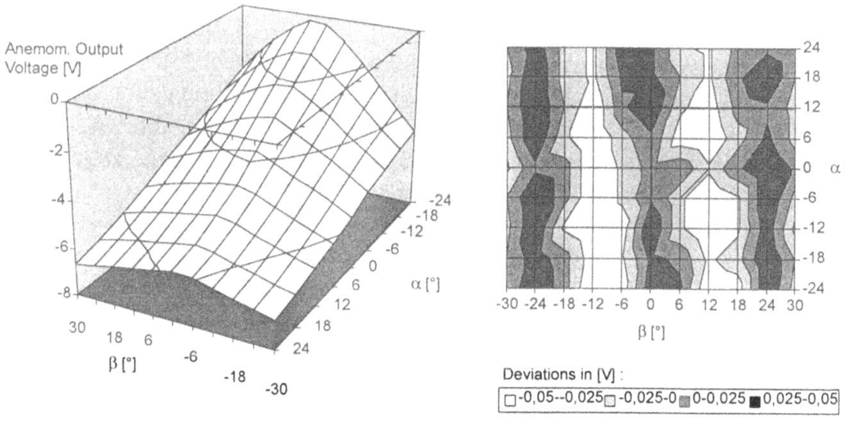

Figure 3.1. Calibration function of a wire of a four-wire probe at 10 m/sec; measurement (left) and deviations between measurement and trivariate approximation (right)

Eq. (3.4) describes a linear equation system consisting of (l+1)·(m+1)·(n+1) equations with equal number of unknown Bézier-factors $b_{w_{ijk}}$, of the form

$$|A| \cdot \underline{b}_w = \underline{y}_w, \; w=1\text{-}4. \tag{3.5}$$

Matrix $|A|$ on the left hand side of eq. (3.5) is symmetrical and positive definite. Thus, eq. (3.5) can be solved efficiently by Cholesky factorization. $|A|$ depends on the parameterization of the calibration axes only, hence its calculation is only once required, since the axis parameterization is common for all wires. Thereafter, eq. (3.5) can be solved consecutively four times, taking the vector $\underline{y}_w$ of the corresponding wire into account each time. *Fig. 3.1* shows the angular calibration surface at Q = 10 m/sec. The deviation between measurement and approximation has an r.m.s. value of typically 0.5 % and does not exceed 1.5 % of the peak-to-peak value of a calibration surface.

Considering a sample of four anemometer output voltages E_1 to E_4, measured in a turbulent flow field by a four-wire probe, each voltage E_w is expressed by means of its approximating polynomial P_w

$$\begin{aligned} E_1 &= \sum_{i=0}^{l}\sum_{j=0}^{m}\sum_{k=0}^{n} b_{1_{ijk}} \cdot B_i^l(x_1) \cdot B_j^m(x_2) \cdot B_k^n(x_3)\,, \\ &\vdots \\ E_4 &= \sum_{i=0}^{l}\sum_{j=0}^{m}\sum_{k=0}^{n} b_{4_{ijk}} \cdot B_i^l(x_1) \cdot B_j^m(x_2) \cdot B_k^n(x_3)\,. \end{aligned} \tag{3.6}$$

This is a system of nonlinear equations consisting of four equations with three unknowns. The redudancy of the system is used to improve the solutions for x_1, x_2 and x_3. Treating the four-wire probe as a combination of four independent „three-wire probes“, equation system (3.6) can be split into four systems of nonlinear equations, each system consisting of three equations with three unknowns. Solving this set of systems of nonlinear equations one gets four different solutions of the three velocity components.

Since all four wires are aligned with respect to the streamwise velocity component u in direction of their maximum sensitivity, it is expected that all four solutions of this component are reliable. For the cross-stream velocity components, v and w, the three-wire combinations containing two wires in the plane parallel to the corresponding component should yield accurate solutions of this component. By contrast, wire combinations containing only one wire in the plane parallel to a cross-stream component yield lower accuracy. This is demonstrated in *Fig. 3.2* at the example of the normal velocity component v. Hence, for every cross-stream component two reliable solutions are obtained, *Fig. 3.3*. This is the significant advantage of the method described, as it yields four (4) different reliable solutions for the streamwise velocity component and two (2) reliable solutions for each cross-stream velocity component.

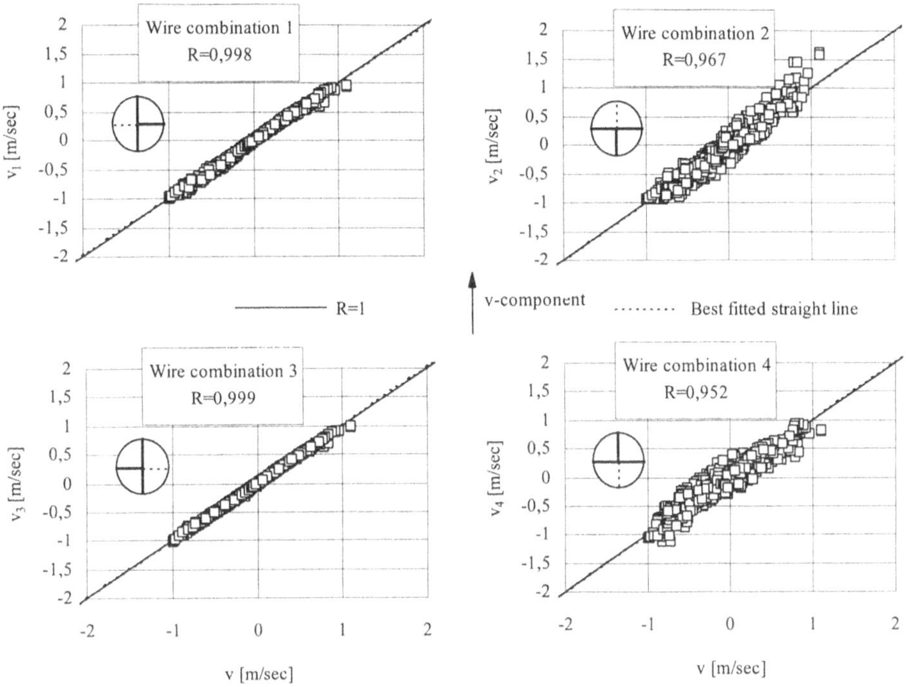

Figure 3.2. Correlations of the normal velocity component v, determined by different 3-wire combinations of a 4-wire probe; R : correlation coefficient

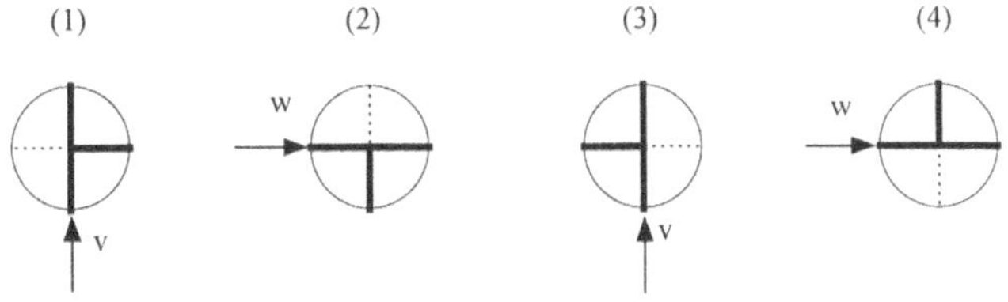

Figure 3.3. Maximum cross-stream sensitivity of different 3-wire combinations of a 4-wire probe

The uniqueness domain of four-wire probes has been investigated analytically by Rosemann (1989). For 45°-slanted wires he found that uniqueness is guaranteed at any pair of pitch and yaw angles satisfying $(\alpha^2+\beta^2)^{0.5} < \phi$, where ϕ was estimated, in agreement with Pailhas & Cousteix (1986), to be at least 30°. This area covers the range of angles of interest in most turbulent flows. In the present work, the uniqueness domain of the four-wire probe used was examined by applying the procedure described above to anemometer voltage data sets $\{E_w\}$ obtained by probe calibration. In the range of -30° to +30° for pitch and yaw angle, and for velocity magnitudes in the range of 1.5 to

15 m/sec, no ambiguities were obtained. The calibrating velocity vectors were recovered within the approximation accuracy.

Assumption-free calibration methods for four-hotwire probes have also been developed by Samet & Einav, (1987), Pompeo, (1990) and Döbbeling (1990). Samet & Einav, (1987) used an orthogonal four-hotwire probe treating the probe as a combination of two V-wire sensors. Each one of the V-wire sensors delivered one solution for the streamwise component and one solution for the cross-stream component parallel to the plane formed by the V-arrangement. This method, however, does not exploit the symmetry of the orthogonal probe type about its longitudinal axis, delivering only two solutions of the streamwise velocity component and one solution for each one of the cross-stream components. Pompeo (1990) used a probe consisting of an X-wire sensor and a perpendicular positioned V-wire sensor. Each one of the two-wire combinations delivered one solution of the streamwise component and one solution of the cross-stream component parallel to the wire combination. Pompeo (1992) reports unsatisfactory results with this probe type, most probably due to the asymmetrical wire arrangement. It is noteworthy that in a probe of this type the wire spacing in the V-sensor must be much larger than the wire spacing in the X-sensor. Döbbeling (1990) developed a 3-D method for orthogonal four-hotwire probes considering that the angular sensitivity of the wires is velocity-independent at high flow velocities. He could then separate the velocity calibration from the angular one and perform an angular probe calibration only at the mean velocity of the velocity range of interest. This method, however, is inappropriate for flow velocities less than ~15 m/sec, as the angular wire sensitivity becomes strongly velocity dependent in this range.

3.2. HOTWIRE PROBE CALIBRATION FACILITY

A satisfactory determination of velocity derivatives from closely spaced velocity measurements necessitates a high accuracy of these measurements. Even if the hardware used is perfect, this accuracy depends to a large extend on the calibration of the probe. Special attention must be paid to the calibration facility used. When measurements in flows with a high degree of turbulence are planned the calibration must yield reliable results also for large pitch and yaw angles. A calibration should furthermore establish unequivocal relations between the anemometer voltage outputs and the velocity components. Hotwire probe calibrations must be performed for different mean stream velocities and different orientations of the probe relative to the mean stream direction. To avoid drifts caused by disconnection of the probe from and reconnection of the probe to the anemometers and by a new probe alignment, it is preferable to calibrate the probe in-situ.

The main requirements to a calibration facility for hotwire probes are :

a) time invariant mean stream velocity,
b) low background turbulence level of the calibrating flow,

c) time invariant flow temperature, equal to the flow temperature during the measurement,
d) high accuracy of the velocity adjustment,
e) high accuracy of the adjustment of the pitch and yaw angles,
f) small perturbation of the flow field by the probe,
g) high degree of automation to reduce the calibration duration.

The probe traversing facility, preferably the same used for the experimental investigations presented in the following sections, should allow a precise positioning of the probe in a plane normal to the flow direction with high repeatability. The probe holder mounted on the traversing facility must have two rotational degrees of freedom, necessary for the adjustment of the pitch and yaw angles during calibration. To guarantee constant incident flow during the calibration the probe must rotate around its tip.

A calibration facility designed to fulfil these conditions was constructed. A jet flow is used for the calibration inside the wind tunnel. The main parts of the jet facility are shown in *Fig. 3.4.* The compressor used is driven by a computer controlled motor through a saw tooth belt drive with a stability of +/- 1 r.p.m and a range from 0 to 6,000

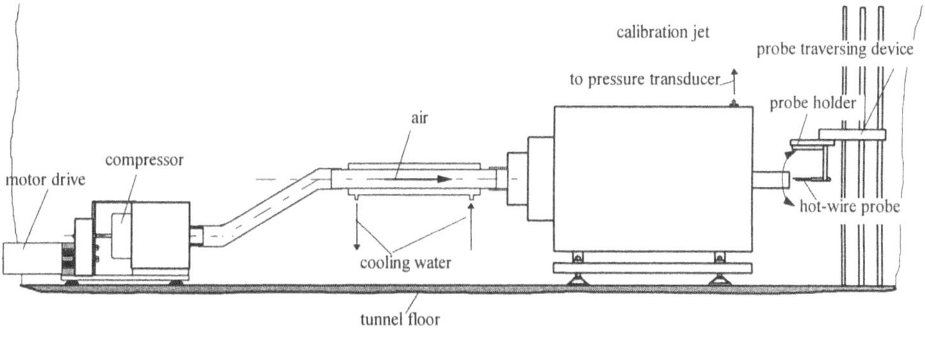

Figure 3.4. Calibration device in wind tunnel

r.p.m. At a belt drive transmission ratio of 1:2 the maximum revolutions of the compressor reach 12,000 r.p.m. A water cooled air conditioner keeps the air temperature in the jet constant within 0.5 °C. Nozzles with diameters from 10 mm to 60 mm are used and the velocity at the exit of the nozzle U_C varies from 1.0 m/s to 100 m/s. With the motor drive used the calibration velocity is adjusted with an accuracy of better than 1 cm/s, and a long term variation of less than +/-0.1 % of U_C is achieved.

The traversing device allows to position the probe with an accuracy of ~0.1 mm in vertical direction. Taking into account that the sensitive diameter of the probe is 3.0 mm this accuracy is considered sufficient. A specially designed probe holder is used for the in-situ calibration of the probe. A sketch of the probe holder is presented in *Fig. 3.5.* The pitch and yaw rotations are driven by computer controlled step motors through spindle gearings. Each step of the motors causes an angular probe deflection of

0.0017 deg. During a probe calibration 9 to 13 pitch and yaw angles are adapted for each of the 7 to 9 velocities adjusted. This yields a total of 567 to 1,521 movements of the holder and the same number of individual measurements for each wire of the probe. The high resolution of the pitch and yaw rotations ensures that at the end of a calibration the initial position of the probe given by $\alpha = \beta = 0$ deg is reached with an accuracy de-

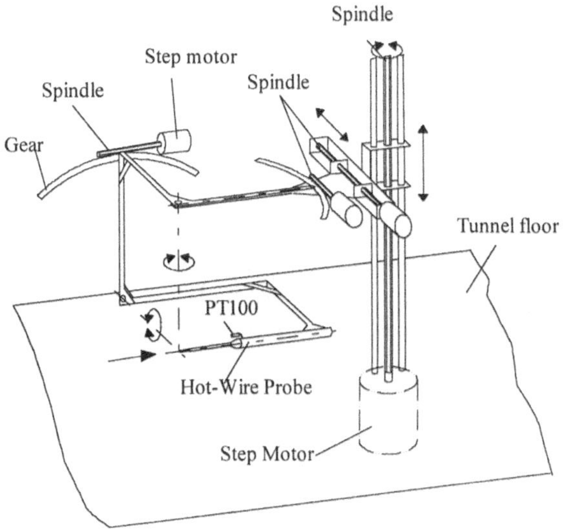

Figure 3.5. Sketch of probe traversing device with probe holder

pending primarily on the play of the gears, which in this case is ~0.1 deg. With this facility an accurate probe calibration run is performed, computer controlled and automatically, in 30 to 60 min.

4. Determination of Velocity Derivatives and Derivative Depending Quantities

4.1. DETERMINATION OF VELOCITY DERIVATIVES

In the following only the 20-wire probes presented in *Fig. 2.6* are considered. The estimations of derivatives presented can be applied to the instantaneous velocity components $\tilde{U}_i$ as well as to their Reynolds decomposition U_i and u_i. The spatial derivatives in streamwise direction at the centroid of the central array A are derived from the time derivatives of U_i (Taylor hypothesis) :

$$\left.\frac{\partial U_i}{\partial x_1}\right|_A = -\frac{1}{U_{1,A}} \cdot \left.\frac{\partial U_i}{\partial t}\right|_A , \; i = 1, 2, 3 . \tag{4.1}$$

For the evaluation of the time derivative a second order central difference scheme is used :

$$\left.\frac{\partial U_i}{\partial t}\right|_A = \frac{U_{i,A}(t+\Delta t) - U_{i,A}(t-\Delta t)}{2\Delta t} \tag{4.2}$$

$\Delta t = 1/f_s$ is the sampling time step and f_s the sampling frequency of the A/D-converter. The second derivatives of U_i are similarly derived from

$$\left.\frac{\partial^2 U_i}{\partial^2 x_1}\right|_A = \frac{1}{U_{1,A}^2} \cdot \frac{\partial^2 U_i}{\partial^2 t}, \tag{4.3}$$

and

$$\left.\frac{\partial^2 U_i}{\partial^2 t}\right|_A = \frac{U_{i,A}(t+\Delta t) - 2\cdot U_{i,A}(t) + U_{i,A}(t-\Delta t)}{4\cdot \Delta t^2}. \tag{4.4}$$

Velocity derivatives in cross-stream directions are estimated using the schemes corresponding to the front view of the four-wire array arrangements of a twenty-wire probes shown in *Fig. 2.6*. When the arrangement of *Fig. 2.6(a)* is used, the first derivatives centered at point A, i.e. the center of array A, are estimated from a second order central difference scheme

$$\left.\frac{\partial U_i}{\partial x_2}\right|_A \cong \frac{U_{i,B} - U_{i,C}}{2\cdot \Delta x_2}, \quad \left.\frac{\partial U_i}{\partial x_3}\right|_A \cong \frac{U_{i,D} - U_{i,E}}{2\cdot \Delta x_3}, \tag{4.5}$$

and the second derivatives from

$$\left.\frac{\partial^2 U_i}{\partial^2 x_2}\right|_A \cong \frac{U_{i,B} - 2\cdot U_{i,A} + U_{i,C}}{\Delta {x_2}^2}, \quad \left.\frac{\partial^2 U_i}{\partial^2 x_3}\right|_A \cong \frac{U_{i,E} - 2\cdot U_{i,A} + U_{i,D}}{\Delta {x_3}^2}. \tag{4.6}$$

For the wire arrangement of *Fig. 2.6(b)* the second order scheme used for the first derivatives is :

$$\left.\frac{\partial U_i}{\partial x_2}\right|_A \cong \frac{1}{2}\cdot\left[\frac{U_{i,B} - U_{i,C}}{2\cdot \Delta x_2} + \frac{U_{i,E} - U_{i,D}}{2\cdot \Delta x_2}\right] = \frac{U_{i,B} - U_{i,C} + U_{i,E} - U_{i,D}}{4\cdot \Delta x_2}, \tag{4.7}$$

$$\left.\frac{\partial U_i}{\partial x_3}\right|_A \cong \frac{1}{2}\cdot\left[\frac{U_{i,B}-U_{i,E}}{2\cdot\Delta x_3}+\frac{U_{i,C}-U_{i,D}}{2\cdot\Delta x_3}\right]=\frac{U_{i,B}+U_{i,C}-U_{i,E}-U_{i,D}}{4\cdot\Delta x_3}\;. \tag{4.8}$$

Second derivatives in directions x_2, x_3 cannot be computed by finite differences in this arrangement. The mixed derivative centered on A is found by applying the scheme :

$$\left.\frac{\partial^2 U_i}{\partial x_2 \partial x_3}\right|_A \cong \frac{-U_{i,C}+U_{i,B}+U_{i,D}-U_{i,E}}{4\cdot\Delta x_2\cdot\Delta x_3}\;. \tag{4.9}$$

For the x_1, x_2 and x_1, x_3 directions Taylor's hypothesis must be used. Mixed derivatives centered at A are then estimated as follows :

$$\begin{aligned}\left.\frac{\partial^2 U_i}{\partial x_1 \partial x_2}\right|_A \cong\; & \frac{1}{2}\cdot\frac{U_{i,B}(t+\Delta t)-U_{i,B}(t-\Delta t)+U_{i,C}(t-\Delta t)-U_{i,C}(t+\Delta t)}{4\cdot\Delta x_2\cdot\Delta t\cdot U_{1,A}}\\ & +\frac{1}{2}\cdot\frac{U_{i,E}(t+\Delta t)-U_{i,E}(t-\Delta t)+U_{i,D}(t-\Delta t)-U_{i,D}(t+\Delta t)}{4\cdot\Delta x_2\cdot\Delta t\cdot U_{1,A}}\;,\end{aligned} \tag{4.10}$$

and

$$\begin{aligned}\left.\frac{\partial^2 U_i}{\partial x_1 \partial x_3}\right|_A \cong\; & \frac{1}{2}\cdot\frac{U_{i,B}(t+\Delta t)-U_{i,B}(t-\Delta t)+U_{i,E}(t-\Delta t)-U_{i,E}(t+\Delta t)}{4\cdot\Delta x_3\cdot\Delta t\cdot U_{1,A}}\\ & +\frac{1}{2}\cdot\frac{U_{i,C}(t+\Delta t)-U_{i,C}(t-\Delta t)-U_{i,D}(t-\Delta t)+U_{i,D}(t+\Delta t)}{4\cdot\Delta x_3\cdot\Delta t\cdot U_{1,A}}\;,\end{aligned} \tag{4.11}$$

with $\Delta x_2 = \Delta x_3 = \Delta t \cdot U_{1,A}$. This determines the A/D sampling frequency, which must be $f_s = U_{1,A}/\Delta x_2$ or an integer multiple of that. One must keep in mind that with the probes and measuring equipment available the accuracy of second derivatives is much inferior to that of first derivatives.

In the numerical expressions of eq. (4.7) to eq. (4.11) there appear quantities measured exclusively by arrays B, C, D and E. This suggests the use of a simplified probe with four arrays as shown in *Fig. 4.1*. The discretization lengths $2\cdot\Delta x_2$, $2\cdot\Delta x_3$ are reduced to the size of these lengths in a twelve-wire probe and its construction is simplified.

The derivatives computed by using eqs. (4.7) to (4.11) are still centered at A, which now is the center of the probe tip. The derivative in flow direction x_1 must be estimated in a different manner :

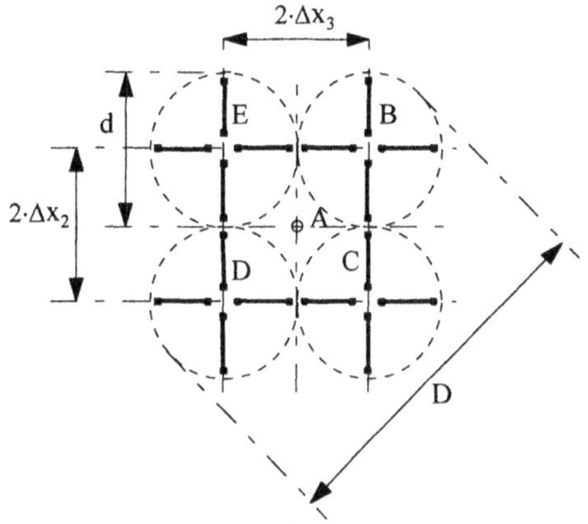

Figure 4.1. Front view sketch of a four-array 16-wire probe.

$$\left.\frac{\partial U_i}{\partial x_1}\right|_A \cong \frac{1}{4}\cdot\frac{U_{i,B}(t+\Delta t)-U_{i,B}(t-\Delta t)+U_{i,C}(t+\Delta t)-U_{i,C}(t-\Delta t)}{\Delta t\cdot U_{1,A}}$$
$$+\frac{1}{2}\cdot\frac{U_{i,D}(t+\Delta t)-U_{i,D}(t-\Delta t)+U_{i,E}(t+\Delta t)-U_{i,E}(t-\Delta t)}{\Delta t\cdot U_{1,A}} \quad . \qquad (4.12)$$

The velocity components in A are given by

$$U_{i,A} \cong \frac{1}{4}\cdot\left(U_{i,B}+U_{i,C}+U_{i,D}+U_{i,E}\right) \quad . \qquad (4.13)$$

The velocity components estimated at point A are averages over an area four times larger than in a twenty-wire probe and the ones used to estimate derivatives with the 16-wire probe. This is also true for the derivatives in flow direction x_1 and is a major disadvantage of such a probe.

4.2. COMPUTATION OF QUANTITIES DEPENDING ON VELOCITY DERIVATIVES

In spite of the improved performance of hot-wire anemometry the accuracy of second and mixed derivatives determined from measurements with multi-hotwire probes is poor. Therefore only quantities containing first derivatives will be considered, i.e. vorticity and rate of strain.

The components of the vorticity vector $\underline{\tilde{\Omega}} = \nabla \times \underline{\tilde{U}}$ are

$$\tilde{\Omega}_1 = \frac{\partial \tilde{U}_2}{\partial x_3} - \frac{\partial \tilde{U}_3}{\partial x_2}, \quad \tilde{\Omega}_2 = \frac{\partial \tilde{U}_3}{\partial x_1} - \frac{\partial \tilde{U}_1}{\partial x_3}, \quad \tilde{\Omega}_3 = \frac{\partial \tilde{U}_1}{\partial x_2} - \frac{\partial \tilde{U}_2}{\partial x_1} \qquad (4.14)$$

or when Reynolds decomposition is performed

$$\Omega_1 = \frac{\partial U_2}{\partial x_3} - \frac{\partial U_3}{\partial x_2}, \quad \Omega_2 = \frac{\partial U_3}{\partial x_1} - \frac{\partial U_1}{\partial x_3}, \quad \Omega_3 = \frac{\partial U_1}{\partial x_2} - \frac{\partial U_2}{\partial x_1} \tag{4.15}$$

for the mean movement and

$$\omega_1 = \frac{\partial u_2}{\partial x_3} - \frac{\partial u_3}{\partial x_2}, \quad \omega_2 = \frac{\partial u_3}{\partial x_1} - \frac{\partial u_1}{\partial x_3}, \quad \omega_3 = \frac{\partial u_1}{\partial x_2} - \frac{\partial u_2}{\partial x_1} \tag{4.16}$$

for the fluctuating part.

For the determination of the rate of strain tensor S_{ij}, which is symmetrical, one needs to compute the following elements

$$\begin{aligned} \tilde{S}_{11} &= \frac{\partial \tilde{U}_1}{\partial x_1}, \quad \tilde{S}_{12} = \tilde{S}_{21} = \frac{1}{2}\cdot\left(\frac{\partial \tilde{U}_2}{\partial x_1} + \frac{\partial \tilde{U}_1}{\partial x_2}\right), \quad \tilde{S}_{13} = \tilde{S}_{31} = \frac{1}{2}\cdot\left(\frac{\partial \tilde{U}_3}{\partial x_1} + \frac{\partial \tilde{U}_1}{\partial x_3}\right), \\ \tilde{S}_{22} &= \frac{\partial \tilde{U}_2}{\partial x_2}, \quad \tilde{S}_{23} = \tilde{S}_{32} = \frac{1}{2}\cdot\left(\frac{\partial \tilde{U}_3}{\partial x_2} + \frac{\partial \tilde{U}_2}{\partial x_3}\right), \quad \tilde{S}_{33} = \frac{\partial \tilde{U}_3}{\partial x_3}. \end{aligned} \tag{4.17}$$

These quantities can again be computed for mean and fluctuating quantities. All derivatives needed can be determined by any of the probes described above.

4.3. ACCURACY AND LIMITATIONS

It is obvious that in some cases measurements with multi-hotwire probes can only give rough estimates of derivatives of velocity components. Between the velocity measurement locations linear velocity distribution is assumed. This is likely to occur when the dimensions of the incoming vortices are much larger than the probe dimensions. But the assumption of linear velocity distribution can still cause large errors when measuring for example in the core of a vortex, *Fig. 4.2*. The only way to improve the measuring accuracy is to manufacture hot-wire arrays with sensing areas much smaller than the Kolmogorov microscale η of the flow investigated. The discretization length of the vorticity sensor has to be a fraction of η as well.

These requirements exceed the capabilities of present day manufacturing techniques. With the production methods and materials available, and using hot-wires with ~1 μm diameter, the diameter of a four-wire array cannot be reduced to less than ~0.4 mm. In boundary layer flows η can be 0.2 mm and less. Hence, the smallest vortex structures are in many cases smaller than the dimensions of the probe and therefore "invisible" to the probe.

The electronic instrumentation is a weak point of the measuring system also. In the case of weak gradients and small discretization distances the signal differences between two adjacent probes would become too small. This condition would reduce the system

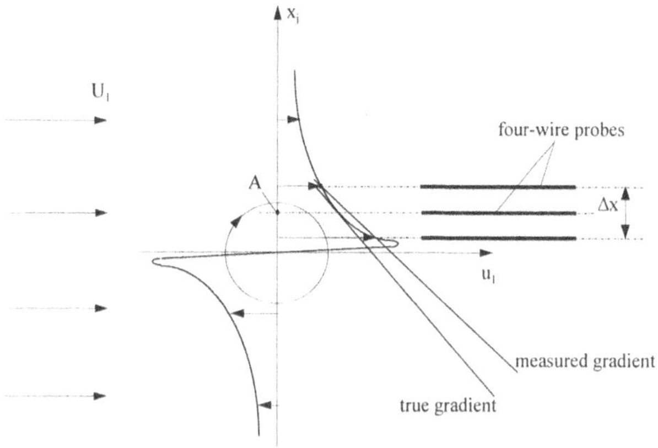

Figure 4.2. Error in the derivative estimation caused by large discretization distances

accuracy inacceptably since the signal-to-noise ratio would decrease significantly in this case. Hence, the discretization increment has to be as small as possible, to allow the resolution of small vortex structures, but it must still remain large enough, to guarantee sufficient system sensitivity at small gradients. Mi & Antonia (1993) suggest a lower bound of the discretization increment of $2\cdot\eta$ and an upper bound of $4\cdot\eta$.

In spite of the limitations and requirements mentioned above, the measuring accuracy of first order velocity derivatives expected for the probes designed is quite high. The array diameter of this probe, which represents today's standards, is 0.8 mm at a discretization distance of 1.4 mm for the probe in *Fig. 2.6(b)* or 2.0 mm for the probe in *Fig. 2.6(a)*. The latter should be 3 to 7 times higher than the Kolmogorov microscale, but 3 to 5 times less than the convective Taylor microscale λ of the flow investigated. Since the essential processes of enstrophy production and viscous dissipation scale with λ (Tennekes & Lumley, 1992), it is expected that, if these conditions are fulfilled, the loss of information due to the lack of resolution of the smallest scales will not be dominant.

5. Measuring Technique Performance Tests

Experimental investigations of turbulent flow fields at moderate Reynolds numbers with a 20-wire probe of the type presented in *Fig. 2.6(b)* (Lemonis et al., 1994, Lemonis, 1995) yield satisfactory results confirming the efficiency of the sensors and the data reduction procedures developed. As will be shown in this section, the data obtained in two basic kinds of turbulent flow, i.e. quasi-isotropic turbulent grid flow and turbulent boundary layer flow, agree well with theory and also with other experimental and numerical data known from literature. This is also true for data obtained from measurements with a 20-wire probe of the type presented in *Fig. 2.6(a)* (Dracos et al., 1989, Tsinober et al. 1992, 1993). This shows that the two arrangements of the wires are nearly equivalent. The principal parameters of the flows investigated are listed in TABLE 5.1 and TABLE 5.2.

TABLE 5.1 : Principal parameters of quasi-isotropic turbulent grid flow

U_{inf} [m/sec]	x/M [-]	λ [mm]	Re_λ [-]	η [mm]
6.65	55	8.7	72	0.52

TABLE 5.2 : Principal parameters of turbulent boundary layer flow

U_{inf} [m/sec]	δ [mm]	Re_δ [-]	Re_θ [-]	u_τ [m/sec]
6.43	155	$6.643 \cdot 10^4$	$6.459 \cdot 10^3$	0.225

5.1. STATISTICS OF VELOCITY FLUCTUATIONS IN TURBULENT BOUNDARY LAYER.

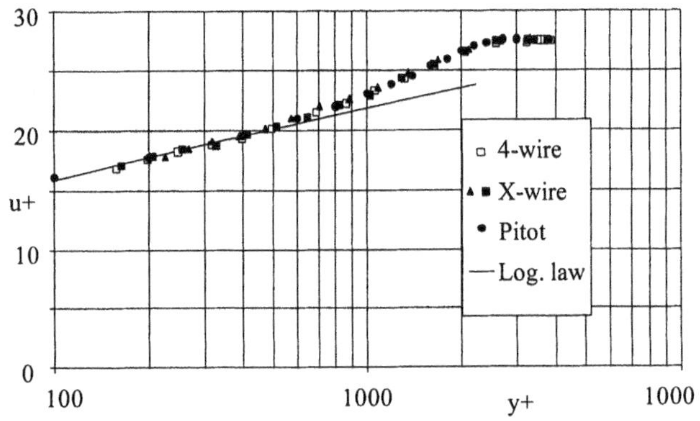

Figure 5.1. Mean velocity profile in boundary layer measured by different probes

The mean velocity profiles in turbulent boundary layer measured with different probes are in mutually good agreement and agree well with the logarithmic law, *Fig. 5.1.* The r.m.s. values of the velocity fluctuations measured by 4-wire probes compare well to measurements with X-wire probes, *Fig. 5.2.* The deviations close to the wall are of the order of max. 3 %, as expected due to the differences in the wire separation between these probe types. The data agree satisfactorily with the measurements of Klebanoff (1954) and Pompeo (1992).

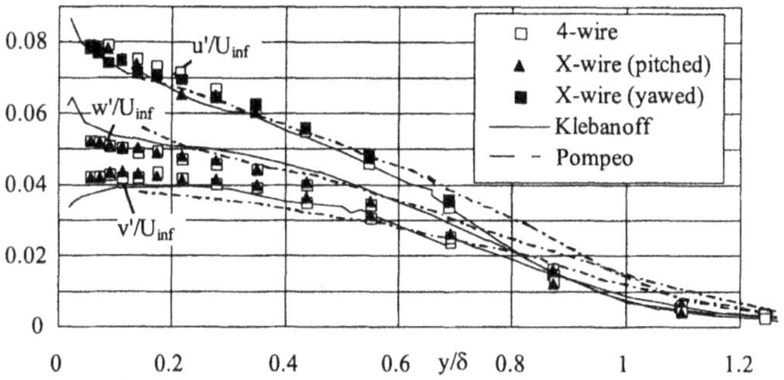

Figure 5.2. Normal stresses measured by X-wire and 4-wire probes

5.2. ISOTROPY RELATIONS IN TURBULENT GRID FLOW

Grid turbulence was generated using a grid with square mesh of M = 6 cm and a solidity factor of 0.44. Grid turbulence produced this way is noticeably anisotropic, with a ratio of the turbulent intensity in the streamwise direction to that normal to it of typically 1.25:1 (Townsend, 1976). However, most of the experimental work known from literature has been analyzed assuming perfect isotropy of grid turbulence.

TABLE 5.3 shows that the results follow generally the theoretical predictions. The correlations involving the streamwise derivatives $\partial u_i/\partial x_1$ agree well with theory as do the correlations involving the spanwise derivatives $\partial u_i/\partial x_2$ and $\partial u_i/\partial x_3$. This is an indication that the application of Taylor's hypothesis for the estimation of streamwise derivatives yields reasonable results. A large part of the deviations is most probably due to the anisotropic nature of the flow. On the average, the deviations from theory are about 8 %, while Tsinober et al. (1992) report deviations between 7 % and 12 % and Kit et al. (1987) of 30%.

TABLE 5.3. Measured relations between r.m.s. of velocity derivatives

$\left\langle\left(\frac{\partial u_i}{\partial x_j}\right)^2\right\rangle^{0.5} : \left\langle\left(\frac{\partial u_1}{\partial x_1}\right)^2\right\rangle^{0.5}$	**The-ory**	**Meas**	**Dev. [%]**	$\left\langle\left(\frac{\partial u_i}{\partial x_j}\right)^2\right\rangle^{0.5} : \left\langle\left(\frac{\partial u_1}{\partial x_1}\right)^2\right\rangle^{0.5}$	**The-ory**	**Meas**	**Dev. [%]**
$\langle(\partial u_2/\partial x_2)^2\rangle^{0.5}$	1.0	1.18	-18.0	$\langle(\partial u_2/\partial x_1)^2\rangle^{0.5}$	$\sqrt{2.0}$	1.45	-2.5
$\langle(\partial u_3/\partial x_3)^2\rangle^{0.5}$	1.0	1.12	-12.0	$\langle(\partial u_2/\partial x_3)^2\rangle^{0.5}$	$\sqrt{2.0}$	1.43	-1.1
$\langle(\partial u_1/\partial x_2)^2\rangle^{0.5}$	$\sqrt{2.0}$	1.50	-6.1	$\langle(\partial u_3/\partial x_1)^2\rangle^{0.5}$	$\sqrt{2.0}$	1.40	1.0
$\langle(\partial u_1/\partial x_3)^2\rangle^{0.5}$	$\sqrt{2.0}$	1.74	-23.0	$\langle(\partial u_3/\partial x_2)^2\rangle^{0.5}$	$\sqrt{2.0}$	1.40	1.0
Mean deviation [%] : 8.1							

5.3. CONTINUITY EQUATION

An important test concerning the reliability of the measurement of the velocity derivatives is the verification of the continuity equation, $\partial u_1/\partial x_1+\partial u_2/\partial x_2+\partial u_3/\partial x_3 = 0$. This check is important, because it employs the derivatives of velocity fluctuations only and thereby characterizes the reliability of measurements of instantaneous velocity gradients.

Measured time series of $\partial u_1/\partial x_1$ were compared to time series of $-(\partial u_2/\partial x_2+\partial u_3/\partial x_3)$ by means of a linear regression model of the form $(\partial u_1/\partial x_1) = C_0 + C_1\cdot(-\partial u_2/\partial x_2-\partial u_3/\partial x_3)$. The intercept was typically between $C_0 = 0.002\cdot(\partial u_1/\partial x_1)'$ and $C_0 = -0.004\cdot(\partial u_1/\partial x_1)'$ and the slope was $C_1 = 1$ +/-0.05. As expected, C_0 is very low while the values of C_1 indicate a lack of the compressibility hypothesis of the order of max. 5 %. The corresponding correlation coefficient

$$R_T = -\frac{\left\langle \left(\frac{\partial u_1}{\partial x_1}\right)\cdot\left(\frac{\partial u_2}{\partial x_2}+\frac{\partial u_3}{\partial x_3}\right)\right\rangle}{\left(\frac{\partial u_1}{\partial x_1}\right)'\cdot\left(\frac{\partial u_2}{\partial x_2}+\frac{\partial u_3}{\partial x_3}\right)'} \tag{5.1}$$

varies between 0.68 and 0.72, in both types of flow investigated, revealing a rather large scatter of the data around the best fit. However, though instantaneously the error in the derivative estimation may be of order 30 %, the reasonable results of intercept and slope obtained indicate that the data are statistically reliable. The values of R_T obtained here are among the highest ever achieved in an experiment. Tsinober et al. (1992) report values of about 0.7 also, while Kit et al. (1987, 1988) obtained $R_T \cong 0.35$ in turbulent grid flow and Balint (1986) $R_T \cong 0.25 - 0.37$ in turbulent boundary layer. Recently, Ong (1992) obtained values of R_T up to 0.75 in turbulent boundary layer. At least in quasi-isotropic turbulence this error is mainly caused by the probe size and can be reduced only by drastically decreasing the probe dimensions. Further investigations are needed to clarify this point.

5.4. STATISTICS OF VORTICITY FLUCTUATIONS IN TURBULENT BOUNDARY LAYER

Measured r.m.s. values of the fluctuating vorticity components, normalized by u_τ^2/ν, are shown in *Fig. 5.3*. They are compared with measurements of Balint et al. (1991) and numerical simulations of turbulent channel flow of Kim et al. (1987). The r.m.s. of the streamwise components are also compared to measurements in turbulent channel flow of Kastrinakis & Eckelmann (1983). The discrepancies in the near-wall region are thought to be caused by the different Reynolds number of the flows. Investigated far from the wall the agreement is better.

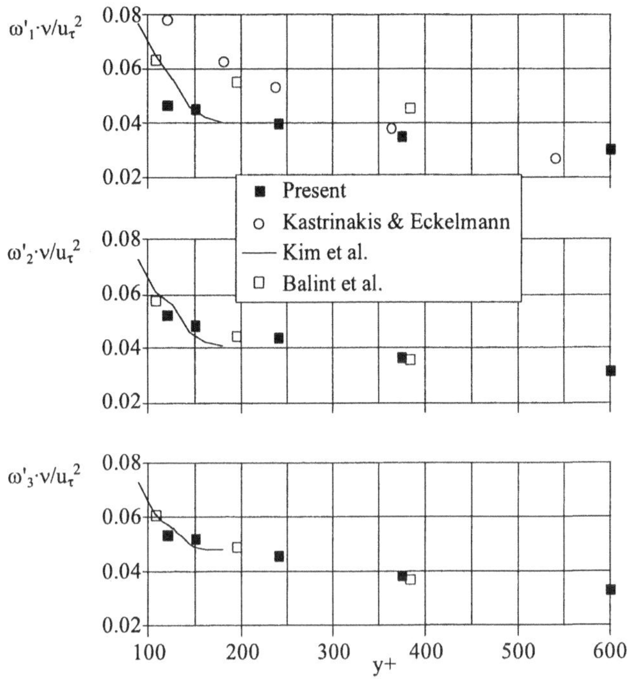

Figure 5.3. Measured r.m.s. of fluctuations of vorticity components normalized with u_τ and ν, compared to measurements of Kastrinakis & Eckelmann (1983, $Re_\theta = 1085$), Balint et al. (1991, $Re_\theta = 2685$) and numerical simulations of Kim et al. (1987, $Re_\theta = 280$)

5.5. VORTICITY ALIGNMENT AND ENSTROPHY PRODUCTION

Our experimental results support the theoretical conclusions and numerical studies concerning the alignment of the vorticity vector with respect to the principal axes of the rate of strain tensor (Betchov, 1956, Ashurst et al., 1987, Jiménez, 1992). Vorticity tends to align with the axis of the intermediate principal strain β and, in accordance with Betchov's assumption, enstrophy production $\sigma = \omega_i \cdot \omega_j \cdot s_{ij} / |\underline{\omega}|^2$ is associated mainly with positive values of β. At negative β's enstrophy is attenuated (*Fig. 5.4*). *Fig. 5.5* shows that the correlation between enstrophy production σ and total strain $|s| = (s_{ij} \cdot s_{ij})^{0.5}$, $s_{ij} = \partial u_i/\partial x_j + \partial u_j/\partial x_i$, is weak, with maximum probability at low values rather than near the extremas. These results are in good agreement with numerical simulation results of isotropic, homogeneous turbulent flow of Jiménez et al. (1992).

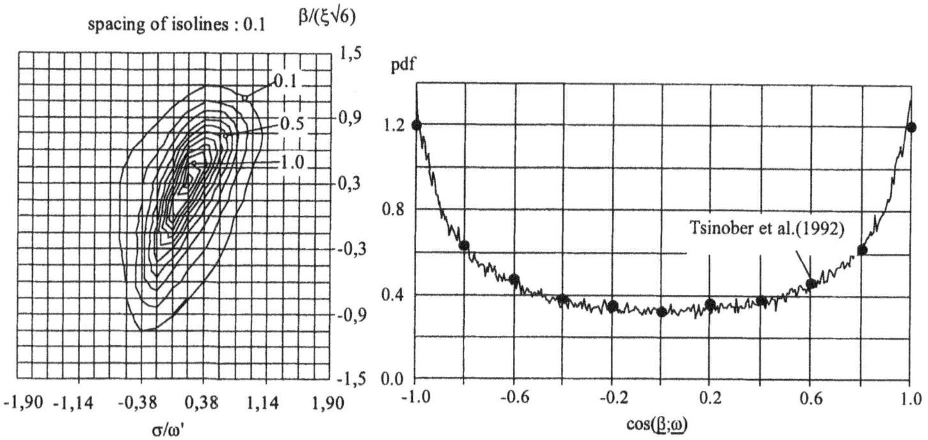

Figure 5.4. Measured joint p.d.f. of enstrophy production versus normalized β-strain and p.d.f.'s of the direction cosine between the vorticity vector and the β-eigenvector of the rate of strain tensor; $\xi = (\alpha^2+\beta^2+\gamma^2)^{0.5}$; $\alpha > \beta > \gamma$: eigenvalues of the rate of strain tensor.

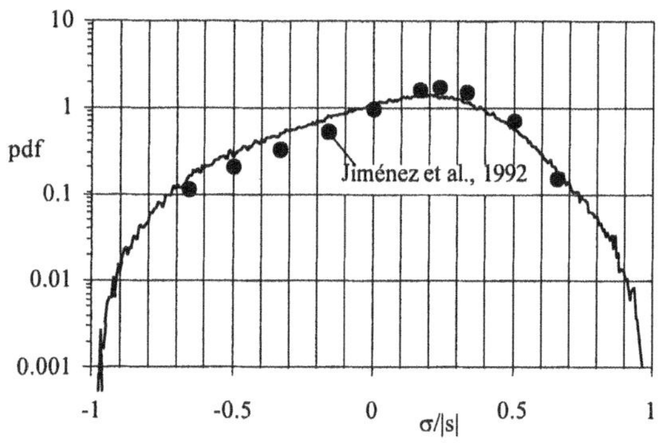

Figure 5.5. Measured probability density function of $\sigma/|s|$

5.6. LENGTH SCALES AND LOCAL DISSIPATION IN TURBULENT BOUNDARY LAYER

The local dissipation rate is calculated from $\langle\varepsilon\rangle = 2\cdot\nu\cdot\langle s_{ij}\cdot s_{ij}\rangle$ and the dimensionless local dissipation rate from $\varepsilon^+ = \langle\varepsilon\rangle\cdot\nu/u_\tau^4$. The convective Taylor microscale is estimated by $\lambda = (\langle u_1^2\rangle/\langle(\partial u_1/\partial x_1)^2\rangle)^{0.5}$ and $\lambda^+ = \lambda\cdot u_\tau/\nu$, and the Kolmogorov microscale by $\eta = (\nu^3/\langle\varepsilon\rangle)^{1/4}$ and $\eta^+ = \eta\cdot u_\tau/\nu$. It is useful to study the dependence of these quantities on the microscale Reynolds number Re_λ , *Fig. 5.6* and *Fig. 5.7*. Here, we could perform a direct comparison to numerical simulations results of Jiménez et al. (1992). It is noteworthy that the data of Jiménez et al. were obtained in a numerical simulation of homogeneous, isotropic turbulence. The results of the grid turbulence measurements have also been plotted in these figures. We see that in all cases the data agree well, especially at high Reynolds numbers. Evidently, these quantities are, at least at high values of Re_λ, independent of the origin of turbulence, an indication of local isotropy. Generally, we recognize asymptotic trends at large Reynolds numbers.

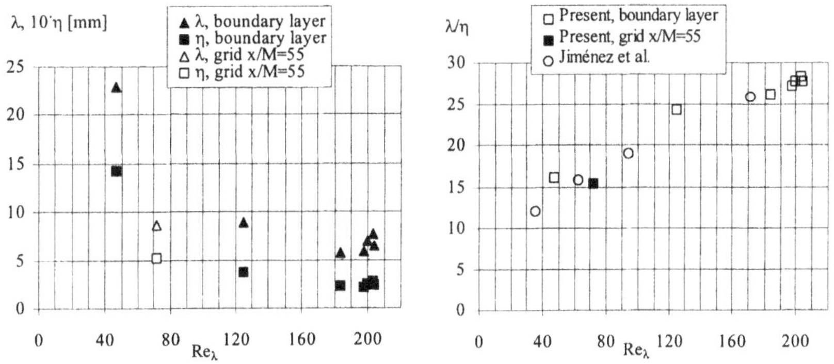

Figure 5.6. Dependence of Taylor microscale λ, Kolmogorov microscale η and of their ratio on the microscale Reynolds number, compared to numerical simulations of Jiménez et al. (1992)

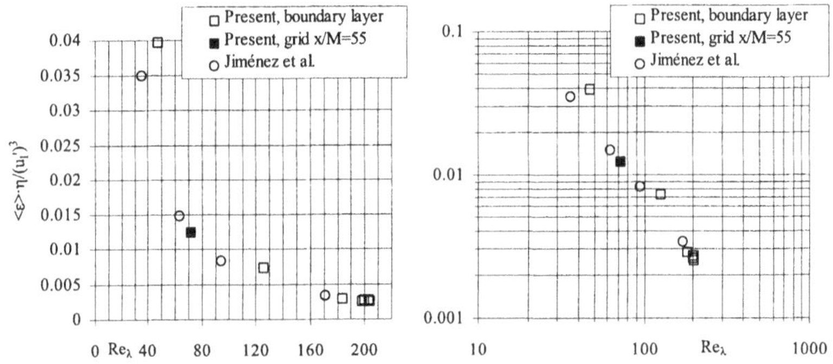

Figure 5.7. Dependence of normalized local dissipation on the microscale Reynolds number, compared to numerical simulations of Jiménez et al. (1992)

6. Concluding Remarks

Recent developments in hot-wire anemometry make possible the measurement of the three components of the velocity vector and of all their first derivatives with satisfactory accuracy. The construction of miniature probes with up to 20 wires is however a major undertaking. With the appropriate equipment and some experience a 20-wire probe can be constructed in two or three weeks. If properly constructed, such a probe is very robust and can even withstand operation in the free atmosphere.

Measurements of quantities depending on first derivatives of the velocity components are mostly in good agreement with theoretical results or with results of direct numerical simulations. This shows that the accuracy with which the derivatives are determined with such probes is better than any reached before. However, further improvements are needed.

Although a big effort was made to reduce as much as possible the dimensions of the probe, its size is still to large for reliable derivative measurements of the velocity in high velocity and Reynolds number flows. These probes are also too large for measurements in the vicinity if a wall.

The experience made with 2.5 μm diameter sensing wires makes one hope that a further reduction of the probe size is possible. If wires of 1.25 μm diameter are used, the dimensions of a probe can be halved, provided the diameter of the prongs can also be reduced to half their actual diameter.

The increase in spatial resolution of the probe must be combined with an increase in the time resolution, i.e. with an increase of the sampling frequency. High quality electronic equipment with very low signal to noise ratio, good calibration facilities and careful calibration of the probes are to much higher degree prerequisites for precise measurements of velocity derivatives than for precise velocity measurements.

At present multi-channel hot-wire anemometer equipment is commercially available from companies like TSI, DANTEC, A&A LAB SYSTEMS and others. The general trend is to make use of digital techniques and simplify and automate the tuning of the bridges and speed up the calibration of the probes. Crucial properties of anemometers are, however, good stability and a very low noise-to-signal ratio. An example of the recent developments in this field will be presented in the next contribution.

References

Abdel-Rahman, A., Tropea, C., Slawson, P. & Strong, A. (1987). On temperature compensation in hot-wire anemometry. J. Phys. E: Sci. Instrum. **20**, pp. 315.

Antonia, R. A., Shah, D. A. & Browne, L. W. B. (1987). Spectra of velocity derivatives in a turbulent wake. Phys. Fluids **30** (11), pp. 3455.

Antonia, R. A., Shah, D. A. & Browne, L. W. B. (1988). Dissipation and vorticity spectra in a turbulent wake. Phys. Fluids **31**, pp. 1805.

Ashurst, W. T., Kerstein, A. R., Kerr, R. M. & Gibson, C. H. (1987). Alignment of vorticity and scalar gradient with strain rate in simulated Navier-Stokes turbulence. Phys. Fluids **30** (8), pp. 2343.

Balint, J.-L. (1986). „Contribution á l'étude de la structure tourbillonnaire d'une couche limité turbulente au moyen d'une sonde a neup fils chauds mesurant le rotationnel". PhD Thesis, University Claude Bernard-Lyon.

Balint, J.-L. , Wallace, J. M. , & Vukoslavĉević, P. (1991). The velocity and vorticity vector fields of a turbulent boundary layer. Part 2. Statistical properties. J. Fluid Mech. **228**, pp. 53.

Batchelor, G. (1960). "An Introduction to Fluid Mechanics", Cambridge Univ. Press, U.K.

Batchelor, G. (1967). "The Theory of Homogeneous Turbulence", Cambridge Univ. Press, U.K.

Bearmann, P. W. (1971). Corrections for the effect of ambient temperature drift on hot-wire measurements in incompressible flow. DISA Inf. **11**.

Betchov, R. (1956). An inequality concerning the production of vorticity in isotropic turbulence. J. Fluid Mech. **1**, pp. 497.

Bruun, H. H. & Tropea, C. (1985). The calibration of inclined hot-wire probes. J. Phys. E: Sci. Instrum. **18**, pp. 405.

Chew, Y. T. & Ha, S. M. (1988). The directional sensitivities of crossed and triple hot-wire probes. J. Phys. E: Sci. Instrum. **21**, pp. 613.

Collis, D. C. & Williams, M. J. (1959). Two-dimensional convection from heated wires at low Reynolds numbers. J. Fluid Mech. **16**, pp. 357.

Döbbeling, K. (1990). "Experimentelle und theoretische Untersuchung an stark verdrallten, turbulenten isothermen Strömungen". PhD Thesis, TU Karlsruhe.

Dracos, T., Kholmyansky, M., Kit, E. & Tsinober, A. (1989). Some experimental results on velocity-velocity gradients measurements in turbulent grid flows. Proc. of the IUTAM Symp. on Topological Fluid Mechanics, Cambridge Univ. Press.

Foss, J. F. (1976). Accuracy and uncertainty of transverse vorticity measurements. Bull. of the Am. Phys. Soc. **21**, pp. 1237.

Foss, J. F. (1981). Advanced techniques for transverse vorticity measurements. Proc. of the 7th Biennial Symp. of Turbulence, Univ. of Missouri-Rolla.

Hinze, J. O. (1975). „Turbulence". McGraw Hill.

Idelchik, I. E. (1986). „Handbook of Hydraulic Resistance". Springer-Verlag.

Jiménez, J., Wray, A. A, Saffman, P. G. & Rogallo, R. S. (1992). The structure of intense vorticity in homogeneous isotropic turbulence. CTR, Proc. of the Summer Progr. 1992.

Jiménez, J. (1992). Kinematic alignment effects in turbulent flows. Phys. Fluids A **4** (4), pp. 652.

Jørgensen, F. E. (1971). Directional sensitivity of wire and fiber-film probes. DISA Inf. **11**.

Kastrinakis, E. G. & Eckelmann, H. (1983). Measurement of streamwise vorticity fluctuations in a turbulent channel flow. J. Fluid Mech. **137**, pp. 165.

Kastrinakis, E. G., Nychas, S. G. & Eckelmann, H. (1984). Vorticity and velocity measurements in a fully developed turbulent channel flow. Turbulence and Chaotic Phenomena in Fluids (ed. T. Tatsumi), IUTAM.

Kavence, G. & Oka, S. (1973). Correcting hot-wire readings for influence of fluid temperature variations. DISA Inf. **15**.

Kim, J., Moin, P. & Moser, R. (1987). Turbulence statistics in fully developed channel flow at low Reynolds number. J. Fluid Mech. **177**, pp. 133.

King, L. V. (1914). On the convection of heat from small cylinders in a stream of fluid. Determination of convective constants of small platinum wires with application to hot-wire anemometry. Phil. Trans. A **214**, pp. 373.

Kit, E., Tsinober, A., Balint, J.-L., Wallace, J. M. & Levich, E. (1987). An experimental study of helicity related properties of a turbulent flow past a grid. Phys. Fluids **30** (11), pp.3323.

Klebanoff, P. S. (1954). Characteristics of turbulence in a boundary layer with zero pressure gradient. NACA TN 3178.

Kovasznay, L. S. G. (1954). Turbulence measurements. Physical Measurements in Gas Dynamics and Combustion (**10**), Princeton Univ. Press, pp. 213.

Kramers, H. (1946). Heat transfer from spheres to flowing media. Physica **12**, pp. 61.

Lasser, D. (1987). „Bernstein-Bézier-Darstellung trivariater Splines". PhD. Thesis, TU Darmstadt.

Lemonis, G., Dracos, T. & Tsinober, A. (1994). Velocity gradient depending quantities in turbulent grid flow. Proc. of the 5th European Turbulence Conference (ed. E. Benzi), Kluwer Acad. Publ.

Lemonis, G. & Dracos, T. (1995). A new calibration and data reduction method for turbulence measurements by multi-hotwire probes. Exp. Fluids **18**, pp. 319.

Lemonis, G. (1995). "An Experimental Study of the Vector Fields of Velocity and Vorticity in Turbulent Flows". PhD Thesis, ETH Zurich.

Lomas, C. G. (1986). „Fundamentals of hot-wire anemometry". Cambridge Univ. Press.

Mi, J. & Antonia, R. A. (1994). Vorticity characteristics of the turbulent intermediate wake. Proc. of the 5th European Turbulence Conference (ed. E. Benzi), Kluwer Acad. Publ.

Monin, A.S. & Yaglom, A.M. (1971). "Statistical Fluid Mechanics", Vol. **1.2**, MIT Press, Cambridge Mass., U.S.A.

Müller, U. R. (1987). Comparison of turbulence measurements with single, X- and triple hot-wire probes. Berichte VDI **121**, pp. 62.

Ong, L. (1992). „Visualization of Turbulent Flows with Simultaneous Velocity and Vorticity Measurements". PhD thesis. Univ. Maryland, College Park.

Pailhas, G. & Cousteix, J. (1986). Method for analyzing four-hot-wire probe measurements. Rech. Aerosp. 1986-2, pp. 79.

Perry, A. E. (1982). „Hot-wire anemometry". Clarendon Press, Oxford.

Piomelli, U., Balint, J.-L. & Wallace, J. M. (1989). On the validity of Taylor's hypothesis for wall-bounded turbulent flows. Phys. Fluids **A1**, pp. 609.

Pompeo, L. (1992). „An experimental study of three-dimensional turbulent boundary layers". PhD. Thesis, ETH Zurich.

Rosemann, H. (1989). Einfluß der Geometrie von Mehrfach-Hitzdrahtsonden auf die Meßergebnisse in turbulenten Strömungen. DLR-FB 89-26.

Samet, F. & Einav, S. (1987). A hot-wire technique for simultaneous measurement of instantaneous velocities in 3-D flows. J. Phys. E: Sci. Instrum. **20**, pp.683.

Taylor, G. I. (1935). Statistical theory of turbulence. Proc. Roy. Soc. London, A **151**, pp. 421.

Tennekes, H. & Lumley, J. L. (1992). „A First Course in Turbulence". MIT Press.

Townsend, A. A (1976). „The Structure of Turbulent Shear Flow". Cambridge Univ. Press.

Tsinober, A., Kit, E. & Dracos, T. (1992). Experimental investigation of the field of velocity gradients in turbulent flows. J. Fluid Mech. **242**, pp. 169.

Tsinober, A., Kit, E. & Dracos, T. (1993). Velocity gradients in a turbulent jet flow. Applied Scientific Research **51**. Advances in Turbulence IV (ed. F.T.M. Nieuwstadt). Kluwer Acad. Publ., pp. 185.

Vukoslavĉević, P. , Wallace, J. M. & Balint, J.-L. (1991). The velocity and vorticity vector fields of a turbulent boundary layer. Part 1. Simultaneous measurement by hot-wire anemometry. J. Fluid Mech. **228**, pp. 25.

Wagner, T. C. & Kent, J. C. (1988). On the directional sensitivity of hot-wires: a new look at an old phenomenon. Exper. Fluids **6**, pp. 553.

Wallace, J. M. (1986). Methods of measuring vorticity in turbulent flows. Exper. Fluids **4**, pp. 61.

Wallace, J. M. & Ong. L. (1994). Local isotropy of the vorticity field in a boundary layer at high Reynolds numbers. Proc. of the 5th European Turbulence Conference (ed. E. Benzi), Kluwer Acad. Publ.

Wyngaard, J. C. (1969). Spatial resolution of the vorticity meter and other hot wire arrays. J. Phys. E: Sci. Instrum. **2**, pp. 983.

THE MULTICHANNEL CONSTANT TEMPERATURE ANEMOMETER

Recent Developments and Use in 2- -and 3-Dimensional Flows.

F.E. JØRGENSEN
Dantec Measurement Technology
Tonsbakken 16-18
DK-2740 Skovlunde
Denmark

1. Introduction

Multichannel measurements are needed in many applications in fluid dynamic research. The constant temperature anemometer (CTA) is often the obvious choice of instrument, when especially time series are of interest, as it provides a continuos signal of high temporal resolution. Basically multichannel measurements are employed, when information about two or more parameters are needed simultaneously in order to describe the flow. Each parameter is represented by a continuos signal from a separate sensor operated from its own CTA-channel. The operation of multichannel anemometers have traditionally been considered a time consuming task, as it includes optimization and setup of a large number of parameters including both velocity and directional calibrations. In recent years computer controlled anemometers have become available that allows automatic setup and calibration of multiple sensor probes. Multichannel CTA measurements have thus become considerably more efficient and reliable.

2. The Computer Controlled Constant Temperature Anemometer

This concept has been implemented in Dantecs StreamLine® system that is a multichannel, high performance research type anemometer with a dedicated application software package. It supports thermal anemometers, both hot-wire and cold-wire anemometers, including necessary calibration- and traversing systems, and works together with other sensor systems for measurement of for example temperature and pressure.

2.1 SYSTEM OVERVIEW

The StreamLine system consists of a Frame with power supplies and a Controller. The interface to the surroundings is very simple. The setup of the CTA is done from the PC via a serial port connected to the Controller, while the analog CTA outputs are

Th. Dracos (ed.), Three-Dimensional Velocity and Vorticity Measuring and Image Analysis Techniques, 43–58.

acquired via an A/D board in the PC. In this way the anemometer may still be considered an analog instrument fully open to any means of data logging or analog Online data analysis.

This open structure has been chosen instead of for example implementing an A/D converter in the system, or transferring data via a bus structure in order to give the user more freedom and make the anemometer system compatible with as many data collection systems as possible.

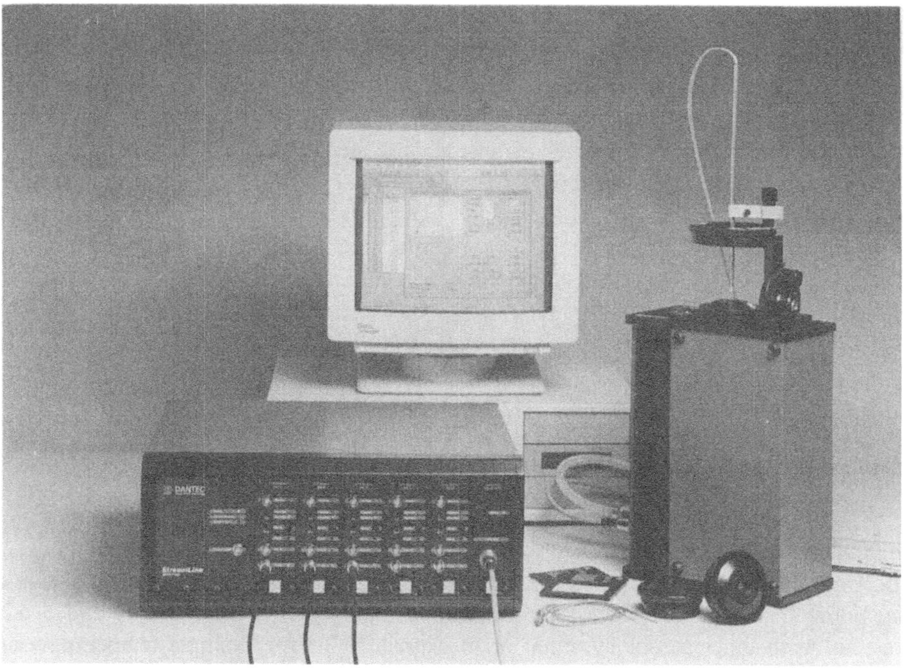

Fig. 1. The computer controlled StreamLine® System with Calibrator.

2.2 HARDWARE IMPLEMENTATION

2.2.1 Frame and Controller

The hardware consists of a Frame with power supplies and mounting slots for 6 anemometer modules. In addition it contains a Controller for communication purposes with the PC. A temperature probe can be connected to the Controller for measurement of ambient temperature, which is a significant influence parameter in CTA measurements. The Frame has 6 analog output bushings from the modules.

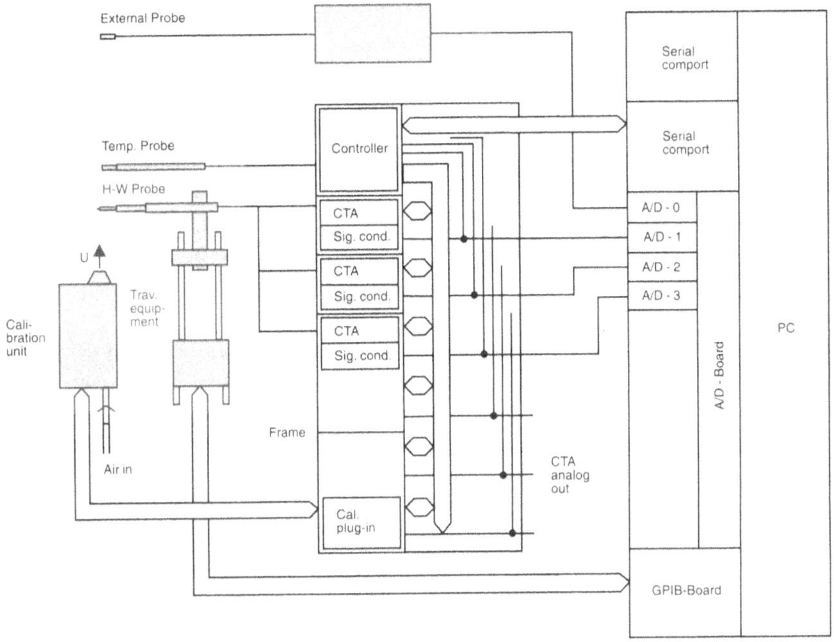

Fig. 2. StreamLine system layout.

The Controller is the gateway from the PC to the anemometer modules installed in the Frame. It communicates with the PC via a standard RS3232D serial interface and it transfers and stores setup data for the modules, samples and returns high speed data for square wave testing of the CTA bridge, samples and returns low speed data for example ambient temperature, and finally controls the configuration and reports errors. The Controller is primarily used for setup and adjustment of the modules and the system can actually run without connection to a PC, once it has been set up via the Controller.

2.2.2 CTA Module with Signal Conditioner

Each CTA module contains the traditional parts: bridge, cable compensation circuits, prefilter, amplifier shaping circuits, servo loop and power source. The bridge can be configured with 1:20 or 1:1 bridge ratio for standard and high frequency work, respectively. The top resistor can be selected to match the power needed by the probe in a specific application. The decade resistance that defines the overheat of the probe is automatically adjusted on basis of a probe resistance measurement. The dynamic bridge balancing is optimized by sampling the probe/bridge response to an imposed square wave signal by means of a fast transient recorder in the Controller.

Before data acquisition low pass filtering of the probe signal must be carried out in order to avoid aliasing, and it is often desirable to amplify the signal in order to utilize the full A/D conversion range. For this purpose the CTA module has a built-in Signal conditioner with programmable highpass and lowpass filters and signal offset and gain. All settings in both the anemometer and the Signal conditioner are carried out by commands from the PC. In most cases it is not needed to perform special optimization of the settings - the factory defaults from the probe database will suffice.

2.2.3 Calibration System.

The calibration system is designed for automatic calibration of probes in air or in other gases. It consists of a Calibration module that plugs into the Frame and is connected via a cable to a Flow Unit that can be placed up to 100 m away from the Frame. It operates from compressed air and creates a free jet with velocities ranging from 0.02 m/s up to Mach 1. The jet profile is very flat and has a very low turbulence intensity, typically around 0.2%.

The velocity in the jet is calculated using the isentropic relations for the expansion through the jet orifice. The parameters are the true differential pressure across the jet and the total temperature and pressure in the stagnation chamber together with the thermodynamic property values for the calibration gas. A high absolute accuracy in calculated velocity is achieved by having he pressure and temperature transducers calibrated against national standards. An individual calibration certificate for these transducers are delivered with each calibrator.

Directional calibration of X-probes and Tri-axial probes can be performed with a pitch-yaw manipulator that allows probe rotation around two axis. The manipulator can be mounted across the jet.

2.3 SOFTWARE IMPLEMENTATION

The application software, StreamWare®, is operating in the Windows environment and is designed to take care of all actions from hardware setup and probe calibration to execution of experiments with acquisition, conversion and reduction of data. In order to setup and run an experiment often more than 100 parameters have to be decided upon and kept track of. This is only possible by utilizing database storage technique.

2.3.1 Software functionality.

When a probe is selected from the probe database and assigned to an anemometer channel connected to an A/D channel in the PC, the anemometer is automatically adjusted accordingly, and the user can immediately set the overheat, read in data and view them Online on the PC screen. Optimum accuracy can be achieved by fine tuning the anemometer and its build-in signal conditioner from the PC screen. An automatic probe calibration procedure creates the necessary transfer functions needed for conversion of voltages into velocity components. After probe calibration velocity components can be presented Online.

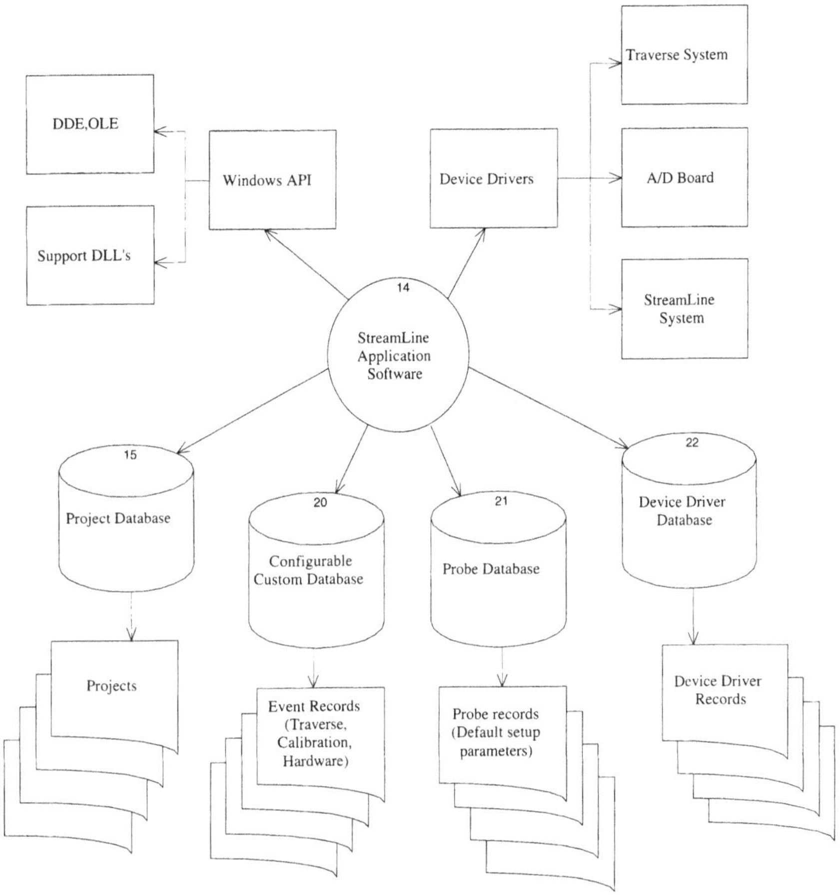

Fig. 3. General structure of the StreamWare Application software.

Raw or reduced data are saved in databases providing easy unpacking and presentation. The presentation is done in data sheets, from where they can be copied via the clipboard to other applications.

Experiments are supported by an experiment layout manager, where the user designs his experiment in the form of block diagrams that define anemometer setup, probe traverse, data acquisition and conversion/reduction routines. The experiment is then performed by a single keystroke that starts the chain of events defined by the block diagrams.

All setups, calibrations, experiments, raw and reduced data are time stamped and saved in a project database. In this way full information about all parameters, actions and data is always present.

2.3.2 Database design.

The software is centered around 4 databases: Probe database, Device database, Function database and Project database. In short a database consists of records of fixed or variable length, and the purpose of the database is to manage a list of such records.

- *Probe database.* Contains a list of standard probes with all the parameters needed to setup and adjust the anemometer. It also contains information about probe geometry's needed for velocity decomposition during data conversion.
- *Device database.* Contains a list of translation drivers that are needed in order for StreamWare to communicate with external devices like A/D boards.
- *Function database.* Contains a list of data reduction programs that allows data to be processed outside StreamWare and to be stored back again into the StreamWare processing environment.
- *Project database.* The first record in the project database is the project definition with the entire physical setup including device configuration and storage of the factory defaults software variables. It is always updated to current status and changes are stored as events following the main record. All records are stamped with identification, type, date and time.

2.2.3 System Configuration

Each project is based on a System configuration that defines the probes and their connections to the Frame. It also defines the connection of the analog outputs from the CTA modules to the A/D board. The configuration is an inherent part of the project

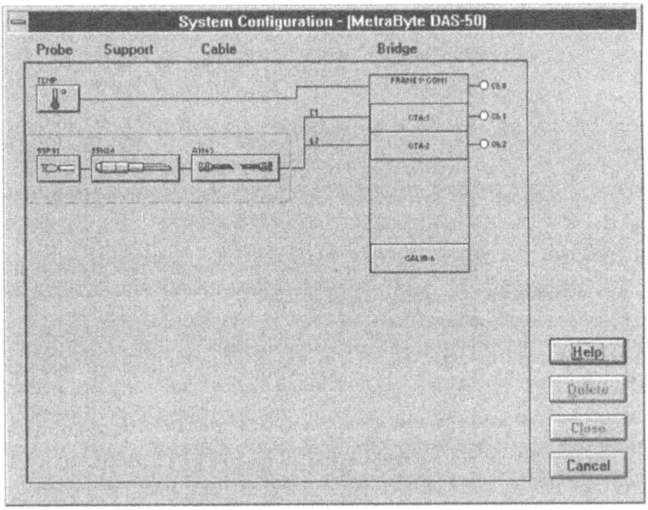

Fig. 4. Configuration of a StreamLine system with an X-probe and an ambient temperature probe for temperature compensation purpose.

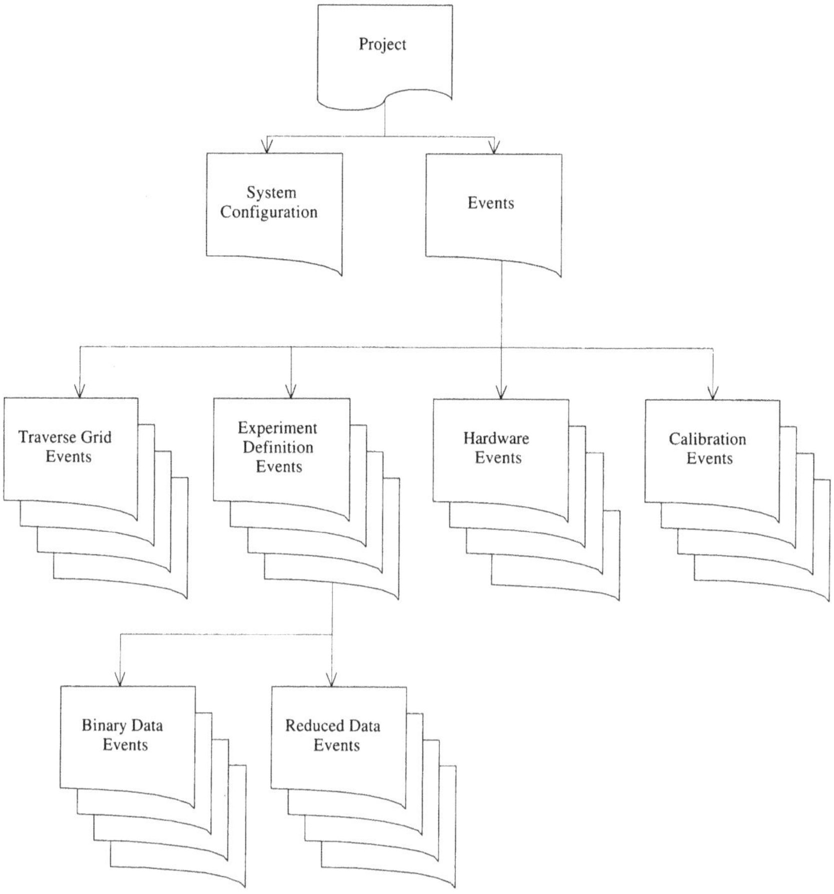

Fig.. 5. Event structure

2.2.3 Event Structure

All actions in a project are named events and contain specific parameters and links to other events. In this way an event structure is created that requires minimum of space and eliminates the need for handling individual files. As the size of a database record is limited raw and reduced data are not placed in databases but in separate files, however organized as databases, linked to the event structure.

Probe specific events	*Experiment specific events*
Hardware setup	Experiment setup
Velocity calibration	Traverse grid
Directional calibration	Raw data
Data conversion/reduction setup	Reduced data

2.2.5 Hardware setup.

The Hardware setup adjusts the overheat and checks the dynamic balancing of the bridge by means of a square wave test. All setup parameters are taken as defaults from the Probe library, and only if the maximum specifications of the system, for example with respect to bandwidth, have to be reached, individual tuning is required. The Signal conditioner can be optimized to the A/D board input range by applying an offset and a gain to the bridge output.

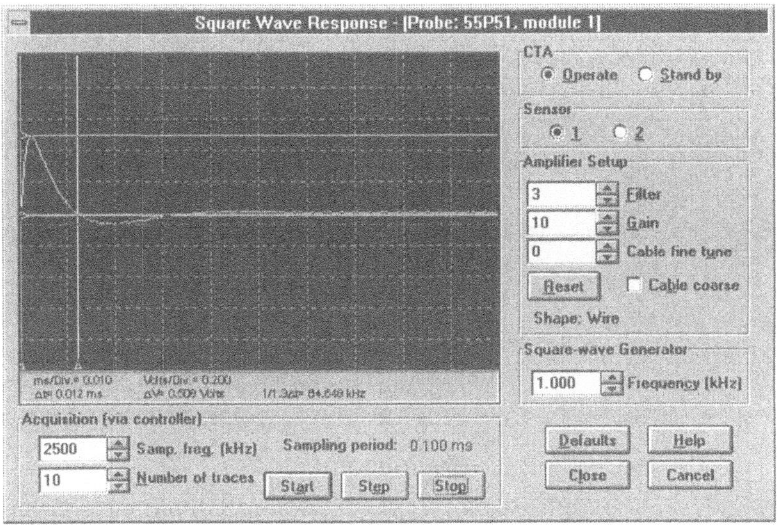

Fig. 6. Square wave test for a gold-plated probe at 20 m/s. The optimum settings provide a bandwidth for the probe and anemometer of approx. 55 kHz.

2.2.6 Experiment layout

The Experiment layout feature provides the user with a tool for designing and documenting his experiment. It is a subprogram that sets up the anemometer, moves a probe around, acquires data, converts them to physical units and saves them in files in raw and reduced form with reference to the experiment. The data acquisition is part of a data flow sequence during which data are exchanged with the surroundings, this means it is possible not only to acquire data but also to transmit data to an external device or make the data acquisition dependent on an external parameter.

2.2.7 Data presentation and export.

Data are presented in datasheets as raw or reduced data in volts or converted to physical units. A graphics editor provides plotting routines. Data and graphs can be pasted into other Windows applications via the clipboard. An optional export routine unpacks the data files and exports them for non-Windows applications, when further processing is needed.

3. Multisensor Probes

Probe Selection and Calibration, Conversion and Decomposition of Probe Signals.

3.1 PROBE SELECTION.

There exist a wide variety of probes for measurement in 2- and 3-dimensional flows. The most common design provides one cylindrical sensor (wire or fiber-film) per velocity component. These probes, however, have a limited useful angular range due to the directional ambiguity of cylindrical sensors. The angular range can be extended by using probes with extra wires, split-fiber probes or by applying other measuring techniques as for instance pulsed wires or flying hot-wires.

Table 1. Multisensor Probe selection versus application

2-D flows	0 - 90°	0 - 180°	0 - 360°
	X-probe	Split-fiber probe* Flying X-probe*	Triple-split fiber probe* Pulsed wire*
3-D flows	70° cone	90° cone	full space
	Triaxial probe	Special 4-wire array*	Split-fibers on Triaxial probe *

Probes denoted with an (*) are not supported by commercially available data conversion software.

3.2 X-PROBES

X-probes have two wires mounted perpendicular to each other normally separated by a wire length in order to avoid thermal interference between the wires. Standard probes have a spatial resolution of about 1x1 mm, special versions down to 0.2x0.2 mm. They are available as miniature probes with the full length of the wires acting as sensors and as gold-plated probes with the wire ends copper -and gold-plated in order to reduce the prong interference.

3.2.1 X-Probe Velocity Calibration

Velocity calibration of X-Probes are carried out for the two wires simultaneously. The probe is placed in a known low-turbulence flow parallel with the probe prongs, i.e. both wires form 45° with the flow vector. The velocity is adjusted to normally 10 to 20 values over the expected range to occur in the experiment. If high turbulence intensities are expected, i.e. the velocity vector moves over the entire useful angular range, it is advised to select the lowest calibration velocity as 10% of the expected minimum velocity. This velocity will correspond to the situation where the velocity vector is parallel with the air.

The calibration points should be selected with a logarithmic distribution, as this will give a better accuracy of the curve fit than an even distribution will.

When the calibration is finished a curve fitting is carried out. A vast number of curve fitting algorithms have been proposed during the history of Anemometry starting with the power law proposed by L.V. King in 1914, later expanded with more terms to increase the accuracy over a wider velocity range. Power laws were especially suited for analog processing in Linearizers and provided a real time linearization of probe voltages. With the introduction of PC's polynomials (4th or 5th order) and look-up tables have become quite popular. Polynomials should however be used with care, as they may oscillate and behave strange outside the calibration range. Normally the

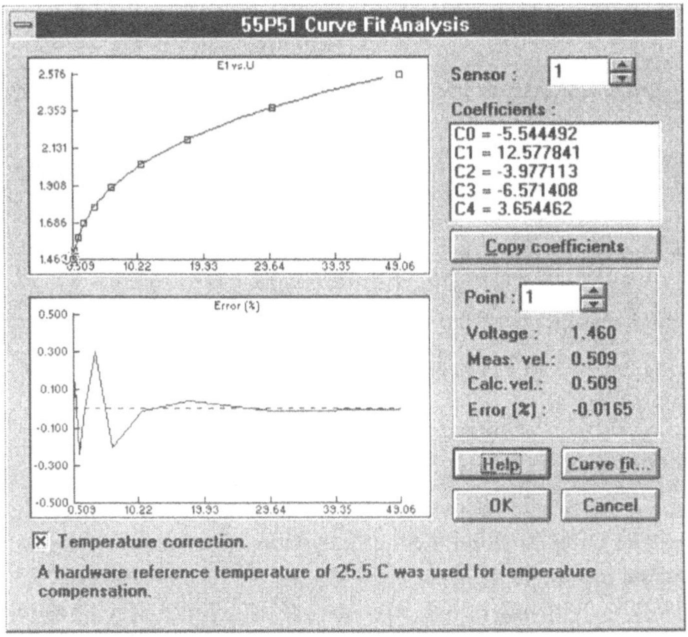

Fig. 7. Velocity calibration of X-probe. Plot of calibration points, transfer function and linearization errors.

difference in linearization accuracy is not significant and the selection of transfer function is often a matter of taste.

The StreamLine application software offers 2 curve fits:

Power law (King): $E^2 = A + B \cdot U^n$

Polynomial: $E = C0 + C1 \cdot U + C1 \cdot U^2 + C1 \cdot U^3 + C1 \cdot U^4 + ..$

and a Look-up table.

3.1.2 X-Probe Directional Calibration

The directional response of X-probes can be described by means of the effective cooling velocities U_{eff1} and U_{eff2} sensed by the two wires, the actual velocities U1 and U2 in the wire coordinate system and the yaw factors k1 and k2:

$$U_{eff1}^2 = k1^2 \cdot U1^2 + U2^2$$
$$U_{eff2}^2 = U1^2 + k2^2 \cdot U2^2$$

The directional calibration is carried out by rotating the probe in a flow of known velocity. In each angular position the probe voltages are measured and converted

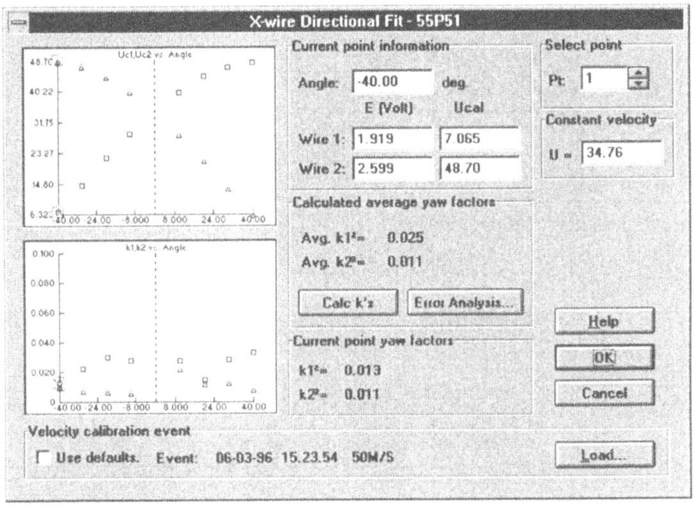

Fig. 8. Directional calibration of an X-probe showing the response of the two wires with respect to +/-40° yaw angle variations. The k-factors are plotted as function of the yaw angle.

directly to effective cooling velocity. If the velocity calibration is carried out at α_1 and α_2 equal to 45, as recommended, the effective cooling velocity can be expressed by means of the calibration velocity Ucal from the transfer function:

$$U_{eff}^2 = U_{cal}^2 \cdot (1+k^2) \cdot \cos^2 45$$

U1 and U2 are the projection of the actual velocity U0 onto the sensors:

$$U1 = U0 \cdot \cos \alpha_1$$
$$U2 = U0 \cdot \cos \alpha_2$$

The effective cooling velocity and the projections of the actual velocity are inserted into the response equations, which are then solved with respect to k1 and k2. This gives one set of k-values for each angular position. Normally the average over the entire angular range is used for k1 and k2. The k-factors varies to some extent with velocity. Only if very high accuracy is required, directional calibrations should be carried out over the entire velocity range. Often individual directional calibration may be omitted and the default values for k1 and k2 used instead.

3.1.3 Conversion and Decomposition of X-probe Data.

Conversion into calibration velocity is done by inserting the voltages from the two sensors into their respective velocity transfer functions.

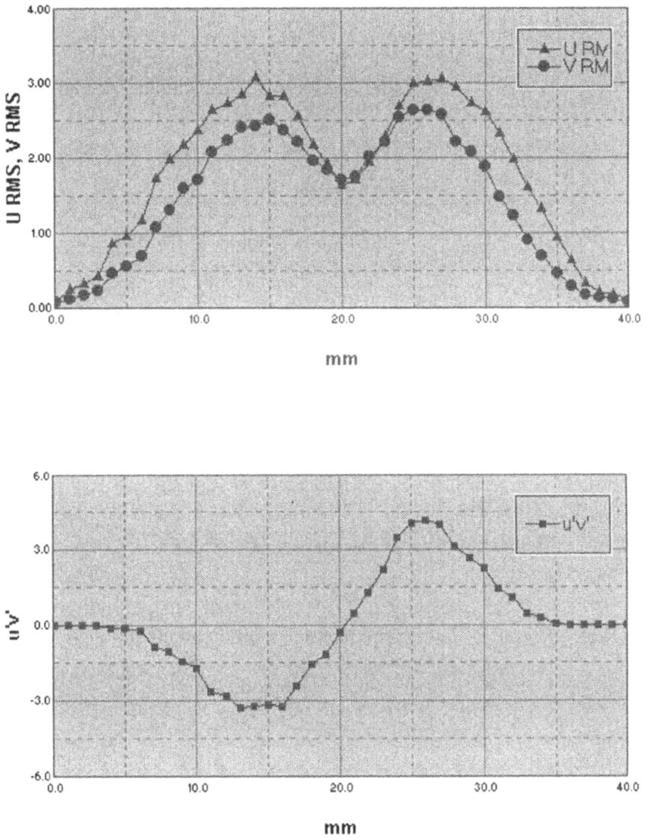

Fig. 9. Rms and uv-profiles in a free jet measured with an X-probe.

Decomposition is the process of finding the velocity components in the direction of the two sensors (wire coordinate system) and finally the components in the probe coordinate system, which is rotated 45° with respect to the wire coordinate system.

The directional response equations are first solved with respect to U1 and U2 for known values of Ueff1, Ueff2 and k1, k2. The velocity components U and V are then calculated as:

$$U = U1 \cdot \cos 45 + U2 \cdot \cos 45$$
$$V = U1 \cdot \sin 45 - U2 \cdot \sin 45$$

When the raw data have been acquired, each set of samples is linearized and decomposed into U and V. They are then ready for reduction into statistical quantities, for example urms, vrms and the cross product uv as shown in Fig. 9.

3.2 TRIAXIAL-PROBES

Triaxial-probes have 3 sensors placed in an orthogonal system. They are used for simultaneous measurement of the 3 velocity components U, V and W in 3-dimensional flows. Due to the directional ambiguity of the sensors the probes are restricted to flows, where the resulting velocity vector stays within the cone of approx. 70° defined by the 3 sensors. Standard configurations have gold-plated wire sensors with a spatial resolution around 2 mm.

3.2.1 Triaxial-Probe Velocity Calibration

Velocity calibration of Triaxial probes are carried out as in the case of X-probes. The probe is placed with its axis parallel to the flow (sensor inclination is 35.3°) and the voltages from the 3 sensors are acquired simultaneously. Related values of flow velocity and probe voltages form basis for curve fitting.

3.2.2 Triaxial-Probe Directional Calibration

The directional response of Triaxial-probes can be described by means of the effective cooling velocities U_{eff1}, U_{eff2} and U_{eff3} sensed by the 3 wires, the actual velocities U1, U2 and U3 in the coordinate system formed by the wires and the yaw and pitch factors k1,2,3 and h1,2,3:

$$Ueff1^2 = k1^2 \cdot U1^2 + U2^2 + h1^2 \cdot U^3$$
$$Ueff2 = h2^2 \cdot U1^2 + k2 \cdot U2^2 + U3^2$$
$$Ueff3 = U1^2 + h3^2 \cdot U2^2 + k3^2 \cdot U^3$$

The pitch factor h describes the effect, when the velocity vector attacks the wire perpendicular to the wire/prong plane. In this situation the heat transfer to the prongs are increased whereby the wire senses an effective cooling velocity bigger than the actual velocity. The pitch factor h is therefore normally bigger than 1.

A straightforward method for directional calibration is to rotate the probe around it axis in a plane velocity field with the axis inclined at an angle θ (normally 30°) with the flow. The probe voltages are then acquired as function of roll angle ω.

The effective cooling velocity is related to the calibration velocity U0 from the transfer function by means of the yaw and pitch factors and the inclination angle 35.3° at which the wires were calibrated:

$$Ueffi^2 = Ucal^2 \cdot (1 + ki^2 + hi^2) \cdot \cos^2 35.3°, \qquad i = 1,2,3$$

The velocities U1, U2 and U3 are the projection of the velocity U0 on the wires:

$$U1 = U0 \cdot \cos\theta,\ U2 = -\ U0 \cdot \sin\theta \cdot \cos\omega,\ U3 = U0 \cdot \sin\theta \cdot \sin\omega$$

The values for Ueff and U1, U2 and U3 are inserted into the response equations, which are then solved with respect to k1, k2, k3 and h1, h2 and h3 for each roll angle. The average of the calculated values of k and h are then used in the conversion routines.

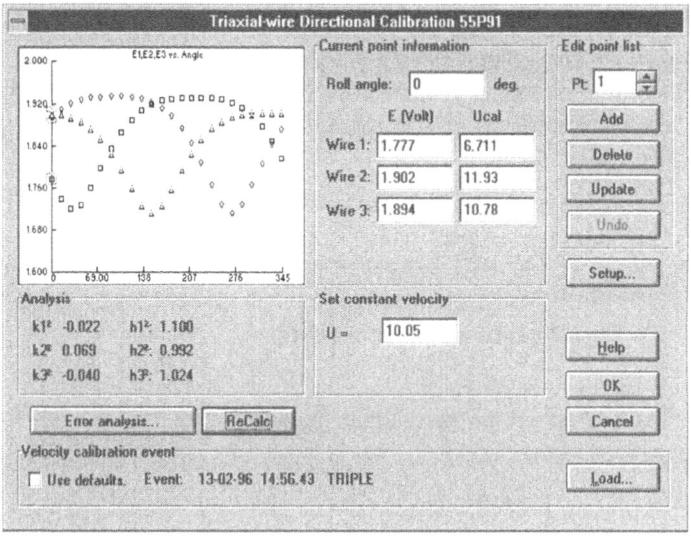

Fig. 10. Directional calibration of a Triaxial probe showing the response of the 3 wires with respect to a rotation of the probe under an inclination angle of 30°.

This will give the yaw and pitch factors that are related to one inclination angle only, which is normally selected close to the maximum acceptance angle of the probe. If optimum accuracy is required the probe should be calibrated at inclination angles that covers the entire expected range. These calibrations should then be repeated over the full velocity range. In such cases it is normally preferred to carry out a global calibration and create a 3-dimensional look-up table with the probe voltages E1, E2 and E3 as parameters.

3.2.3 Conversion and Decomposition of Triaxial-probe Data.

Conversion to calibration velocities Ucal1, Ucal2 and Ucal2 are done by inserting the probe voltages into their respective transfer functions. The effective cooling velocities are calculated from x() and inserted into (y). The final equation system for calculation of the velocities in the wire coordinate system then becomes:

$$\begin{matrix} U1^2 & & k1^2 \;\; 1 \;\; h1^2 & ^{-1} & Ucal1^2\cdot(1 + k1^2 + h1^2)\cdot\cos^2 35.3 \\ U2^2 & = & k1^2 \;\; 1 \;\; h1^2 & & Ucal2^2\cdot(1 + k2^2 + h2^2)\cdot\cos^2 35.3 \\ U3^2 & & k1^2 \;\; 1 \;\; h1^2 & & Ucal3^2\cdot(1 + k3^2 + h3^2)\cdot\cos^2 35.3 \end{matrix}$$

The velocity components U, V and W in the probe coordinate system are found by projection:

$$U = \cos 45\cdot\cos 35.3\cdot U1 + \cos 45\cdot\cos 35.3\cdot U2 + \cos 54.7\cdot U3$$
$$V = -\cos 45\cdot U1 + \cos 45\cdot U2$$
$$W = -\cos 45\cdot\cos 35.3\cdot U1 - \cos 45\cdot\sin 35.3\ 250U2 + \cos 35.3\cdot U3$$

In practice the probe is oriented so that the probe coordinate system is identical with the laboratory coordinate system. Accurate measurements of the velocity components by means of a Triaxial-probe require very accurate alignment of the probe both during calibration and measurement in order to avoid systematic errors. It is therefore recommended to use alignment procedures that involve optical devices. If proper care is taken, velocity components may be determined with an accuracy better than 5%.

4. Conclusion.

By utilizing modern computerized Constant Temperature Anemometers measurements in two -and three-dimensional flows have become easier and faster to carry out. The setup and calibration of the CTA bridges are highly automized with data conversion and reduction routines at hand. Traversing and data acquisition can be preprogrammed and carried out on a single command.

An automatic calibration facility yields fast and accurate velocity and directional calibrations of multisensor probes. The accuracy of X-probes and Tri-axial probes can be improved by individual directional calibration. Build-in conversion routines provide on-line display of the velocity components in two- and three-dimensional flows. Examples show that the total time for configuration, setup and calibration (velocity plus directional) of X-probes and Tri-axial probes is reduced to between 10 to 15 minutes. Measurements of for example a shear stress profile in a jet with data reduction and presentation of uv in a graph can thus be done in less than half an hour, even when system configuration is included. The dedicated application software keeps track of all actions in the project and makes the experiment traceable.

5. References

Bruun, H.H. (1995), *Hot-Wire Anemometry*, Oxford Science Publications, Oxford, New York, Tokyo.

Jørgensen, F.E. (1994), *A new concept for an automated hot-wire anemometer with fully integrated experiment manager*, Dantec Information 13, 2-9.

Stannov, T. (1995), *An accurate, easy to use low-turbulence calibrator for hot-wire anemometers*, Dantec Information 14, 6-11.

CHAPTER II

PARTICLE IMAGE VELOCIMETRY

PARTICLE IMAGE VELOCIMETRY :

A NEW APPROACH IN EXPERIMENTAL FLUID RESEARCH

WING T. LAI
TSI Incorporated
Fluid Mechanics Instrument Division
St. Paul, Minnesota, USA

1. Introduction

Although many flow visualization techniques have been employed to study the dynamics and topology of coherent structures in turbulent flow, none of the currently available techniques can provide instantaneous vorticity fields. Vorticity is one of the fundamental characteristics which is important for the understanding of the physics of turbulence phenomena. While numerical simulation allows us to explore sophisticated aspects of vortex dynamics, it is generally limited to simple geometries and low Reynolds numbers. The new approach with particle image velocimetry (PIV) has shown promise in the measurement of vorticity fields. In recent years, PIV has been used very extensively in fluid flow research. Some of the work can be found in references [1,2,3]. The most salient characteristic of the technique is its ability to measure flow in a two-dimensional plane, providing the two components of velocity vectors. The technique also makes it possible to capture instantaneous turbulent flow fields for the purpose of identifying fundamental mechanisms of turbulence and extracting their spatial structure. Due to its capability of getting an instantaneous measurement of velocity field accurately, PIV can also provide the vorticity and rate of strain fields. A PIV measurement of the vortices in an air jet is shown in Figure 1, with the velocity vector field overlaid on the image. Figure 2 gives the corresponding vorticity field.

The fundamental principle of PIV is simple. It measures the displacements ΔX and ΔY of the particle images as shown in Figure 3. The displacements must be small enough so that $\Delta X/\Delta t$ is a good approximation of velocity of U. That is, the trajectory must be nearly straight and the speed along the trajectory should be nearly constant. These conditions can be achieved by choosing Δt to be small compared to the Taylor microscale of the Lagrangian velocity field, subject to the constraints of accuracy.

Th. Dracos (ed.), Three-Dimensional Velocity and Vorticity Measuring and Image Analysis Techniques, 61–92.

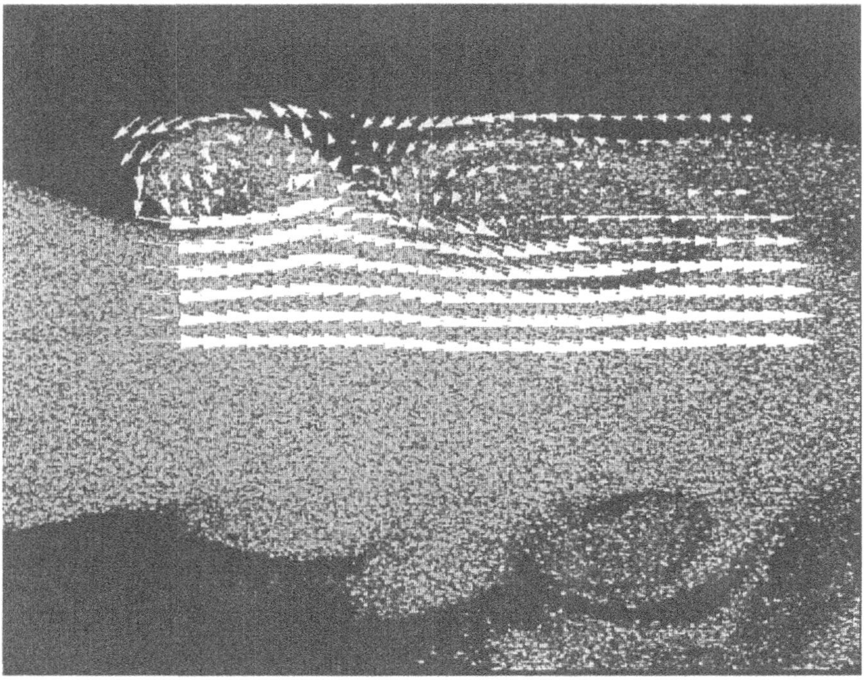

Figure 1. PIV measurement of an air jet

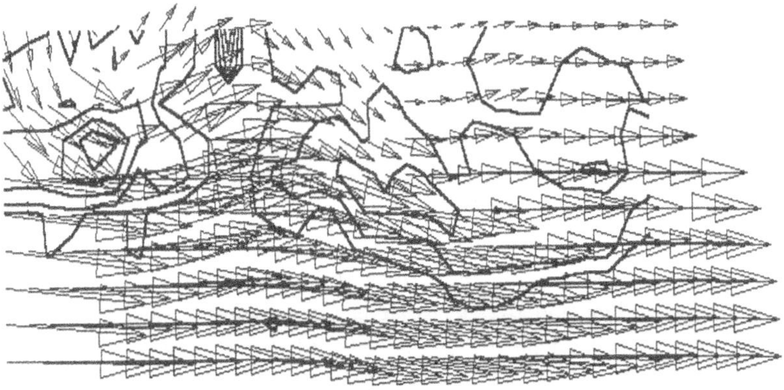

Figure 2. Vorticity field calculated from the PIV velocity measurement

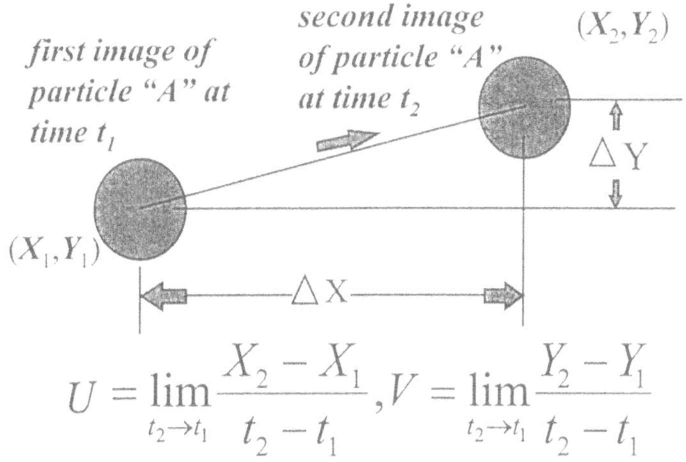

Figure 3. The fundamental principle of PIV

A typical PIV system is comprised of two systems : image capture and image analysis. In an image capture system, a pulsed laser is generally used for illumination due to the high energy in each laser pulse. Either a photographic camera or a CCD video camera can be used as recording medium. Obviously a CCD video camera provides more instantaneous feedback of the results, but the pixel resolution for most commercial video cameras is generally less than the resolution offered by photographic film. Consequently it may reduce the imaging size and the accuracy of the measurement. The effect of pixel resolution on the accuracy and the imaging size will be discussed.

A problem for PIV is the directional ambiguity associated with the measurements. To resolve such problem, an image shifting device is normally needed in image capture. Some of the common image shifting techniques include rotating mirror, polarization optics and translation of pixel lines. All the different techniques are widely used for PIV measurements but each may only be applicable to a certain velocity range. The directional ambiguity in a flow field can also be resolved when the images are captured into two video or photographic frames. Since the sequence of the frames is known, the crosscorrelation analysis scheme can be applied to the image frames to obtain the correct velocity magnitude as well as the direction. The major advantage of such image capture method and crosscorrelation analysis scheme is that no image shifting device is required.

Several PIV experiments were performed to illustrate the robustness of such method and their results will be discussed in a later section.

Time history is of great interest in fluid flow measurements for ensemble statistics. To measure a time-resolved flow, it is necessary to capture the flow at the frame rate of the video camera. High resolution video images contain a massive amount of data so that it was previously impossible to capture and store more than a few frames at a time. Nowadays new high speed camera interfaces are available to transfer data at a sustained rate of over 90 MBtyes per second into a storage media, like the host computer memory. This allows effectively real-time rata acquisition at the full camera frame rate, even from several high resolution cameras simultaneously. A automated processing scheme is suggested to allows the flow time history to be reconstructed, with a temporal resolution limited only by the camera frame rate.

Most of the particle image velocimetry studies are confined to two-dimensional flow fields, however the need for three-dimensional measurements has been emerging for many applications. A stereoscopic PIV system can be extended from a conventional system to measure a three-dimensional flow field. Generally the additional hardware required for the extension is a second camera and a translation stage for the two cameras to form a stereoscopic configuration. The concept is to use the two images captured from the cameras to resolve the out-of-plane motion. A more elaborate method for three-dimensional measurement is the holographic PIV method. The recording phase for holographic PIV is almost identical to the conventional PIV method, except the image is captured on a hologram. The reconstruction phase to extract the particle displacements requires a very different optical arrangement from the normal PIV system. The advantages of the two methods for three-dimensional measurements will be discussed.

2. Pixel resolution and the resolution of particle displacement

It is recognized that the pixel resolution in a recording medium can affect the accuracy and the imaging size in PIV measurements. To analysis such effect, the relationship between particle image size and the pixel size is inspected. Comparison of a number of recording media in their pixel resolutions and dynamic ranges is also performed.

The accuracy of the particle displacement is determined by the finite uncertainty that exists in locating the center of each particle image, and it directly affects the spatial and temporal resolution of a PIV experiment. If the errors associated with particle response were negligible, it is conceivable to make the finite errors vanish by decreasing the time between the two laser pulses, Δt, in image capture. However Δt cannot be made arbitrarily small because the displacement must be larger than the uncertainty in its measurement in order to make accurate velocity measurements. Therefore the value of Δt depends upon the displacement accuracy, and the displacement accuracy determines the spatial resolution. This result can be given as that the product of the uncertainty in the velocity and the uncertainty in spatial location is a constant of the PIV system that is determined by its pixel resolution.

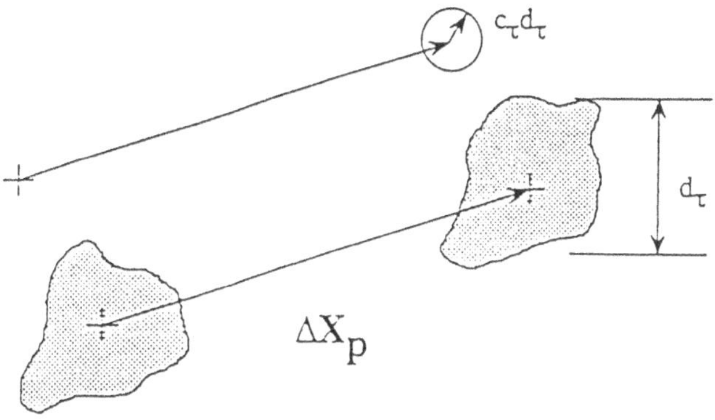

Figure 4. Displacement of the particle images and its uncertainty

To relate the pixel resolution with the particle displacement resolution, it is needed to derive a set of criteria. Consider the measurement of the displacement between two images of the same particle, as shown in Figure 4. Due to noise from the recording medium, and effects from background speckle and aberrations of lens, the images of the particles are usually irregular in shape. Regardless of the source, the noise makes even the first and the second exposures of the same particle appear different, and there is an rms error in determining the displacement of the centroid of the particle images which can be designated as $\sigma_{\Delta X}$, with ΔX_p as the particle displacement. The precise relationship between the rms displacement error and the properties of the image is complex, however, on the basis that it is easier to locate the center of a small image than a large image, Adrian [1] stated that the rms error can be represented as

$$\sigma_{\Delta X} = c_\tau \, d_\tau$$

where d_τ is the mean diameter of the image, and c_τ is a constant that depends on the ability of the analysis procedure to determine the displacement between the images. Typically the constant is 1 - 10% of the image diameter. The recorded image diameter is given approximately as

$$d^2_\tau = d^2_e + d^2_R$$

where d_R represents the resolution of the recording medium and d_e is the mean diameter of the optical image prior to being recorded. The mean diameter of the optical image is related to the actual particle diameter and its effect by diffraction limited imaging. According to Adrian [4], two quantities can be used to define the capability of a PIV system, especially the recording medium. The dynamic velocity range (DVR) is the ratio of the maximum velocity to the minimum resolvable velocity and the dynamic spatial range (DSR) is given as a ratio between the largest observable length scale to the smallest observable length scale. The combination of the two quantities is given as

$$(DSR)\,(DVR) = L_x / c_\tau\, d_\tau$$

where L_x is one of the dimensions to define the format of the recording medium. The other dimension is given as L_y. The above equation shows that the capability of a PIV system to have both a large dynamic velocity range and a large dynamic spatial range is determined by a dimensionless constant $L_x / c_\tau\, d_\tau$ which is a characteristic of the system. The format length L_x characterizes the physical size of the recording medium, which determines its capacity to hold information. The recorded image diameter, d_τ, is a measure of the performance of the optical imaging system and the recording system; and the constant c_τ depends on the ability of the interrogation algorithm to measure displacements accurately. It can be easily observed that PIV systems having large values of $L_x / c_\tau\, d_\tau$ are well-suited for turbulence research which normally has large velocity fluctuations.

According to the above equation, one important consideration for configuring a optimum PIV system is to match the size of an image with the resolution of the recording medium, given as d_R as defined earlier. The resolution elements of photographic film are the grains in the emulsion, and their resolution, in mm, is conventionally given as d_R = 2 /(number of resolvable line pairs per millimeter). The resolution elements of a video device are the pixel elements, whose resolution d_R is characterized by the spacing. The representation of the particle image in the recording medium is like sampling an analog image at discrete points in space. If the optical image diameter d_e is too small, the image is under-sampled according to Nyquist's criterion, and information of the image will be lost. Further under-sampling leads to both mean bias errors in locating the particle and a random error. If the image is over-sampled with too many of the discrete points, the capacity on the recording medium is wasted. Over-sampling has very little effect in reducing the random error once the number of samples per diameter c_R exceeds a critical value of 2 pixels per diameter in a video array as suggested by Wernet and Pline [5]. If the image diameter satisfies the relationship given below

$$d_e > c_R\, d_R$$

then the finite resolution of the recording medium has minimal effect on the accuracy of measuring the displacement of the particle image. If the number of samples per diameter c_R is set at 2, the magnification of the imaging lens can be optimized to be equal to $c_R d_R$.

This makes d_e as small as possible, given the resolution limits of the recording medium, the dynamic range product is then maximized. Table 1 shows a comparison of various recording media and their characteristics for a PIV measurement. All the characteristics given the in the table were calculated based on particle size of 1 micron which is typical in an air flow measurement. Again the resolution of the film is taken to be 2/(line pairs per mm), and the effective pixel format is given as L_x/d_R by L_y/d_R. The dynamic spatial range was derived assuming the dynamic velocity range of 100:1.

The general trend given in the table is that the field of view increases as the format size and the number of pixels are increased. The number of pixels varies over a 200:1 range between the standard Sony camera and a large 200 mm x 250 mm sheet of high resolution black-and-white film, and the major effect is the change of the dynamic spatial resolution (DSR). The standard resolution Sony video camera operating to fully resolve the particle images is not able to cover a field of view much larger than 10 mm. If the cameras are used for larger field of view by decreasing the magnification, the diameter of the images will decrease, and the rms and the random errors will increase. High resolution cameras with up to 2K x 2K pixels may be comparable to 35 mm cameras, giving fully resolved images on fields of view roughly 30 mm which is adequate for some of the laboratory-scale experiments. However the dynamics spatial range of around 100 may not be good enough for high Reynolds number experiments.

The most impressive capabilities for the study of high Reynolds number turbulence measurements happens with the use of the high resolution, large format photographic recording medium. The field of view of more than a meter can be achieved with spatial resolution of 300 microns full scale. This corresponds to more than 3000:1 dynamic spatial range, good enough to measure three decades of turbulence spectrum.

A similar conclusion on the recording medium performance can be applied to water flow applications which require larger seed particles, 10 to 50 microns, to compensate for the scattered light lost due to the smaller ratio of refractive indices. The recorded image of a 20 micron particle is less sensitive to the effects of parameters other than the geometric diameter, so the optimum magnification will vary in lesser amount for applications in water flow.

3. Directional ambiguity in PIV measurements

In any double-pulsed PIV measurements, the magnitude of the displacement can be determined correctly, but its direction is ambiguous unless the first and second images can be identified. Various techniques can be used to distinguish the second images from the first images. Images of the particles can be marked by different intensity as shown in Figure 5a; or by different colors as given in Figure 5b. Separation using intensity is generally not a reliable method because the image intensity is also dependent upon the location of the particle in the light sheet. In Figure 5c a coded sequence is used to provide the distinction as suggested by Marko and Rimai [6], and the technique of using fluorescent or phosphorescent particles whose images leave decaying streaks was

Table 1. Formats and characteristics of various recording media

Recording medium	Format L_x x L_y (mm)	Pixel size (microns)	Resolution (line pairs/mm)	Effective pixel format	Optimun magnification	Field of view (mm)	Dynamic spatial range (DSR)
Sony XC-75 (RS-170)	6.45 x 4.84	8.4 x 9.8		768 x 494	0.59	10.8	68
Dalsa IA-D2	10.24 x 10.24	10.0 x 10.0		1024 x 1024	0.71	14.4	91
Kodak MegaPlus 1.4	8.96 x 7.04	6.8 x 6.8		1317 x 1035	0.48	18.7	118
Kodak MegaPlus 4.2	18.26 x 18.40	9.0 x 9.0		2028 x 2044	0.64	28.5	184
TMAX 400 film	35 x 24	5.0 x 5.0	100	7000 x 4800	0.35	100	626
TMAX 400 film	125 x 100	5.0 x 5.0	100	25000 x 20000	0.35	360	2236
TMAX 400 film	250 x 200	5.0 x 5.0	100	50000 x 40000	0.35	715	4472
Technical Pan film	125 x 100	1.5 x 1.5	320	83333 x 66666	0.11	1140	7450
Technical Pan film	250 x 200	1.5 x 1.5	320	166666 x 133333	0.11	2270	14900

Optimum magnification, $M_{opt} = (c_R d_R) / (1.5 \;\delta z\; \lambda + d_p^2)^{1/2}$

$c_R = 2.0$, d_p = particle diameter = 1.0 micron, δz = depth of field = 1000 microns,
λ = wavelength of laser light = 532 nm

Field of view = L_x / M_{opt}

Dynamic velocity range (DVR) = 100: 1 and $c_\tau = 0.05$

employed by Gharib et al.[7], is given in Figure 5d. These methods work for low image density PIV, but aside from the two-color technique, they are not generally useful for high image density PIV. It is because statistical analysis by autocorrelation always forms a function which is symmetric in its argument, regardless of the arrangement of the original pulse configurations.

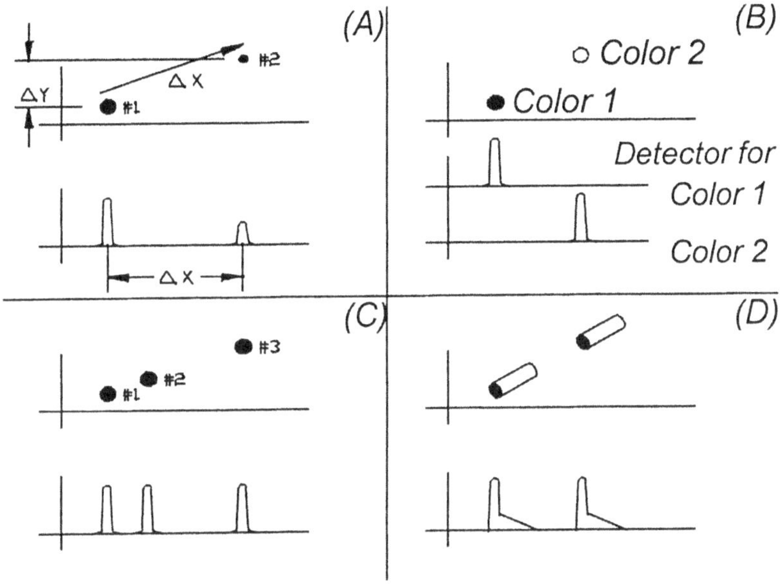

Figure 5. Image shifting techniques : (a) by intensity, (b) by color, (c) with code sequence, (d) with fluorescent or phosphorescent particles

For resolving the directional ambiguity in high and low particle density PIV measurements, the most recognized technique is image shifting. The concept of the image shifting is very similar to the method of frequency shifting used in laser Doppler velocimetry, at which negative Doppler frequencies are avoided by imposing a positive bias on the frequency of the signal using an acousto-optic cell, normally a Bragg cell. In the same token, the image shifting technique for PIV imposes a bias or image shift displacement, X_s, to the second particle image with respect to the first so that all the second images are located on the same side from the first images. The selection of X_s requires the knowledge of the maximum negative velocity components in the x and y directions, similar to the magnitude of frequency shifting employed in laser Doppler velocimetry. It is obvious that the bias, X_s, has to be large enough to cover for all positive and negative displacements, but should not be too large such that the resolution of the actual physical displacements will be reduced. Once the resultant displacements are measured, the bias is then subtracted to give the actual displacements. Figure 6 shows that the second image is moved to the positive quadrant when a shift vector, X_s, is added to the actual displacement.

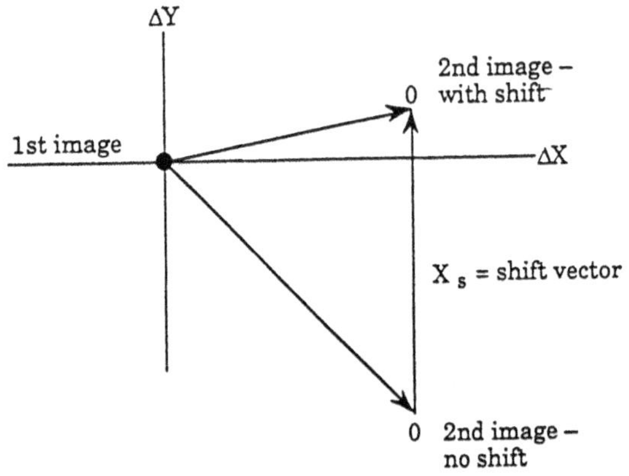

Figure 6. Image shifting moves the second image to the positive quadrant

There are several image shifting techniques in PIV. Employment of a rotating mirror or a polarization optics, or special design in the video camera to translate pixel lines between laser pulses are a few of the image shifting techniques. The rotating mirror image shifting method has been commonly used, given in Figure 7, were introduced by Landreth et al. [8] and Kompenhans et al.[9]. With rotating mirror image shifting, a rotating mirror is placed between the illuminated plane in the flow and the recording medium. Mirror rotation in the time between the laser pulses at T_1 and T_2 adds the image shift displacement to the flow displacement. For the stationary particle shown in the figure, the artificial displacement corresponds to the image shift displacement, determined by the angular movement of the mirror. It can be realized that the rotational speed of the mirror may pose a limit on the amount of image shift displacement. For high speed measurement, time between the two pulses needs to be reduced in order to confine the particle displacement. Therefore the rotational speed of the mirror is required to be sufficiently high to provide the appropriate amount of the image shift displacement. Most of the commercial rotating mirror image shifter devices can be operated in the hundred revolution per second range, thus applicable to velocity up to 500 m/sec. The investigation of flow under a helicopter rotor by Raffle and Kompenhans [10] employed a rotating mirror to provide image shifting for the measurement which had a mean velocity of about 250 m/sec.

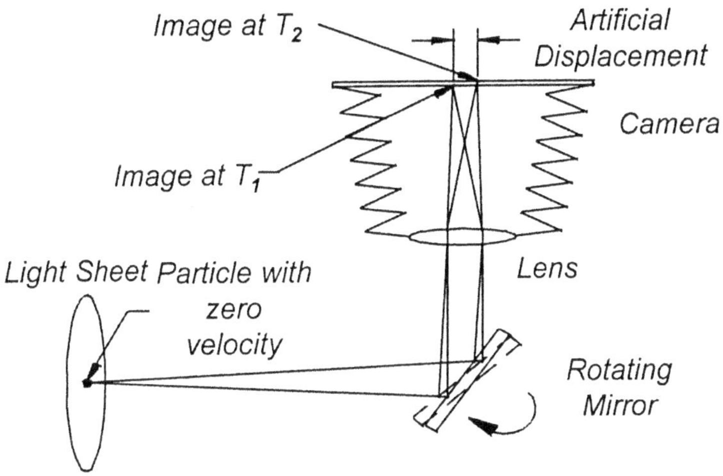

Figure 7. Image shifting technique with rotating mirror

The image shifting method using a polarization optics or calcite crystal is mostly applicable to supersonic flows. As shown in Figure 8, a calcite crystal is placed between the recording medium and the image plane. The calcite crystal allows light polarized in the direction parallel to its axis to pass through unshifted (as indicated by the particle image at T_1) while shifting the light whose polarization axis is perpendicular to its axis (shown as the particle image at T_2). The magnitude of the shift is dependent on the thickness of the crystal. For this technique to be feasible, the polarity of the second laser is required to be rotated by 90 degrees. Such requirement can be met with a laser system which employs two separate laser heads with polarization of the two laser beams at 90 degrees apart. In general a laser system with two laser heads is needed to measure high speed flow so that small pulse separation can be obtained. Nowadays most of the commercial PIV laser systems employ two separate laser heads to provide flexible pulse separation and maximum pulse energy, hence the calcite crystal technique is very feasible. The work by Landreth and Adrian [11] employed a double refracting crystal to provide image shifting in their high speed flow measurements.

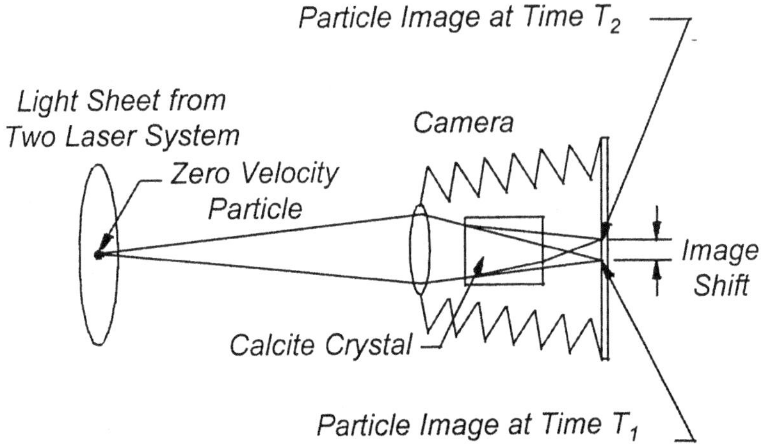

Figure 8. Image shifting technique with polarization optics

Image shifting using the movement of pixel lines in the CCD array of a video recording medium has been employed recently. The general concept of the method is given in Figure 9. Essentially the first laser pulse expose the CCD array at T_1, and before the second laser pulse arrives at T_2, a number of pixel lines on the CCD array are moved downward so that the entire images on the array are shifted. The number of pixel lines to be moved is then equivalent to the amount of image shift. In order to resolve the directional ambiguity, four or more number of pixel lines are normally needed to be shifted. Depend on the total time required to shift the required number of pixel lines, it may pose a certain constraint on the minimum pulse separation that such method can be applied.

Capturing the PIV images on two separate video frames from a video recording medium and analyzing the two frames with crosscorrelation analysis technique is becoming more widely used to resolve the directional ambiguity in PIV measurements. Such method has been shown to be very robust and more convenient to be used when compared with the image shifting techniques described above. Further discussion of the crosscorrelation is given in the next section.

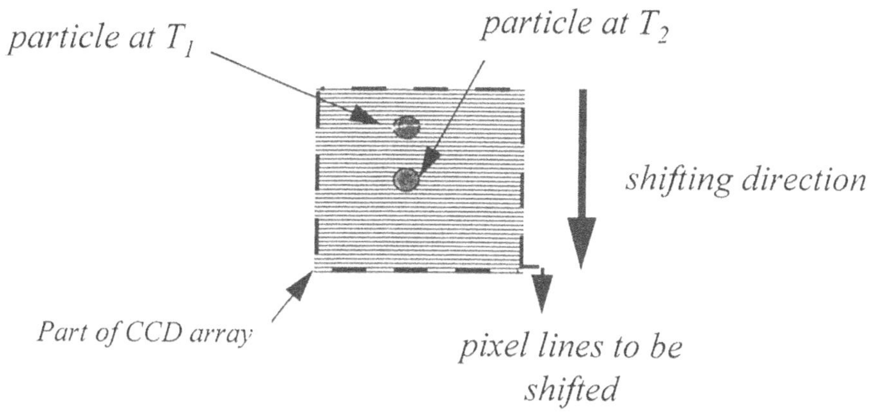

Figure 9. Image shifting using translation of pixel lines in a CCD array

4. Crosscorrelation analysis technique in PIV measurements

For high image density, it is known that the autocorrelation and the crosscorrelation are the two statistical approaches for image analysis. The principles of the approaches are described in details by Keane and Adrian [12,13]. In general, the autocorrelation approach requires a double or multiple pulse image for the analysis. Further, due to symmetry in the correlation function, some kind of image shifting techniques or image tagging techniques are normally required to resolve the direction of the displacement. On the other hand the crosscorrelation analysis approach calculates the displacements by correlating two image frames. Since the sequence of the image frames is known as a priori, the direction of the displacement is resolved.

It is shown in Keane and Adrian [14] that the crosscorrelation approaches of interrogation of single-exposure and double-exposure double-pulse are superior to the autocorrelation approach using double pulse image for a number of reasons. The versatility in selecting the size and offset of the interrogation windows permits a higher level of spatial resolution by more efficient pairing of particle images. Ideally there are no loss pair in the optimal implementation of the crosscorrelation approach, the spatial resolution can be improved or the seeding density can be reduced while maintaining acceptable detection probabilities. Since the need for image shifting is not necessary for crosscorrelation, the dynamic range of velocity measurement is also increased.

Despite the superiority of the crosscorrelation approach, there was a practical limitation of the technique. Since the correlation is performed on two separate video frames, the frame rate of the image capture hardware imposes a limit on the time separation between the laser pulses which subsequently restricts the range of the flow to be measured. For a standard RS-170 format CCD video camera, the time between frames is 33 msec, or 40 msec for the standard CCIR format CCD video camera. Assuming the size of an interrogation area of 1 mm, and the displacement of the particle to be 1/4 of the interrogation size, at 0.25 mm, the maximum velocity that will result with the pulse separation of 33 msec is 0.0075m/sec which is too low for a majority of flow applications. Although the interrogation size can be increased to accommodate for higher velocities, it is obvious that the velocity range cannot be increased without substantial sacrifice of the spatial resolution, which is not desirable for PIV.

Alternate techniques have been attempted to shorten the pulse separation on two video frames to increase the velocity range. One of the techniques is to simply locate the laser pulse as close as possible at the end of one video frame and the second pulse at the beginning of the next frame, a technique known as frame straddling. Such technique can generally reduce the pulse separation to about 3 msec, which corresponds to the time required for a typical RS-170 camera to discharge the image from the CCD array. However the velocity range can only be improved by an order of magnitude to 0.075 m/sec, but it is still not in any practical range. Another technique to shorten the pulses is to locate the two pulses between the odd and even fields in an interlaced CCD video camera. Usually the pulse separation can be reduced to less than 100 μsec with such technique, but the interlaced images generally require the particle images to be large in order to compensate for the incomplete particle image due to the interlacing effect. As a result a large magnification may be required, meaning a small field of view for the image capture. Further the interlaced particle images usually have detrimental effect to the accuracy of the measurements.

To take full advantage of the crosscorrelation approach, the combination of the frame straddling technique and a design of fast image transfer from the CCD array was investigated. The result was a specialized CCD video camera which allows image capture in subsequent frames with pulse separation at a few microsecond apart. The characteristics of the camera is that the charge in the photosensitive pixels can be transferred very quickly into memory banks residing next to the photosensitive pixels, hence allowing the transfer to the charge to be completed on the order of a few microseconds. By employing the frame straddling technique, the time between the two pulses on the two video frames is then reduced to 2 μsec. As a result the velocity range is substantially increased. Assuming the same 0.25 mm particle displacement for a 1 mm interrogation size, the maximum velocity to be measured becomes 125 m/sec when the pulse separation is 2 μsec. It is easily seen that such velocity range is more applicable for fluid flow measurements.

To show the applicability of the specialized camera, a number of experiments were performed, utilizing the frame straddling technique and the crosscorrelation approach to obtain the velocity fields. The experiments include (1) convection flow in an enclosure,

(2) interaction of a jet to free stream water flow, and (3) water spray flow. The experimental setup for all the experiments was almost identical except the laser. A diagram showing the experimental setup is given in Figure 10. Two versions of the camera were employed for the experiments, one with low resolution of 768 by 484 pixels and the other with 1K x 1K pixels. The timing sequence for the laser pulsing and image capture was controlled by a synchronizer electronics. Whenever necessary an articulated arm was utilized to position the laser sheet to the measurement regions. A Windows™ based software package, Insight™, was employed to control the image capture and to perform image analysis. To ensure proper frame straddling, a certain pulse delay was set in the synchronizer to ensure the first laser pulse to be positioned at the very end of a video frame. Depending on the pulse separation for each experiment, the second pulse was recorded on the subsequent video frame.

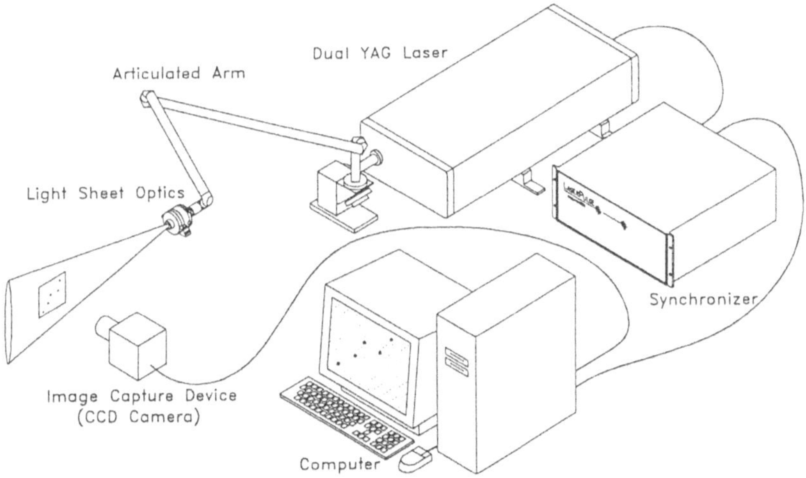

Figure 10. Experimental setup for the PIV experiments

1. Convection flow in an enclosure

The experiment was a study of convection flow in an enclosure with a size of 5 cm at all dimensions.. A drawing of the test setup is given in Figure 11. As indicated in the figure three surfaces of the enclosure were cooled at 10°C while the top surface was maintained at room temperature of 25°C. The fluid in the enclosure was salt water. A 2W argon-ion laser was employed as the illumination source for the PIV measurements. Silver coated hollow spheres with mean diameter of about 15 μm and density of 1.02 gm/cc were employed as seed particles. Since the velocity of the flow was extremely small, on the order of mm/sec, the laser sheet was continuously on for illumination and the frame rate of the video camera was used to "pulse" the laser beam. The low resolution version the camera was employed as the image capture device.

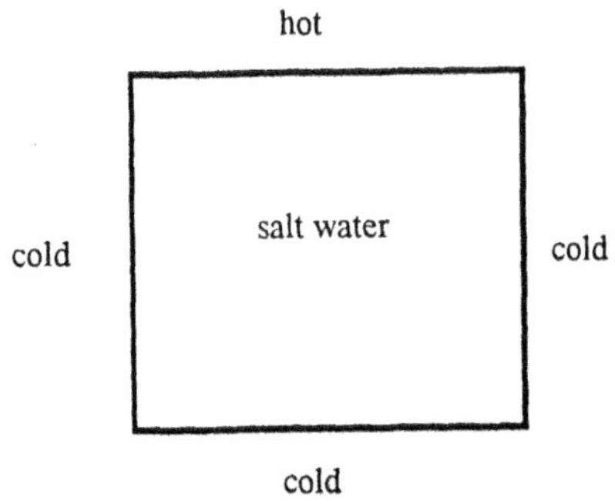

Figure 11. Convection flow in an enclosure

Since the velocity was on the order of millimeter per second, the crosscorrelation analysis was performed between the first and the fourth frame, giving an effectively pulse separation of 132 msec. The velocity vector field shown in Figure 12 is one of the measurements. As expected, two circulation regions caused the density gradient was observed in the enclosure and solidification of the salt crystals was also apparent on the interior surfaces. The two circulation regions have almost identical velocity magnitudes, but opposite in direction. The dynamic range of the flow was quite high, as indicated by the stagnant flow at the center of the circulation and the large velocity at the top of the enclosure.

2. Interaction of a jet to free stream water flow

Interaction of a water jet to flow in a water tunnel was the second experiment which also employed the low resolution version of the camera. A water jet at 2 m/sec was impinging to a water tunnel which had a free stream velocity of 50 cm/sec as shown in Figure 13. Measurements were taken at regions where the interaction of the jet flow to the free stream took place. A 2W argon-ion laser and the a beam modulator were used to provide the laser pulses. Seed particles employed were the silver coated hollow spheres with mean diameter of 15 μm and density of 1.02 gm/cc. Typical pulse separation between the two consecutive frames was 2 msec.

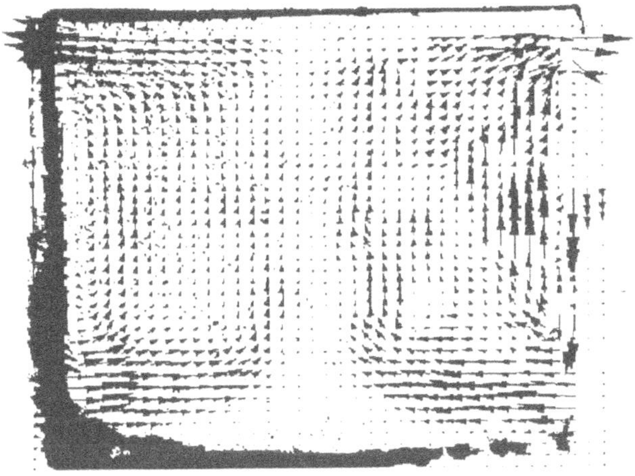

Figure 12. Velocity vectors in the convection flow

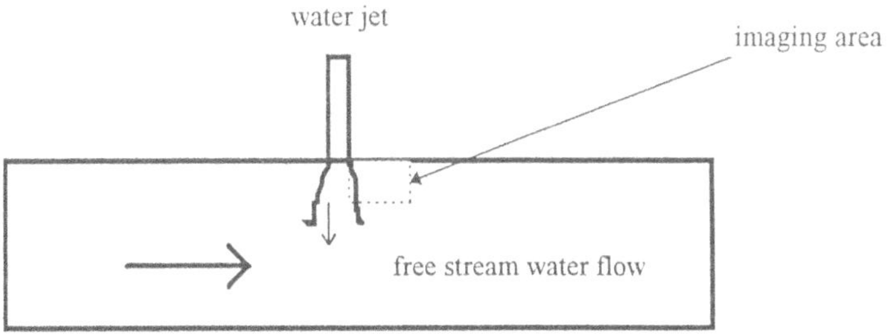

Figure 13. Interaction of a impinging water jet to free stream water flow

The two images used for the crosscorrelation are given in Figure 14. Size of the imaging area was about 4 cm by 2.6 cm. The results from the figure show that the impinging jet generated a shear layer to the free stream flow and formed a recirculation region next the layer. The dynamic range between the jet and the stream was large, as shown by the velocity vectors as well as the elongated particle images in the jet region. The stretched particle images in the jet region was caused by the long pulse duration during illumination.

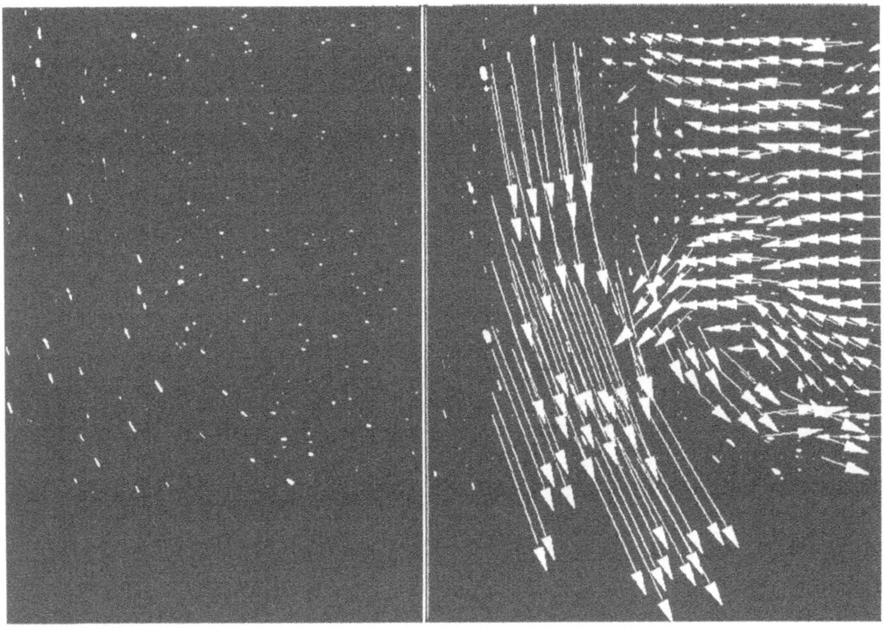

Figure 14. Velocity vectors of the impinging jet experiment

3. Water spray flow

The third experiment which utilized the specialized camera with 1K by 1K pixels resolution was a water spray experiment. Illumination was provided with a dual Nd:YAG laser with about 50 mJ energy per laser pulse. The seed particles were water droplets in the spray, and velocity at the center portion of the spray was about 50 m/sec. The pulse separation was set at 10 μsec. Region of measurement was the exit of the spray as indicated in Figure 15. Imaging area was about 10 cm by 10 cm. Two of the images captured for the crosscorrelation analysis are given in Figure 16, together with the velocity vector field.

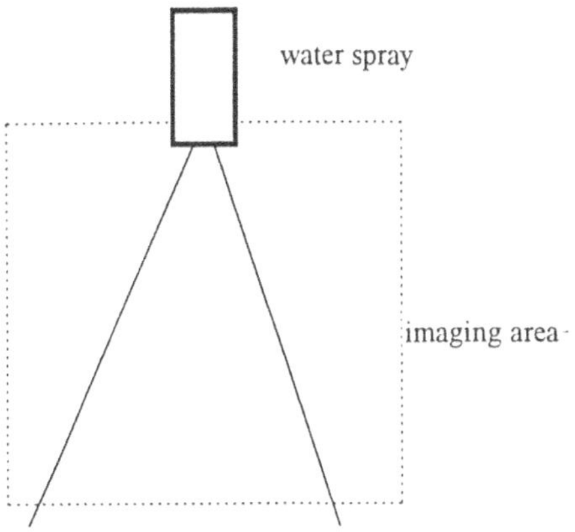

Figure 15. Experiment of the water spray flow

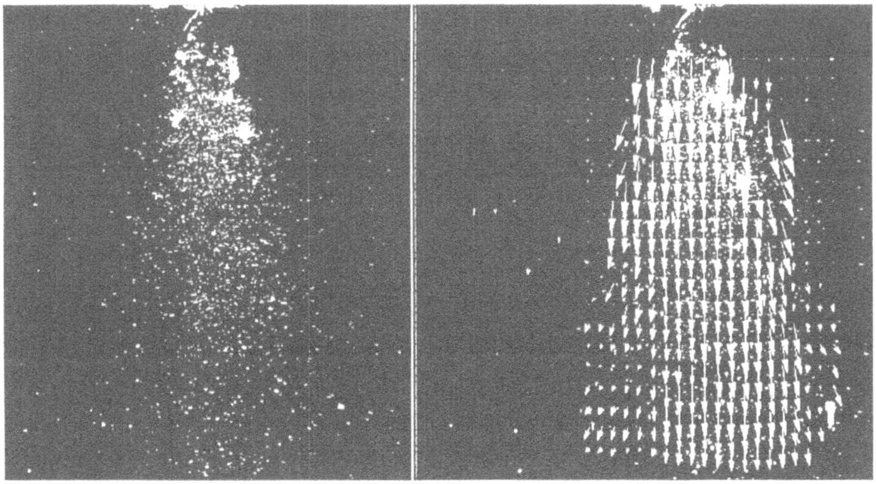

Figure 16. Velocity vectors of the water spray experiment

The superiority of combining frame straddling for image capture and crosscorrelation for image analysis is well proven with the results from the experiments discussed above. The specially developed camera was an important component because it allowed image

5. Real-time image capture and batch image processing

For measurement in a time-resolved flow, it is necessary to capture the flow at the frame rate of the video camera. In order to capture the flow at the frame rate, an important aspect is the ability to transfer the image to the host computer memory in a very short time, typically in less than 1 μsec, before the next video frame is ready for image capture. A new camera interface was developed to allow data transfer at a speed of 60 to 90 Mbytes/sec. This effectively provides real-time data acquisition at the full frame rate of the camera, and even from several high resolution cameras simultaneously. Such data transfer rate is actually necessary for real-time image capture in three-dimensional PIV measurements which generally employed two video cameras.

A water jet experiment is used to show the effective data transfer rate for real-time flow image capture. The PIV components employed for the experiment were same as those given in Figure 10. An argon-ion laser together with a beam modulator were used to provide the laser pulses. The 1K by 1K version of the specialized camera described above was used as image capture device. Since the frame rate for the camera is 30 frames/sec, 15 pairs of images can be obtained with the camera. The Insight™ package was used for image capture and image analysis. The high data transfer camera interface installed in the computer which was equipped with 128 Mbytes RAM and 2 Gbtye hard disk. The test setup was a water jet flow with speed of about 1 m/sec. The region where the measurements were made is given in Figure 17. The time resolved flow fields are provided in Figures 18a to 18e. Each flow field plot is 66 msec apart because of the image pair needed for cross-correlation. It is noticed that the plots provide a good indication of the real-time effect of the flow field.

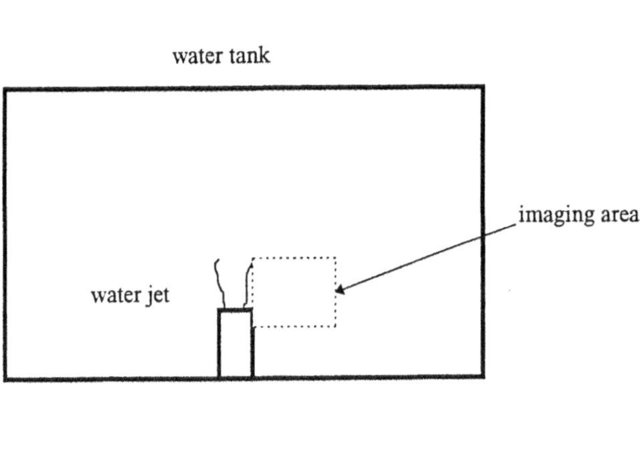

Figure 17. Measurement region for the water jet experiment

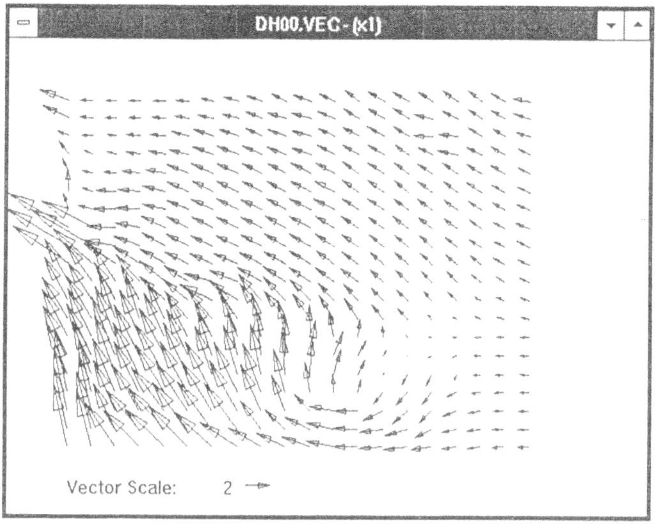

Figure *18a*. Velocity results at t = 0 sec

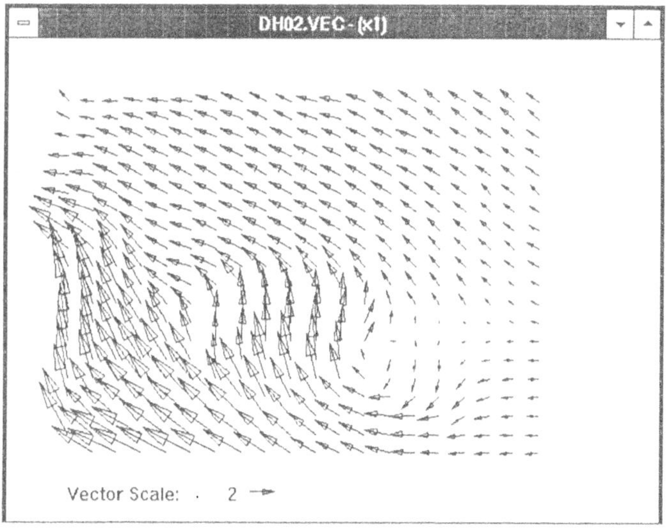

Figure *18b*. Velocity results at t – 66 msec

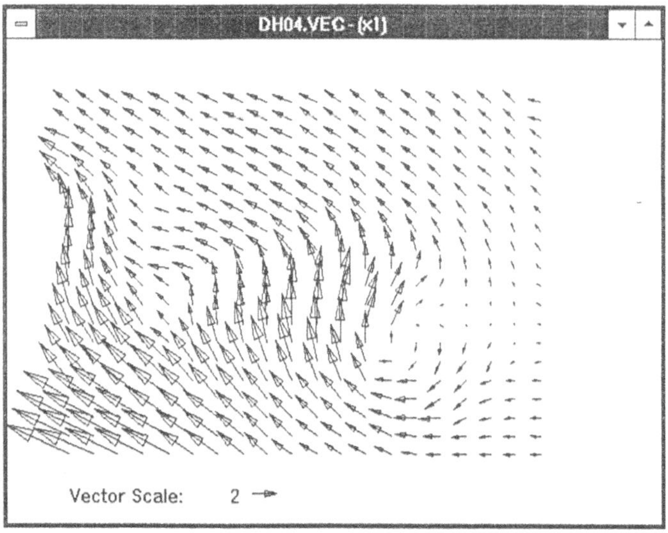

Figure *18c*. Velocity results at t = 132 msec

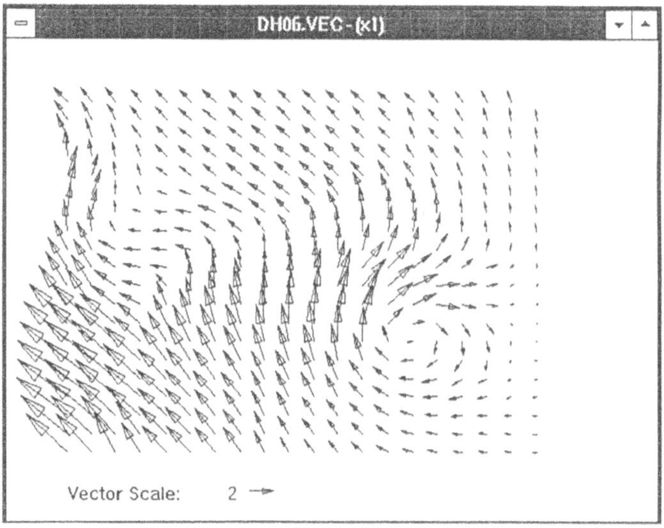

Figure *18d*. Velocity results at t = 198 msec

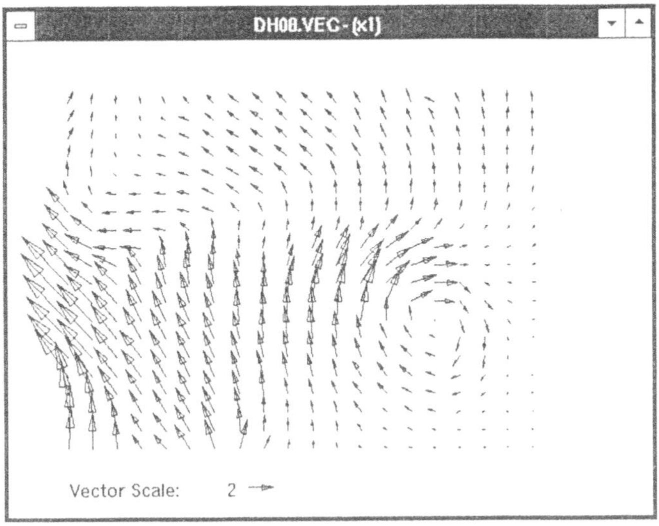

Figure *18e*. Velocity results at t = 264 msec

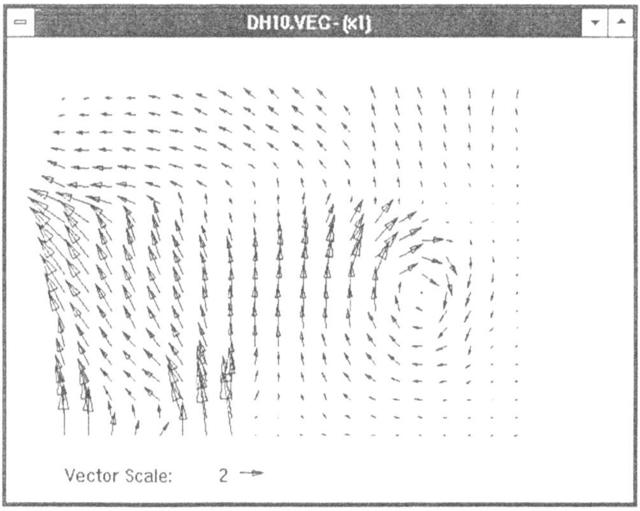

Figure *18f*. Velocity results at t = 330 msec

The ability to automate the sequence of image capture and analysis is also an important aspect for PIV measurements. To complement with the real-time image capture, a batch processing has been developed to allow a large number of images to be analyzed automatically. With the capability of real-time capture and batch processing, time history of a flow field can truly be measured by PIV with a temporal resolution limited only by the video camera frame rate.

6. Three-dimensional measurements with PIV

It is well known that three-dimensional measurements in fluid flow research are of great interest. Naturally the attempt to use the PIV technique to obtain a three-dimensional flow filed is very attractive. The extension of the PIV technique for three-dimensional flow has received considerable amount of research effort from researchers worldwide. In general two categories of three-dimensional PIV systems have been developed, the stereoscopic PIV system and the holographic PIV system. The two types differentiate substantially in the interrogation or reconstruction phase, while the imaging phase can be very similar in the hardware requirements. A direct extension from a conventional two-dimensional PIV system to measure three-dimensional velocities has also been suggested and developed.

A simple attempt to extend a standard PIV system with a single camera to measure the three components of velocity was suggested by Arroyo and Greated [15]. Their technique was based on combining the standard PIV with stereoscopy. A special mirror system was constructed to allow the flow field to be captured onto two stereoscopic images on the same negative. The three velocity components were then obtained by combining the in-plane measurements from both images. The technique was applied in a study of a Rayleigh streaming flow in a rectangular tube with velocity in the few millimeter per second range. Other measurements with a single camera configuration were also performed by Stolz and Kohler [16], Halloin and Jottrand [17] and Raffel et al. [18]. The studies by Stolz and Kohler [16] and Halloin and Jottrand [17] were based on particle tracking to estimate the out-of-plane motion. The techniques seemed to work well in low density particle situations. The approach applied by Raffel et al. [18] was to analyze the height of the correlation peak of the out-of-plane velocity component. Two parallel light sheet were formed and the out-of-plane velocity estimation was achieved by analyzing images of particles within the parallel light sheets by spatial crosscorrelation. The initial conclusion of the study was shown to be applicable to flows with out-of-plane component in one direction only.

1. Stereoscopic approach

A more common approach to the three-dimensional measurements is to employ two cameras. Such approach is often referred as the stereoscopic approach. With the stereoscopic system the three velocity components in the plane illuminated by the light sheet are measured. In general the out-of-plane component is not measured directly but is derived from the planar measurements of the two cameras.

For flow field with low particle density, particle tracking can be used with the stereoscopic setup. The experimental work by Robinson and Rockwell [19] was based on tracking particle images on contiguous planes where the out-of-plane velocity component was obtained with a spectral interpolation method. Guezennec et al. [20] also developed a particle tracking scheme with images taken from two CCD video cameras for the three-dimensional measurements. A special software algorithm was developed to match the particles and an adaptive Gaussian window filtering technique was used to validate the results. The algorithm was shown to be accurate and robust.

For flow field with high particle density, particle tracking may not be feasible, then the correlation analysis method may be needed. A twin-photographic camera stereoscopic system was developed by Prasad and Adrian [21] for three-dimensional high image density PIV measurements. The system was applied to a measurement of a rotating disk which was submerged in a cylindrical tank filled with glycerin Further a set of stereoscopic equations was developed to resolve the aberrations due to the air-liquid interface. Image shifting by physically moving the camera system during recording was implemented in the experiment to resolve directional ambiguity. The system enabled accurate measurements of up to 0.2% of full scale for the in-plane displacement and 0.8% of full scale of out-of-plane component for the liquid flow. However the system seemed to be applicable only to low velocity measurements primarily due to the requirement to move the camera system as the image shifting method.

Recently the stereoscopic system developed earlier by Prasad and Adrian [21] was modified to employ CCD video cameras as the image recording medium. The CCD video cameras used with the system are two Kodak MegaPlus 4.2 cameras each one has a pixel resolution of close to 4.2 million pixels. As given in Table 1, the resolution of the Kodak camera is much lower that the photographic film (100 mm x 125 mm) employed with the liquid flow measurements. The convenience and the instant feedback with the video camera, however, more than compensate for the discrepancy on the pixel resolution. The camera system was applied to the measurements of a boundary layer flow in a wind tunnel as described in Lai et al [22]. Some of the details are presented here.

The arrangement of the stereo camera setup is given in Figure 19. Separation of the cameras was set at 25 degrees to obtain a object distance of 1617 mm and a field of view of 80 mm. Illumination of the flow field was provided by two Nd:YAG lasers with a wavelength of 532 nm and at pulse frequency of 10 Hz. Each laser was capable of delivering 200 mJ per pulse of energy with a pulse duration of 8 nsec. The laser beam was generated into a sheet of about 1 mm in thickness in the test section. Timing sequence of the laser pulses and the image capture with the cameras was coordinated with a synchronizer. Image capture and analysis was provided by the Insight™ software. The software controlled the two video cameras for simultaneous image capture, and image analysis was performed with the autocorrelation technique. Corrections for the distortion of the image plane, misregistration of the images, and the variation of the magnification factors were also performed in the software. A layout of the laser optics and computer hardware for the measurements is given in Figure 20.

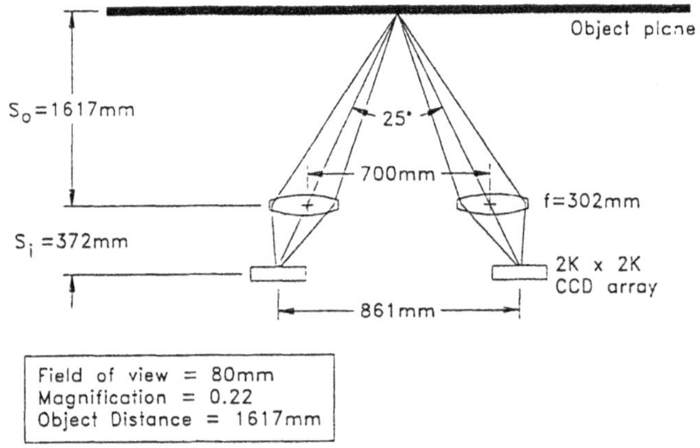

Figure 19. The arrangement of the stereo-camera system

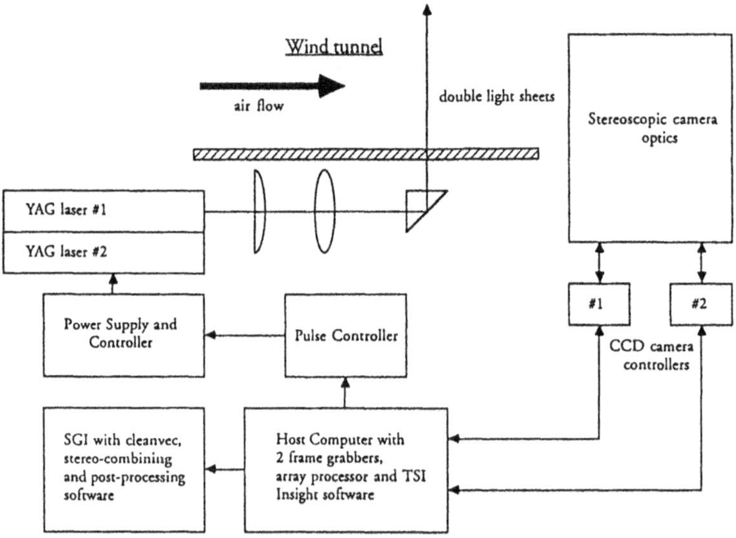

Figure 20. Layout of the laser optics and computer hardware for boundary layer measurements

The measurements were taken in an eiffel-type low turbulence wind tunnel. The free stream velocity was 3.4 m/sec and the measurement area was about 80 mm by 80 mm. Figure 21 is a plot of the in-plane velocity with a velocity of 3.0 m/sec subtracted from all the vectors. The purpose was to reveal the structure of the flow field. It can be observed that a few distinct vortices were formed in the layer. Interestingly the plot of the three-dimensional vectors given in Figure 22 showed that the vortices had fairly strong out-of-plane motion, as indicated on the plot at the left side of the figure. Some of out-of-plane components had a magnitude as much as one-third of the free stream velocity. Figure 23 is a contour of the out-of-plane component, denoted as W, revealing the strong three-dimensional vortices in the boundary layer.

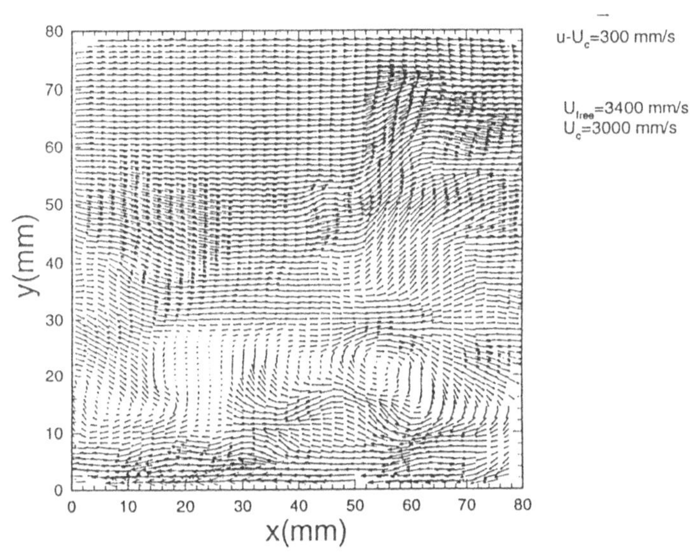

Figure 21. In-plane velocity vectors of the boundary layer measurement

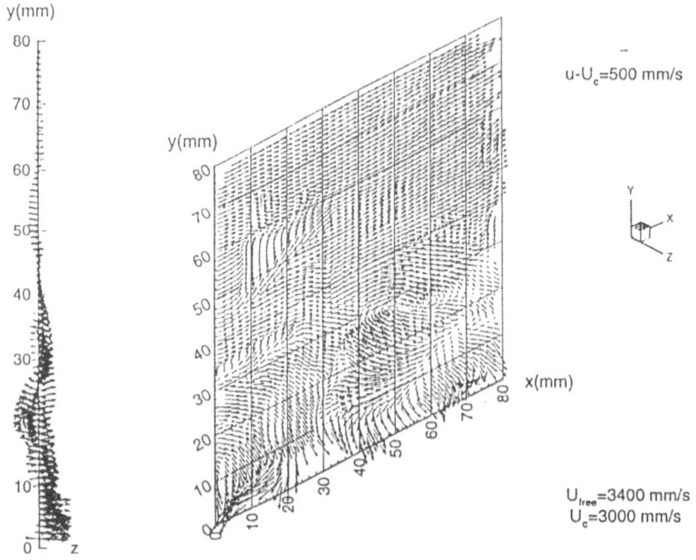

Figure 22. Plot of the three-dimensional vectors

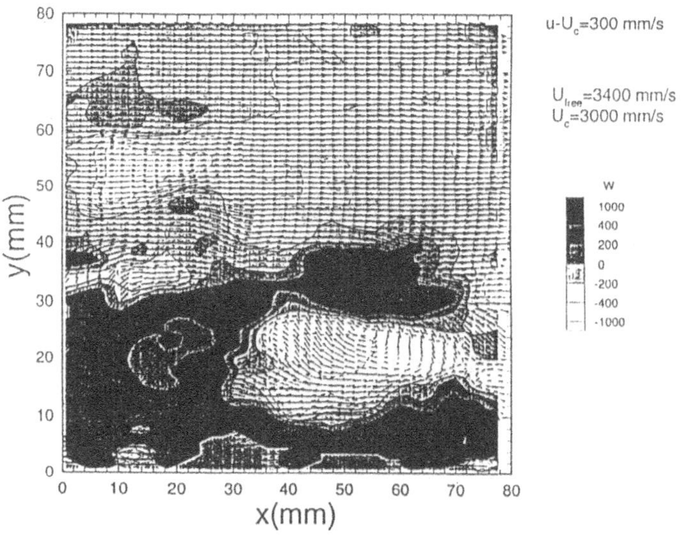

Figure 23 . Contours of the out-of-plane component with the x-y vectors as background

2. Holographic approach

As different from the stereoscopic approach described above, the holographic approach is a true three-dimensional measurement technique for a volumetric domain. The complexity of such approach is also increased substantially. A typical holographic system is divided into two sub-systems, an imaging sub-system and a reconstruction sub-system. Similar to a conventional PIV system, the illumination of the measurement volume is provided with a high energy pulsed laser such as the Nd:YAG laser. The images of the particles are then recorded on a hologram. To reconstruct and interrogate the flow information from the volumetric domain, a CCD camera is normally used to scan through the hologram to record the image field. The image field is digitized in a computer and correlation analysis is used to compute the velocities.

Despite the complexity holographic PIV has received a lot of attention. A few studies in holographic PIV have been taken by Barnhart et al. [23], Bernal and Scherer [24] and Meng and Hussain [25]. The holographic PIV system used by Barnhart et al. [23] employed a phase conjugate play-back geometry to compensate for aberrations in the imaging sub-system. The system also employed a reference multiplexed, off-axis geometry for the determination of velocity directions with crosscorrelation analysis technique, and a stereo camera geometry for the determination of the three velocity components. The system was applied to measure air flow in a turbulent pipe with a Reynolds number of about 6000. A total of 448,362 velocity points were measured in a measurement volume of 24.5 mm x 24.5 mm x 60 mm. Bernal and Scherer [24] constructed a two-axis holographic PIV system for the measurements of a free surface water jet. The system offered a spatial resolution of 2 mm and a spatial dynamic range of about 100. The measurement accuracy with the system was determined to be 2% of the maximum flow speed. The study by Meng and Hussain [25] showed that a system with in-line recording and off-axis viewing for recontructions offered improved signal-to-noise and depth resolution over a conventional in-line holography by an order of magnitude. The particular system was applied to measure a vortex ring.

It is obvious that the benefit and the amount of information obtained from a holographic PIV is substantially higher than a stereoscopic PIV system, primarily due to the fact that the holographic system provides a true three-dimensional measurement in a volumetric domain. However the complexity of such system also prohibits the generalization of the technique to be used in every experimental environment, hence the stereoscopic system provides a good compromise for three-dimensional PIV measurements.

With the specialized camera using frame straddling and crosscorrelation emerging as the important component to PIV measurement, the extension to stereoscopic PIV system will be simplified because the requirement for image shifting will not be necessary. In additional advances in computer hardware, camera interfaces and analysis software will allow image analysis to be performed efficiently and accurately. Therefore three-dimensional PIV systems may be eventually become a common tool for fluid flow research.

7. Conclusions

The PIV technique has become a viable tool for fluid flow research. The ability of the technique to provide a velocity field allows many other flow properties to be measured. Some of the properties are vorticity and strain rates. Flow dynamics and turbulence structures can also be easily investigated from the instantaneous field measurements. In general photographic camera with large format film still provides the best resolution and can be adapted for very large dynamic spatial range, and should be the preferred recording medium for turbulent flow measurements. However the convenience and the instantaneous feedback offered by CCD video cameras overshadow the pixel resolution benefit from photographic cameras, thus making the video cameras as the more popular choice as recording media for PIV.

Image shifting is required to resolve flow direction in PIV measurements. Some of the current image shifting techniques were described and each one may be applicable for a certain velocity range. The most promising technique to resolve flow direction, however, is the use of the crosscorrelation analysis method. Employing the concept of frame straddling, a specialized camera was shown to simplify PIV measurements substantially. The real-time image capture can be achieved with a camera interface which can transfer image data at a speed of up to 90 Mbytes/sec. Batch processing in the PIV software automates the analysis process, making it feasible for time-resolved flow measurements.

Finally three-dimensional PIV using the stereoscopic arrangements may become more attractive with the use of the frame straddling in image capture and crosscorrelation for analysis. Together with the advances in computer hardware, analysis software and camera interfaces, three-dimensional PIV systems will be the tool for fluid flow research in the near future.

8. References

1. Adrian, R.J.: Multi-point optical measurements of simultaneous vectors in unsteady flow -- a review, *Intl. J. Heat and Fluid Flow* (1986), 127-146
2. Kompenhans, J. and Hocker, R.: Application of particle image velocimetry to high speed flows, VKI series , *Particle Image Displacement Velocimetry,* Bruxelles, 1988
3. Humphreys, W.M., Bartram, S.M., and Blackshire, J.L.: A survey of particle image velocimetry applications in Langley aerospace facilites, *AIAA* 93-0411, 1993

4. Adrian,R.J.: Statistical properties of particle image velocimetry measurements in turbulent flow, In *Laser Anemometry in Fluid Mechanics* (1988), pp.115-129. Ladoan: Lisbon, Protugal.
5. Wernet,M. and Pline,A.: particle displacement tracking technique and Cramer-Rao lower bound error in centroid estimates from CCD imagery, *Expts. Fluids* **15**, 295-307
6. Marko,K.A. and Rimai,L.: Video recording and quantitative analysis of seed particle track images in unsteady flow, *Applied Optics* **24** (1985), 3666
7. Gharib,M., Hernan,M.A., Yavrouian,A.H., and Sarohia,V.:Flow velocity measurement by image processing of optically active tracers, *AIAA* Paper 85-0172 (1985)
8. Landreth,C.C., Adrian,R.J., and Yao,C.S.: Double pulsed particle image velocimeter with directional resolution for complex flows, *Experiments in Fluids* **6**, 119-128
9. Kompenhans,J., Raffel,M., Vogt,A., and Schafer,H.J.: Instantaneous flow field measurements on pitching airfoils by means of Particle Image Velocimetry, *Proc. Optical methods and data processing in heat and fluid flow* (1994), London, 117-121
10. Raffel,M., and Kompenhans,J.:PIV measurements of unsteady transonic flow fields above a NACA 0012 airfoil, SPIE Vol. 2052 *Laser Anemometry Advances and Applications* (1993), 527
11. Landreth,C.C., and Adrian,R.J.: Electrooptical image shifting for particle image velocimetry, *Applied Optics* **27** (1988), 4216-4220
12. Keane, R.D., and Adrian, R.J.: Optimization of particle image velocimeters. Part I: Double pulse systems, *Meas. Sci. Technol.* **1** (1990)
13. Keane, R.D., and Adrian, R.J.: Optimization of particle image velocimeters. Part II: Multiple pulsed systems, *Meas. Sci. Technol.* **2** (1991)
14. Keane, R.D., and Adrian, R.J.: Theory of cross correlation analysis of PIV images, *Meas. Sci. Technol.* **3** (1992)
15. Arroyo,M.P., and Greated,C.A.: Stereoscopic particle image velocimetry, *Meas. Sci. Technol.* **2** (1991), 1181-1186
16. Stolz,W., and Kohler,J.: In-plane determination of 3D-velocity vectors using particle tracking anemometry (PTA), *Experiments in Fluids* **17** (1994), 105-109
17. Halloin,V.L., and Jottrand,R.: A new technique for the experimental study of 3D velocity fields, *Experiments in Fluids* **17** (1994), 115-118
18. Raffel,M., Gharib,M., Ronneberger,O., and Kompenhans,J.: Feasibility study of three-dimensional PIV by correlating images of particles within parallel light sheet planes, *Experiments in Fluids* **19** (1995), 69-77
19. Robinson,O., and Rockwell,D.: Construction of three-dimensional images of flow structure via particle tracking techniques, *Experiments in Fluids* **14** (1993), 257-270
20. Guezennec,Y.G., Brodkey,R.S., Triqui,N., and Kent,J.C.: Algorithms for fully automated three-dimensional particle tracking velocimetry, *Experiments in Fluids* **17** (1994), 209-219

21. Prasad,A.K., and Adrian,R.J.: Stereoscopic particle image velocimetry applied to liquid flows, *Experiments in Fluids* **15** (1993), 49-60
22. Lai,W.T., Liu,Z., Adrian,R.J., and Mienhart,C. : Measurement of turbulent boundary layer using stereoscopic particle image velocimetry, to be presented at the Eighth International Symposium on Applications of Laser Techniques to Fluid Mechanics, Lisbon, Portugal, 1996
23. Barnhart,D.H., Adrian,R.J., and Papen,G.C.: Phase-conjugate holographic system for high-resolution particle image velocimetry, *Applied Optics* **33** (1994), 7159-7170
24. Bernal,L.P., and Scherer,J.: Holographic velocity measurements in a turbulent free-surface jet, *AIAA* paper 95-0642 (1995)
25. Meng,H., and Hussain,F.: In-line recording and off-axis viewing technique for holographic particle velocimetry, *Applied Optics* **34** (1995), 1827-1840

REAL-TIME SIGNAL PROCESSING IN PARTICLE IMAGE VELOCIMETRY

The importance of real-time image processing in optimizing data collection and analysis in Particle Image Velocimetry.

D.R. McCLUSKEY AND E. HAYES
DANTEC MEASUREMENT TECHNOLOGY A/S
2740 SKOVLUNDE
DENMARK

Abstract

Many researchers require whole field velocity data with high spatial and temporal resolution to obtain new experimental information about the flow fields of interest. Thus PIV, Particle Image Velocimetry, is becoming a preferred technique. In the literature, many implementations of PIV systems are described. However, a key point in which systems differ is in the time required from image capture to presentation of the corresponding vector map - the processing time. For practical use this potential time bottleneck can be reduced substantially by very fast processing. Thus the focus of the paper is on the camera-illumination control-processing combination for real time PIV measurements. The seeding, laser and illumination light sheet are only briefly mentioned as these are components that are generally available.

In this paper, key parameters for successful PIV measurements will be related to the benefits of using a real time processor that allows on-line access to the PIV vector maps. Possibilities for quickly optimizing a set-up are shown. The immediate benefits are shorter experimental time, reduced running costs and the possibility of quickly analyzing long data series for comparison of results with conventional point measurements such as CTA and LDA as well as with CFD.

1. Introduction

1.1 BASIC PRINCIPLE

With the PIV measuring technique, the velocity is measured by recording the displacement of small particles carried with the flow and subsequently analyzing the particle displacements recorded. Particles embedded in a region of the flow are

Th. Dracos (ed.), Three-Dimensional Velocity and Vorticity Measuring and Image Analysis Techniques, 93–112.

illuminated by two or more short light pulses fired with a known time separation. The particle positions on each exposure are recorded by a camera.
The particle displacements in the flow field are found from the displacements in the image plan. Knowing the magnification of the imaging and the time separation between the light pulses, the velocity projections on the measuring plane are found. Mostly 2-D flows are recorded.

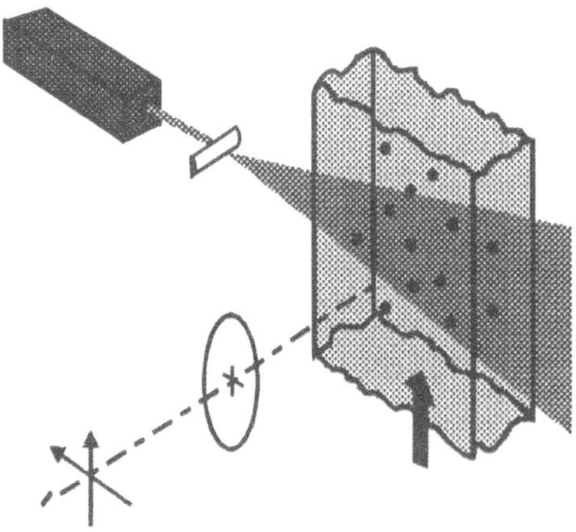

Fig. 1: Outline of a basic PIV system

The results, the primary velocity data, are usually calculated by identifying the average displacement of particle pairs on the recorded images by using a correlation technique. The deduced velocities are presented in a vector map with typically from 1,000 to 10,000 vectors per map.
Using conventional processors such as PC's or PC-array processor combinations, the processing time for a single image map can vary from several seconds to minutes or longer. While this may, at first sight, seem reasonable for a single measurement, it is actually very inconvenient as there are many parameters that must be adjusted and fine tuned during an experiment. The need for fast processing becomes much more obvious as the number of image maps to be processed increases.

1.2 PIV MEASUREMENT PROCESS

A complete PIV experiment can be divided into the five basic procedures, all of which require control or user interaction creating the need for efficient experimental procedures. There are five key experimental steps:

1. Seeding of the flow.
2. Illumination of cross section of the flow.
3. Recording of scatterers in the flow field.
4. Processing of the records.
5. Analysis for presentation

These topics have been discussed generally in the literature. However some comments on these key steps in the context of efficient experimental procedures and real rime measurements are appropriate.

1.3 SEEDING REQUIREMENTS

The size and type of the particles whose positions are recorded depends on the actual application. For flow studies, the general requirement is that the particles must be small enough to follow fluctuations in the flow field and at the same time be good scatterers.

Typical examples of seeding particles are microfeather spheres (Ikeda *et al.*, 1994) or TiO2 particles for measurements in IC-engines, smoke and water/glycerine particles for air flows and latex particles or pollen in water flows (Mengel *et al.*, 1993).

For good PIV data it is normally important to obtain a constant and uniform particle distribution as well as a reproducible and known particle size distribution Compared to LDA, the flow area that must be seeded is much larger. Also the seeding level must be adjusted so that the particle image distribution on the image map is fairly uniform and sufficiently plentiful for flowmapping. Since the particles are often generated using dry powder dispersers, atomizers, fog- and smoke generators, adjustment of the seeding level and distribution must be done experimentally by evaluating the image maps and vector maps recorded.

Thus, suitable seeding levels can be quickly achieved with a real-time processing system that gives immediate feedback on the effect of changes.

1.4 ILLUMINATION

A light sheet is formed to illuminate a cross section of the flow field. A light sheet can be less than 1 mm thick so a narrow slice of the flow is illuminated. For experiments with cross-flows or large particles, adjustment of the light sheet thickness or the time between light pulses may be needed.

The amount of light available must be sufficient for the recorder to detect the particles. The illumination system must be controllable so that the appropriate pulse time intervals for the recordings must be selectable easily. The exposure time should be determined so that the particles do not move significantly during each illumination pulse. Preferably the displacement of the particle during illumination should be less than 0.1 x diameter of image size. The time interval between pulses should be chosen such that the particle displacement on the recorder image plane is within the dynamic range of the PIV analysis system.

1.5 ILLUMINATION SOURCES

1.5.1 CW Lasers

When studying low speed flows, a CW Ar-ion laser will often give sufficient power. CW Argon-ion lasers from 300mW to 10 W are the most commonly used. Electronic shuttering must be used to pulse the light to give multiple pulses for auto-correlation or double pulses for cross-correlation.

A standard light sheet with cylindrical lenses can be used A more sophisticated light sheet is a top hat form. In both cases, the light intensity in the illuminated area will normally be low.

1.5.2 Polygon Scanner Light Sheet

A novel use of a rotating polygon mirror as a beam scanner scanner (Grey *et al.*, 1991) makes better use of the power in a continuous laser as the full beam power is scanned across the flow area in a stroboscopic fashion to make a double or multiple pulse.

The swept light sheet is created by the laser beam being directed to a Polygon Scanner directly. The Polygon Scanner can be fitted with a different number of mirror facets which together with the rotation speed of the Polygon scanner determines the maximum velocity measurable. The combination of 18 facets at a selection of rotational speeds up to 15,000 rpm enable sweep speeds as high as 4.5 kHz and light sheet sweep angles of up to 40 deg.

A Polygon scanner can further be fitted with an electro-mechanical shutter with computer controlled driver for pulsing the incoming light beam. The scanner motor and shutter can be controlled to set the rotational speed and the shutter time needed for the measurement.

1.5.2 Pulsed Lasers

When studying large flow areas or higher speed flows, a high energy intensity in the illumination is required. A pulsed Nd:YAG laser is used in combination with high power optics. This type of laser emits pulses of laser light where both the pulse separation and repetition frequency can be varied. The pulse duration (in the order of 10 nanoseconds) is usually fixed by the laser specification. A light sheet optics using high power glass components must be used with the more powerful Nd:Yag lasers.

1.5.3 Laser Sheet Pulse Rate

The light sheet repetition pulse rate depends on the type of illumination source used.

For continuous lasers, the pulse rate is variable, depending on the shuttering device used.

For Nd:Yag lasers, it is the laser specification that sets the limit, typically at 10 Hz. Double cavity Nd:Yag lasers are very flexible in timing while maintaining power in the pulses. Single cavity double pulsed lasers allow a limited range of pulse intervals with decreasing power per pulse as the pulse interval increases.

1.6 RECORDING A FLOW FIELD

An image of the light scattering particles in the illuminated cross section of the flow field must be captured at distinct time intervals determined by the illumination pulses on a medium suitable for later analysis. The most common ways of making PIV recordings are by CCD-camera or by photographic film.
Recent improvements in CCD technology have made today's CCD recorders almost comparable to photographic film in pixel resolution, pixel density, sensitivity and dynamic range, making the PIV technique much more user friendly.

Once the PIV images have been recorded, the image maps have to be converted into useful vector maps. CCD technology is necessary to implement real-time processing and presentation of vector maps. However, for data recorded on photographic film, a real time processor can still be used with advantage to analyze the developed file images.

2. The Need for a Real Time PIV Processor

The whole measurement process from illumination, cameras, processing through to data storage and analysis needs to be considered as an integrated process. In practice, this requires selection and synchronization of illumination and camera instrumentation as well fine tuning processing parameters to suit measurement of the particular flow field under investigation (McCluskey, 1995, McCluskey *et al.*, 1995).

2.1 SYNCHRONIZATION OF PULSED ILLUMINATION AND RECORDING

2.1.1 Continuous Lasers

For cross-correlation PIV vector processing, it is necessary to have two images of the flow field each of which have been created by exposure to a single pulse of illuminating laser light. Thus, for a measurement system including a trigger event, a polygon scanner, a laser shutter and a cross-correlation camera, the following sequence of signals is required to capture image maps (Stages 3 to 8 are shown in Fig. 2):

1. The polygon scanner should be set to rotate at the frequency necessary to obtain the required time interval between pulses. (In practise, there is also an output tachometer signal from the polygon scanner which is continually monitored. In this way the rotation frequency can ce checked).
2. To prevent exposure of the camera to stray/non-signal laser light, the laser shutter should be set to the closed position.
3. A trigger should initiate the measurement sequence.
4. The free-running camera should be re-set so that the time between image pairs is the same as the required time interval between pulses and the double-frame rate is set to the required vector map rate.
5. The integration of the light falling on the CCD chip, for the first frame of the image-pair, should occur at the start of a sweep of the polygon scanner.

6. A laser shutter should be opened for the duration of the sweep of the polygon scanner so that the camera can be exposed to laser light scattered by the seeding particles in the flow field.
7. The camera frame should be advanced (when the laser shutter is closed) and the image transferred to the register in the camera.
8. The laser shutter should be re-opened to expose the second frame of the image pair to scattered laser light from another sweep of the scanning beam and then closed.
9. Both the first image in the register in the camera followed by second image from the CCD chip should be transferred for cross-correlation vector processing.
10. Stages 5 to 9 should be repeated at the required vector frame rate

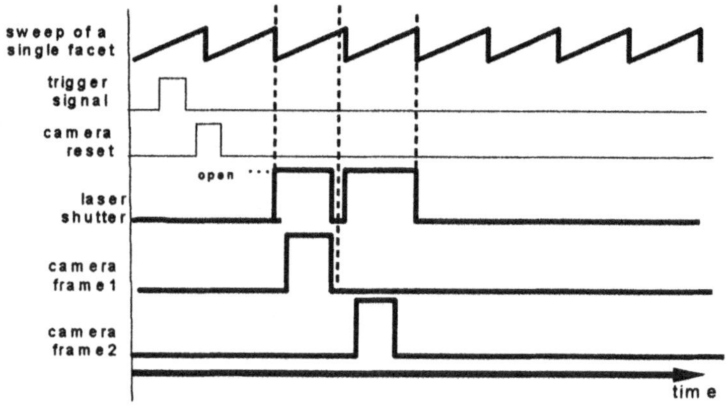

Fig. 2. Automatically generated system synchronisation and control sequence.

2.1.2 Pulsed Lasers

The synchronization timing sequence for a typical double cavity Nd:Yag laser is shown in Figure 3. The laser can operated in a free-run mode, where the laser informs the rest of the system that it is ready to emit laser light. Alternatively, the laser cavities are instructed to fire their flash lamps and when to open their Q-switches to emit laser radiation by asynchronization signals. The camera synchronization signals are as shown above, except that the laser shutter signals can be replaced by the Q-switch / Flash lamp signals.

Manually generating such synchronisation and control sequences is a rather laborious and time consuming task, particularly since the time between pulses and pulse frame rate should be modified to optimise the data acquisition and processing parameters for the particular flow field under investigation. Hence, automatic generation of synchronisation and control signal sequences and their implementation are important requirements for a real time processing.

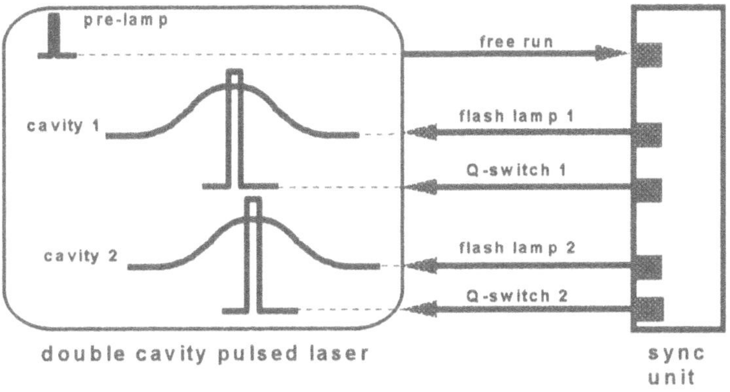

Fig. 3. Communication signals with a double-cavity, Q-switched pulsed laser for synchronisation.

2.2 OPTIMIZATION OF PIV RECORDS

Assuming that the laser pulse / camera have been synchronized, fine adjustment is usually necessary so that the captured images of the particle pairs are suitable for analysis. Obtaining good PIV images for analysis depends on the interactive selection of several other parameters depending on:

- Min. velocity
- Max. velocity
- Max. cross velocity
- Spatial resolution
- No. of pixels in CCD
- No. of vectors in array

These requirements have been fully described in the general literature on PIV and will not be discussed here.

2.3 IMAGE MAP ANALYSIS

Several methods are available to extract the information about the displacement of scatterers between two exposures. In general the particle displacement is found by calculating how similar, or correlated, a recorded exposure is with a second exposure after a certain interval.

When the PIV image is analysed pixel by pixel, the usual procedure is to divide the picture into interrogation areas. A specified interrogation area of a suitable small size is chosen and an average displacement for the particle images is computed. The size of the interrogation area is selected small enough that the corresponding section of the flow may be assumed to have a constant velocity but large enough so that it will contain sufficient particle images pairs for processing.

The PIV image is then analysed by alternately analysing an interrogation area to determine the velocity at the point and stepping to the neighbouring interrogation area, until the complete PIV recording has been covered.

An efficient method of analysing an interrogation area in a PIV image is to use the auto correlation or cross-correlation function to determine a dominating displacement trend in each interrogation area in the image. Converting the image map into a vector map is an integral part of the experimental procedure. The need for real time processing is obvious here as the vector map gives an immediate feedback on the correct selection of the various experimental parameters.

3. An Innovative Real Time Processor for PIV

The novel real time processor for PIV described below forms the basis of the remainder of the paper. The FlowMap PIV 2000 Processor described is capable of outputting a vector map in less than 200 msec., approximately 100 to 1000 times faster than a conventional frame grabber based / pc system.

3.1 PIV 2000 PROCESSOR

The FlowMap PIV Processor 2000 is a high speed dedicated processor for PIV measurements. With the basic configuration the processing time is up to 6 Mpixel per sec. For a CCD recorder with 1k x 1k pixel array and an interrogation area of 64 x 64 pixels, the vector array of 32 x 32 vectors is processed in less than 200 msec in the basic configuration. Using options, the speed can be increased to less than 50 msec per vector map.
The significant speed superiority is achieved by replacing the conventional frame grabber by a specially designed high speed input buffer and correlator unit.

The main processor components are shown in Figure 4.
1. Main frame with power supply
2. Synchronisation Board
3. Input Buffer
4. Vector Evaluation and Validation (Correlator)
5. 4-ch. Analog Input for User'sDevices
6. PC-interface

The Synchronisation Board consists of a multi-channel programmeable signal generator for controlling lasers and cameras. Further, the synchronisation board includes a 4 channel A/D converter for external analog sources (temperature, pressure, etc.).

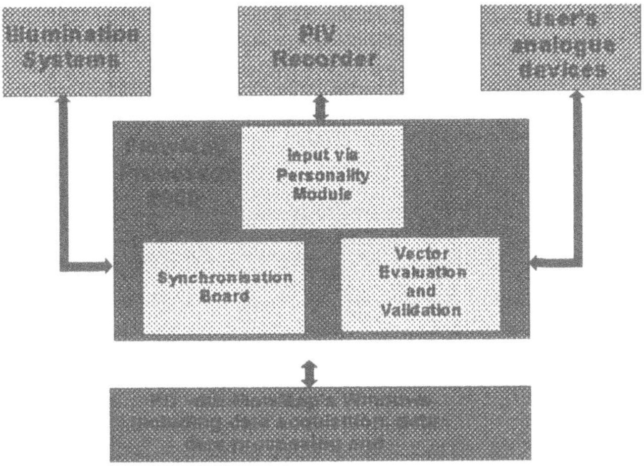

Fig. 4: Block diagram of PIV 2000 Processor

The Input Buffer has an on board memory of 4 Mega pixels and can be expanded to 128 Mega pixels for recording of image sequences. Thus, with a typical 1k x 1k camera, up to 128 image maps can be captured without the need for processing. Also, the input buffer supports down load from and up load to PC and interrogation area read out to the correlator.

Depending on the actual application, the Vector Evaluation and Validation is performed at high speed using either auto and/or cross correlation. In the correlator, post processing, peak finding and validation is performed, on-line (Host-Madsen *et al.*, 1994).
Various criteria are selectable as shown in Figure 5. The most important point is that for an array of approximately 1,000 vectors (32 x 32 vectors), manual evaluation of validity criteria is not practical for on-line presentation of vector maps.
A built-in IEEE card interfaces the FlowMap processor to a PC forcomplete control via the application software.

In the dedicated hardware the processing stages from obtaining a CCD image map input to obtaining a vector map output for auto-correlation are illustrated in Fig. 5 and is as follows.

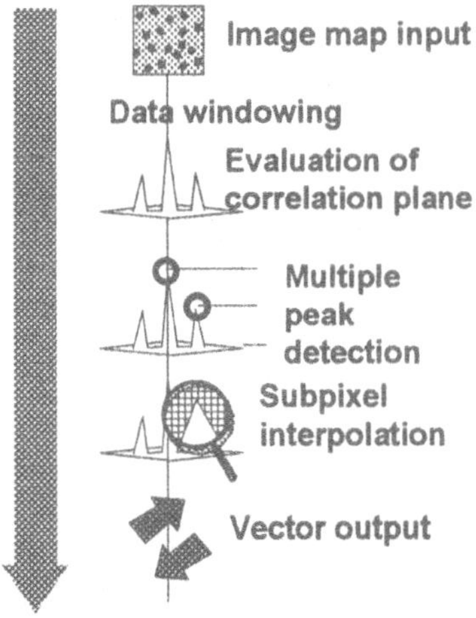

Fig. 5. Stages involved in processing a CCD image map into a velocity vector map using auto-correlation.

1. A whole CCD camera image is subdivided into interrogation regions of selectable dimensions: a range of interrogation area sides is supported because the size of the interrogation area influences the size of the velocity structures that can be resolved as follows

$$\lambda_{struct} = \frac{N_{int} \times d_{pixel}}{M}$$

where N_{int} is the number of pixels in the interrogation area
d_{pixel} is the diameter of the pixel elements on the CCD chip
M is the object:image magnification

Furthermore, it constrains the upper limit of the velocity dynamic range as follows

$$\mathbf{d}_{max} = M\mathbf{v}_{max} t \leq 0.25 N_{int} d_{pixel}$$

where $\mathbf{d}_{max}$ is the maximum measurable displacement vector
$\mathbf{v}_{max}$ is the maximum velocity vector to be found in the flow field
t is the time interval between successive illumination pulses.

This stage takes place in a purpose designed input buffer which replaces conventional frame grabbers which generally can only either read or write. Using frame grabbers limits the total number of frames that can be recorded in a given time series as well as interrupting the measurement process. Therefore, the input buffer, unlike frame grabbers, can simultaneously read image maps from a CCD recorder, store image maps in a memory area and either divide stored image maps into interrogation units and send these for vector processing or send the

entire image to the PC. Thus, together with the correlator unit, it is in principle possible to acquire an infinitely long time sequence.
The following stages take place in the correlator unit.

2. A data window can be applied to the image map. The purpose of this window function is two fold. Firstly, calculation of the correlation plane by FFTs produces cyclic noise. This can be reduced by a technique known as zero padding, where the grey scale value of the pixels at the outer edges of the interrogation area are set to zero. That is, the window function W(*x*,y)

$$W(x,y)=\begin{cases} 1 \quad when & \begin{cases} \dfrac{-kX}{2}\le x\le\dfrac{kX}{2} \\ and \quad \dfrac{-kY}{2}\le y\le\dfrac{kY}{2} \end{cases} \\ 0 \quad when & \begin{cases} \dfrac{-X}{2}\le x\le\dfrac{-kX}{2}\,r \\ or \quad \dfrac{kX}{2}\le x\le\dfrac{X}{2} \quad or \\ \dfrac{-Y}{2}\le x\le\dfrac{-kY}{2} \quad or \\ \dfrac{kY}{2}\le x\le\dfrac{-Y}{2} \end{cases} \end{cases}$$

where *k* determines the width of the top-hat window
x is the distance from the centre of the correlation plane in the x-direction
X is the side of the interrogation are in the x-direction
y is the distance from the centre of the correlation plane in the y-direction
Y is the side of the interrogation are in the y-direction

A Gaussian window is similarly effective in reducing cyclical noise. Furthermore, applying a Gaussian window improves measurement accuracy in cases where particle images straddle the boundaries of interrogation regions: if a window function were not applied, those cut particles would bias the measurement towards zero. In this case, the true displacement vector actually ends outside the interrogation region and, since the calculation is unaware of the existence of the second half of the particle image, it assumes that the displacement vector is actually half a particle image diameter less than its true length. A Gaussian window function effectively filters the grey scale values of the interrogation area so that the data in the centre of the region has the most significant contribution to the following FFT stage and, as one moves radially out from the centre, particle images are attenuated by an increasing amount and so contribute less and less to the velocity measurement. One such window is

$$W(x,y) = \exp\left[-\left[\frac{1}{k}\right]^2\left(\left(\frac{2x}{X}\right)^2 + \left(\frac{2y}{Y}\right)^2\right)\right]$$

where k is the width of the Gaussian window.

Window functions have not been widely discussed in PIV partly because their application on previous PC based systems incurred a significant overhead in processing time. However, with the developed instrumentation, all processing stages are streamlined so that a window function can be applied to, for example, a second interrogation region while the evaluation of the correlation plane of the first interrogation region takes place. In effect, there are no processing overheads. A further feature of the system is that the window function is implemented by a programmable chip, and, as research in this area of PIV processing develops, new window functions can be readily applied.

3. The correlation plane is determined by FFT processing of the image map. To optimise processing speed, the calculation is performed by dedicated hardware.

4. Filtering the correlation plane can improve the signal to noise ratio since it effectively reduces camera and light sheet noise. Again, this is an area of PIV signal processing which has been neglected in the past. As with window functions, there is no time overhead in applying a filter because the processing sequence is streamlined and filters are implemented by a programmable chip.

5. The correlation plane is scanned and, in auto-correlation, the self-correlation peak and the two highest/signal peaks are detected. In cross-correlation, the two highest/signal peaks are detected. The height and width of the self-correlation peak (when using auto-correlation analysis), together with the height and width of the two highest correlation peaks, are then output for up-loading to the PC. The height of the signal peak and second highest peak can then be used as a quality check. This data quality check is particularly useful when optimising data recording and processing parameters.

6. The position of the signal peaks is determined to subpixel resolution using an interpolation routine. Subpixel interpolation effectively means that velocity accuracy can be more than the resolution of the individual pixel elements.

A similar processing sequence is implemented for cross-correlation analysis. In this case, two interrogation regions are required and these can be from successive images from the same camera, different cameras or, in a single frame case, from different interrogation regions.

Obviously, the time taken to evaluate a given vector map is dependent upon the user's selection of analysis options, particularly the size and overlap of the interrogation region as well as the pixel array size of the selected PIV recorder. Nonetheless, multiple vector maps processed from high format CCD cameras are processed virtually in real-time.

3.2 REAL TIME PIV IN PRACTICE

All of the options for data processing, whether to use auto or cross-correlation analysis, the size of the interrogation region, application of window functions, application of filters and so on, are selected via Windows menus in the software. The parameters are then downloaded to the processor prior to the start of the measurement process. This centralises control of the processor unit along with synchronisation and control of CCD cameras and illumination systems, since the PC software also generates these. Figure 6 shows how the software generated signals are implemented by the Synchronisation board in the processor which provides the physical communication links between the processor and other elements in the PIV system.

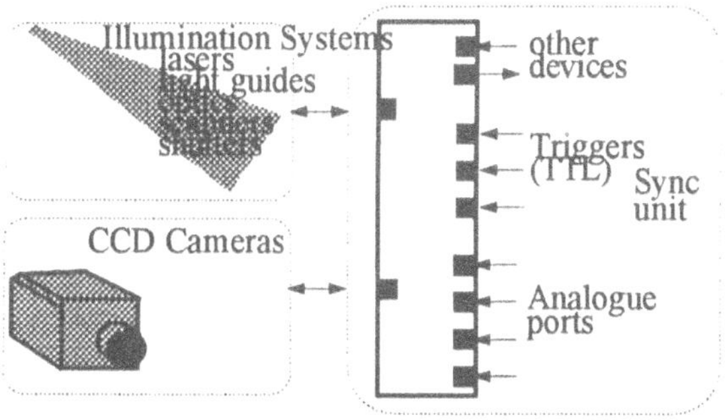

Fig. 6. Overview of synchronisation signals.

3.2.1 Camera Synchronization and Control

In PIV, it is important to control the camera shutter so that the CCD sensor is only exposed to the light from the illumination system since ambient light will reduce the signal-to-noise ratio. Furthermore, since camera shutters are generally rather slow, it is important to ensure that the camera is fully open when the illumination pulses are fired so that the illumination energy is effectively deployed and successive images of a given particle have the same energy density on the recorder. Thus, it is necessary to control the shutter and have a record of its opening and closing times.

With CCD cameras it is also possible to control the integration period of the pixels. The integration period is the interval in time over which the pixels contribute the measurement of detected light. Therefore, the contribution of ambient light to measurement noise can be minimised by ensuring that the integration time only covers the interval between and the duration of the illumination pulses. This is illustrated in Figure 7.

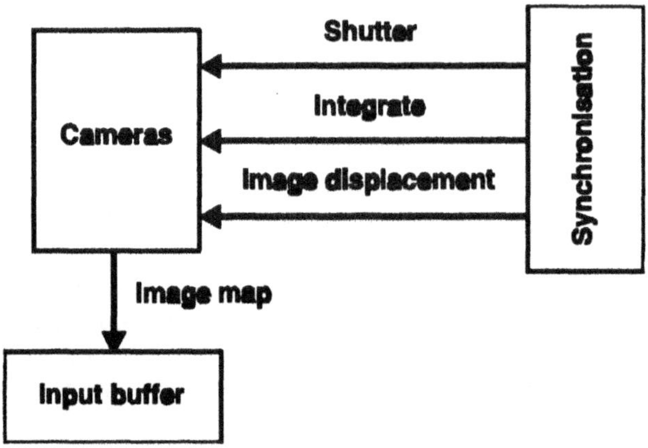

Fig. 7. Key camera synchronisation and control signals.

The CCD camera can be selected to meet the requirement of the application. The cameras are linked to the FlowMap processor with personality modules. The personality module can be upgraded to account for future developments in the CCD technology.
The cameras presently supported are listed in Table 1:

Table 1: CCD cameras presently supported by PIV2000

Camera	Pixel Array	Type
Kodak MegaPlus 4.2I	2029 x 2044	Auto-correlation
Kodak MegaPlus 1.4I	1317 x 1035	Auto-correlation
Dantec PIV 700	768 x 480	Cross-correlation
Dantec FlowSense	1024 x 1024	Auto-correl. with pixel shi

Each CCD camera above uses square pixels for accurate geometry. The camera / lens combination gives an approximate 100% effective fill of the sensitive area.

3.3 REAL TIME Vs CONVENTIONAL FRAME GRABBER SYSTEMS

Implementations of PIV based on FrameGrabber / PC's are time consuming in converting the camera image to the vector map. Table 2 compares the typical delays involved in a short experimental run with 375 vector maps recorded with the real time system.

Table 2: PIV vector processing systems and their overhead

	On-line System	Conventional PIV system
Camera technology	CCD	CCD
Images per run	375	375
Processing	High speed correlator	PC / array processor
Required hard disk space per run	2.1 Mbytes	140 Mbytes
Total number of vectors to be processed	135,000	135,000
Number of vectors processed per run	135 000 (4 correlators)	1000 (40 vectors/sec)
Delay until next run due to off-line processing time requirements	NONE	1 full hour

This time has a significant knock-on effect when initially setting-up a PIV measurement run, resulting in long, costly delays between set-up and results. In contrast, with an integrated real-time PIV system, it should be possible to fine-tune data acquisition and processing parameters while running the experiment. Furthermore, once the system has been optimised, the individual vector maps can be monitored while performing practical experimental measurements. That is, by monitoring the in-coming vector maps in real-time, the experimenter is in complete control of the situation.

3.4 DATA BASE ORGANIZATION

The developed Real Time PIV system includes a comprehensive PC based software package which operates in the Windows™ platform. This provides an icon-orientated graphical interface for integrating data acquisition and control; data validation; and, data analysis and storage. as shown in Figure 8. The software essentially provides an automated interface which integrates all instrumentation in the PIV system as shown by McCluskey *et al.*, (1996).

3.5 SET-UP AND PROCESSING

A key element in the software is the user interface which controls the set-up of all instrumentation in the system as well as the processing options. After connection of the relevant equipment, the relevant device libraries are activated. The key data acquisition and processing parameters are input and the details of how to attain this is handled in the background by the system.

During measurements, the PIV Processor processes the incoming image map from the CCD cameras into a raw vector map, virtually in real-time, and transfers these to be displayed and stored (recorded) on the PC. It is also possible to transfer the image map to the PC directly without processing. The software has a data base which contains an archive of the data acquisition and processing parameters associated with a given data record, automatically gives a unique filename to the incoming data record, and, updates a icon display of the data base. With a virtually on-line vector processing system this is essential to prevent chaos.

3.6 ANALYSIS

There is a wide variety of analysis routines available to extract different information from the data which include data validation, filtering, streamline and vorticity calculations. Therefore, to each PIV measurement there can be several derived data sets. Following analysis, the derived data sets are automatically stored on the fixed disc from which they can be retrieved for subsequent analysis, graphic presentation and/or output either as a hard copy or as an ASCII file. The software database plays an instrumental role by automating good housekeeping of the data and is illustrated in Figure 8:.

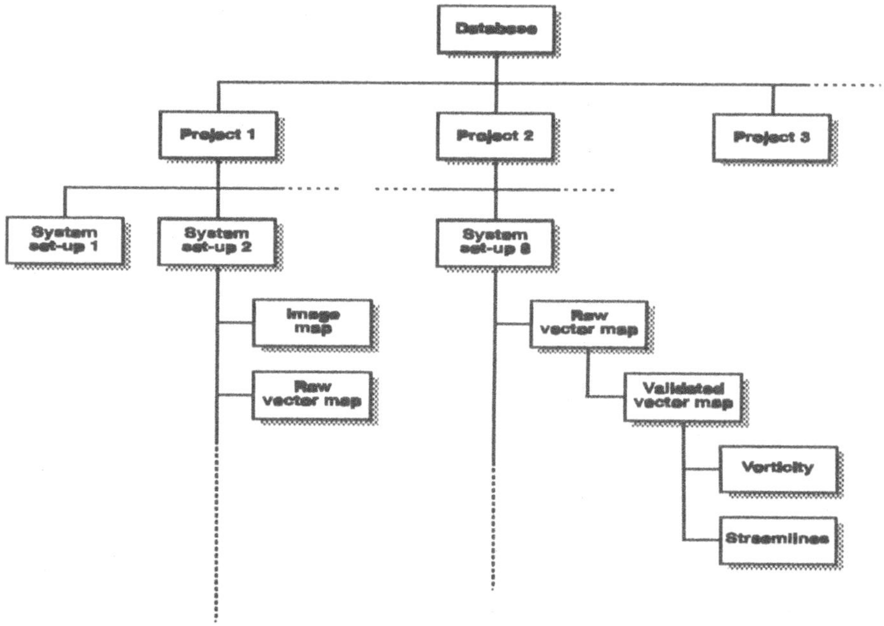

Fig. 8 Software database structure for automatic documentation of real time data

4. Benefits of Real Time Processing

4.1 MULTIPLE ANALYSIS USING VECTOR OVERLAP

In practise, the number of independent vectors in a vector map can be increased by using overlapping interrogation areas so that all the particle images are used, including those near the edges. Overlapping interrogation areas by 25 % of their side is recommended, since it is highly probable that a particle that has an image near the outer edge of the interrogation region may not contribute to a standard vector calculation as its paired image may be in a neighbouring interrogation area. Thus vector overlap increases the need for fast processing.

4.2 MULTIPLE CHOICE VECTOR MAPS, DATA ANALYSIS AND DISPLAY

The experimental set-ups and the raw data are displayed as icons in the icon-orientated graphical interface generated by the software. There is a wide variety of analysis routines available to extract different information from the raw velocity vector map data which include data validation, filtering, streamline calculations, vorticity calculations and so on. Therefore, to each raw PIV measurement there will be several derived data sets, perhaps in a complex chain of calculation options, as shown in Fig. 9. In the database, each of these derived data records is as an icon. Furthermore, the database automatically keeps track of such calculation chains, archiving the calculation method and parameters. Thus, full measurement traceability is ensured.

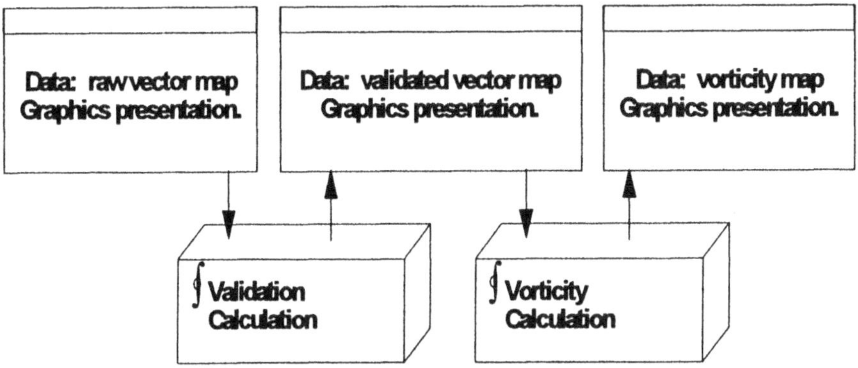

Fig. 9 Example of a calculation chain in the developed software.

Using the calculation chain shown in Fig. 9, the validated vector map and vorticity map shown in Figure 10 were obtained by applying the moving average validation method described by Høst-Madsen *et al.*, (1994).

4.3 MULTI-CAMERA SITUATIONS

The usual configuration of PIV systems is 2-D with a single camera. However, the growing interest in using several cameras simultaneously places greater demand on the need for real-time processing. With the PIV 2000, up to 4 cameras can be installed. Multiple camera options may be useful in complex flow fields for analysis of several different flow areas. A further possibility is to use one camera to obtain a whole field image and another to view some fine-scale details of particular sub-regions of the flow field. Other possible configurations include 3-D using 2 cameras.

Using several cameras increases the requirement for calculation speed.

4.4 LONG SERIES OF VECTOR MAPS

For time history, the typical max. pulse rate of a Nd:Yag laser is about 10 Hz (100 Hz in special cases). Thus a flow can be illuminated at maximum every 0.1 sec. For slowly changing phenomena, this can be sufficient as the development in flow structure can be followed at this sample rate. Additionally, only with real-time vector processing is it feasible and cost-effective to accumulate a significant number of vector maps for statistical analysis.

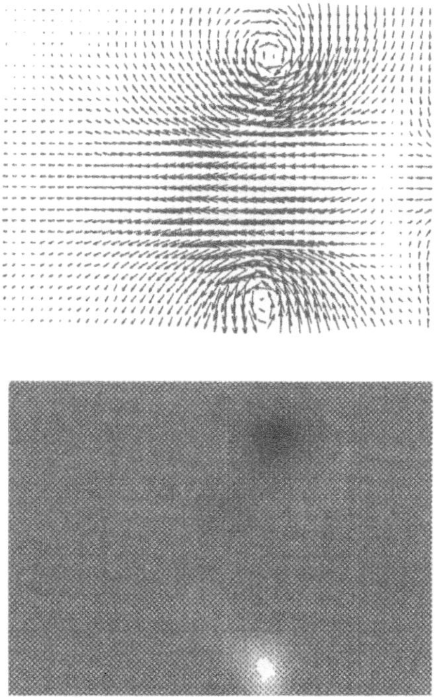

Fig.10: Vector map and corresponding vorticity map evaluated by real time system.

5. Experiments With Real Time Processing

The benefits of the PIV 2000 Real Time PIV Processor described have has been successfully demonstrated in a variety of experimental situations. Flow studies covering an area up to 70 cm x 70 cm inside a car cabin have been described by Mengel et al (xxx) while the evolution of vortical flow fields using extensive time series has been reported by McCluskey *et al.*, (1995). Comparison of time based data series obtained with CTA and LDA have been compared with data series taken with the PIV 2000 processor (McCluskey *et al.*, 1996).

6. Summary

PIV is a particularly powerful measuring tool where the flow field contains turbulent features which are transitory, periodic or an instantaneous event.

The PIV measuring technique, with the introduction of PIV 2000 Processor for synchronization and on-line processing of vector maps, now has become easy-to-use. The range of applications is now expected to be extended as much larger amounts of data can now be recorded and analyzed while reducing the experimental time when compared to conventional PIV systems using phoptographic techniques or CCD techniques with low speed processing.

7. Acknowledgements

The Dantec FlowMap System presented in this paper has been developed by a team at the R&D Department of the Fluid Motion Group, Dantec Measurement Technology A/S, Copenhagen, Denmark.

8. References

Grey, C., Greated, C.A., McCluskey, D.R. and Easson, W.J, (1991), An analysis of the scanning beam PIV illumination system, *Meas. Sci. Technol., vol 2.*

Host-Madsen A. and McCluskey, D.R., (1994), On the Accuracy and Reliability of PIV Measurements, *7th International Symposium on Applications of Laser Techniques to Fluid Mechanics, Lisbon, Portugal*

Ikeda Y., Nishigaki, M., Ippommatsu, M., Hosokawa, S., and Nakajima, T., (1992), Optimum seeding particles for successful LDV measurements, *6th International Symposium on Applications of Laser Techniques to Fluid Mechanics, Lisbon, Portugal*

Mengel, F. And Mørch.T., (1993), Prediction of PIV Recording Performance, *Laser Anemometry Advances and Applications, SPIE Vol. 2052, 331 to 338*

Mengel, F. And Mørch.T., (1995), PIV measurement of air flow inside a car cabin, *Euromech Colloquium, Rome, Italy, June 6-7, 1995.*

McCluskey, D.R., (1995) Obtaining High Resolution PIV Vector Maps in Real Time. *Proceedings of International Workshop on PIV - Fukui 95, Fukui, Japan.*

McCluskey, D.R. and Jacobsen, T.: Instrumentation for Real Time PIV Measurements, *Joint ASME, JSME, EALA 6th international conference on laser anemometry, South Carolina, August, 1995.*

McCluskey, D.R., Laursen, T.S., Rasmussen, J.J., and Stenum, B., (1995), Evolution of Vortical Flow Fields measured by Real Time PIV System. *Joint ASME, JSME, EALA 6th international conference on laser anemometry, South Carolina.*

McCluskey, D.R., Hayes, E.I. and Zhang Jie, (1996), Software for control, acquisition and archival of continuous PIV data, *4th Asian Symposium on Visualization, Beijing, PRC, May 15-18.*

McCluskey, D.R., Jørgensen, F.E., and P. Gjelstrup, (1996), PIV, LDA and CTA Common features and differences. *8th International Symposium on Applications of Laser Techniques to Fluid Mechanics, Lisbon, Portugal..*

INVESTIGATION OF UNSTEADY FLOW FIELDS IN WIND TUNNELS BY MEANS OF PARTICLE IMAGE VELOCIMETRY

J. KOMPENHANS, M. RAFFEL, C. WILLERT, M. WIEGEL,
C. KÄHLER, A. SCHRÖDER, B. BRETTHAUER,
H. VOLLMERS, B. STASICKI
Deutsche Forschungsanstalt für Luft- und Raumfahrt (DLR)
Institut für Strömungsmechanik, Göttingen, Germany

Particle Image Velocimetry (PIV) is increasingly used to investigate unsteady velocity fields instantaneously. For the first time the PIV technique allows the recording of a complete velocity field in a plane of the flow within microseconds. PIV provides information about unsteady flow fields which is difficult to obtain with other experimental techniques. It also helps to save operational time in the wind tunnel. At DLR a set up for PIV has been developed which can be operated under the rough environmental conditions (noise, vibrations) of a large industrial wind tunnel. This system has been successfully applied to very low speed as well as to high speed flows (U = 1 ... 600 m/s). The evaluation and post-processing of the PIV recordings run fully automated on a workstation.

1. Introduction

Today PIV is applied to the investigation of flow fields in many different areas. This means that also a number of special implementations of the PIV technique had to be developed for different applications such as in biology or in turbomachinery. One important application of PIV in industrial research is in aerodynamics. A PIV system for the investigation of air flows in wind tunnels must be operated as well in low speed flows (e.g. flow velocities of less than 1 m/s in boundary layers) as in high speed flows (flow velocities up to 600 m/s in supersonic flows with shocks). Flow fields above solid, moving, or deforming models have to be investigated. The application of the PIV

Updated version of a contribution to the Proceedings of the 2nd Int. Conf. on Exp. Fluid Mechanics (I.C.E.F.M. 94), July 4-8, 1994 - Torino, Italy

Th. Dracos (ed.), Three-Dimensional Velocity and Vorticity Measuring and Image Analysis Techniques, 113–127.

technique in large, industrial wind tunnels poses a number of special problems: large observation area, long distances between the observation area and the light source and the recording camera, time limitations for the measurement, and high operational costs of the wind tunnel.

In spite of all these severe problems, the use of the PIV technique is very attractive in modern aerodynamics, because it helps to understand unsteady flow phenomena as e.g. in separated flows above models at high angle of attack. PIV enables spatially resolved measurements of the instantaneous flow velocity field within a very short time and allows the detection of large and small scale spatial structures in the flow velocity field. Another need of modern aerodynamics is that the increasing number and increasing quality of numerical calculations of flow fields require adequate experimental data for validation of the numerical codes in order to decide whether the physics of the problem has been modeled correctly. For this purpose carefully designed experiments have to be performed in close cooperation with those scientists doing the numerical calculations. The experimental data of the flow field must possess high resolution in time and space in order to be able to compare them with high density numerical data fields. The PIV technique is an appropriate experimental tool for this task, especially, if information about the instantaneous velocity field is required.

The description of the problems as given above leads to the definition of requirements which should be fulfilled when PIV is applied in aerodynamics. First of all, a high spatial resolution of the data field is necessary in order to resolve large scale as well as small scale structures in the flow. This condition directly influences the choice of the recording medium (video or photographical recording). A second important condition is that a high density of experimental data is required for a meaningful comparison with the results of numerical calculations. Thus, the image density (i.e. number of particle images per interrogation spot) must be high. A powerful seeding generator (high concentration of tracer particles in the measuring volume in the flow even at high flow velocities) is needed for this purpose. As the flow velocity is measured indirectly by means of the measurement of the velocity of tracer particles added to the flow, the tracer particles must follow the flow faithfully. This requires the use of very small tracer particles. However, small particles scatter little light. This fact results in a third important condition for the application of PIV in aerodynamics: A powerful pulse laser is required for the illumination of the flow field.

2. Experimental set up for PIV

A summary of the theoretical and experimental aspects of the PIV technique is given by Adrian [1]. The experimental set up of a PIV system for application in aerodynamics typically consists of several subsystems. First of all tracer particles, generated by an aerosolgenerator, have to be added to the flow. These particles have to be illuminated in a plane of the flow at least twice within a short time interval. The light scattered by the particles has to be recorded either on a single frame or on a sequence of frames. The displacement of the particle images between the light pulses has to be determined through evaluation of the PIV recordings. In order to be able to handle the great amount of data, which are collected by employing the PIV technique, a sophisticated post-processing system is required. The general concept for the development of the DLR PIV-system for application in wind tunnels was to have a system with the following features: robust and reliable; easy installation and adjustment; fast, but accurate evaluation; fully automatic evaluation (because of the large amount of data); automatic validation of data; easy adaptation of the PIV system to different wind tunnels and different aerodynamic problems.

Fig. 1 briefly explains the set up for recording. Small tracer particles are added to the flow (olive oil droplets, diameter ≈ 1 μm). A plane (light sheet) within the flow is illuminated two times by means of a Nd:YAG pulse laser (wavelength λ = 532 nm, energy 70 mJ per pulse for the experiments reported in this paper, pulse length 20 ns, time delay between pulses ≈ 20 μs - depending on the flow velocity and the magnification at imaging). It is assumed that the tracer particles move with local flow velocity between the two illuminations. The light scattered by the tracer particles is recorded via a high quality lens either on a single photographical negative ('dual illumination - single frame'

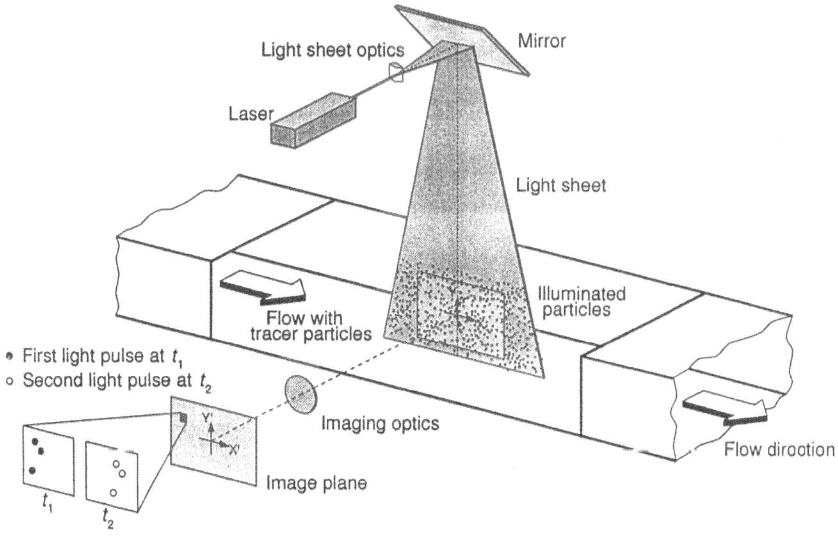

Figure 1. Experimental set up for PIV recording in the wind tunnel.

recording technique) or on two separate frames on a special cross correlation CCD sensor ('dual illumination - dual frame' recording technique). After development the photographical PIV recording is digitized by means of a scanner. The output of the CCD sensor is stored in real time in the memory of a work station.

For evaluation the digital PIV recording is interrogated in small sub areas or 'interrogation spots' (e.g. 32 × 32 pixels or 64 × 64 pixels). The local displacement vector for the images of the tracer particles of the first and second illumination is determined for each interrogation spot by means of statistical methods (auto- and cross correlation), [2]. It is assumed that all particles within one interrogation spot have moved homogeneously between the two illuminations. The projection of the vector of the local flow velocity into the plane of the light sheet (2D velocity vector) is calculated taking into account the time delay between the two illuminations and the magnification at imaging.

The process of interrogation is repeated for all interrogation spots of the PIV recording (e.g. 110 × 75 interrogation spots for 35 mm film = 8200 velocity vectors/recording). At present the spatial resolution at the evaluation of PIV recordings captured by means of a cross correlation CCD sensor with 1000 × 1000 pixels is nearly comparable to that of recordings stored on a photographic film (24 × 36 mm^2 format), digitized by means of a scanner to 2400 × 3600 pixels, due to the fact that smaller interrogation windows are possible. With the 1k × 1k video camera it is possible to capture more than 100 PIV recordings per minute. The evaluation of one video PIV recording with 3600 instantaneous velocity vectors is less than 1 minute. Increasingly video cameras become first choice for the application of PIV in wind tunnels due to their higher flexibility, easier handling and shorter time required to obtain results. Nevertheless, the photographic recording technique will still be used in future as 6 × 6 cm^2 or larger formats allow the recording of larger observation areas at high spatial resolution.

After automatic evaluation of a PIV recording a certain number of obviously incorrectly determined velocity vectors can usually be found by visual inspection of the raw data. In order to remove these 'incorrect data' the raw flow field data have to be validated. For this purpose special algorithms had to be developed, which must also work automatically [3]. These algorithms shall ensure with high level of confidence that no questionable data are stored in the final PIV data set. Questionable data are rejected, if their validity cannot be guaranteed by the validation algorithms. It should be emphasized that for all PIV results presented in this paper only this validation algorithm has been applied after automatic evaluation.

It becomes obvious from the description of the recording and evaluation process as given above, that the sign of the direction of the motion of the particles within each interrogation spot cannot be determined if the 'dual illumination - single frame' recording technique is employed in the way it was described above, since there is no way to decide which image is due to the first and which is due to the second exposure of a particle. Although, for many applications the sign of the velocity vector can be derived from a priori knowledge of the flow, in reverse flows, for example in areas of detached flow and in wakes behind models, additional steps have to be taken to determine the sign of the displacement and hence the velocity vector correctly. The great interest in PIV measurements in many different fields of research requires a flexible technique for ambiguity removal which can be applied to a variety of experimental situations. Especially for aerodynamic investigations it is very important to be able to apply this technique to high speed flows, i.e. with short time intervals of a few microseconds between the two illuminations. One method for ambiguity removal is the image shifting technique as described by Adrian [4], which enforces a constant additional displacement, that must be greater than the maximum displacement due to the reverse flow, on the images of all tracer particles at the time of their second illumination. Only minor modifications of the evaluation software are required when applying the image shifting method for recording at PIV. At present the most widely used experimental technique for image shifting is to utilize a rotating mirror system. The observation area in the flow is imaged onto the recording area in the camera via a rotating mirror. At DLR a high speed rotating mirror system has been developed in order to be able to carry out photographical PIV recordings even in transonic flows [5]. The frequency of rotation ranges from 1 Hz to 100 Hz. This allows to cover a range of shift velocities from 5 m/s to 500 m/s. The high speed rotating mirror system has been successfully applied to different aerodynamic investigations in wind tunnels. A detailed analysis of the errors involved with image shifting by means of a rotating mirror has also been presented in [5].

Ambiguity removal poses no problem for CCD sensors which work in the 'dual illumination - dual frame' recording technique, because the particle images due to each illumination are stored on two separate frames, thus giving information about the temporal sequence at recording.

Flow fields with the following parameters have been investigated by means of the PIV-system as described above: velocity range: U = 0 ... 600 m/s, observation area: up to 300×200 mm^2, spatial resolution: $\approx$ 1-2 mm in the flow field (depending on magnification), dynamic range of velocity (without image shifting): 1 : 10, accuracy of velocity measurement: < 1 % of mean flow velocity.

One statement should be made concerning the accuracy of the PIV technique. The absolute over-all accuracy of the PIV system depends on many different parameters and has to be checked separately for each experiment. There are no simple means by which the accuracy of PIV can be checked against an other measuring technique. Thus, usually it is tried to determine the accuracy of each individual component of the PIV system to derive a statement about the over-all accuracy [6]. One has to distinguish between three different error sources: first, a systematic error which influences all velocity vectors of a PIV recording in the same manner (e.g. the magnification factor at imaging, scale factor of the evaluation system), the second type of errors influences each velocity vector in a different, but steady way (e.g. velocity lag of tracer particles, optical distortion of the lens system, out-of-plane velocity component, errors due to image shifting) - however, neighboring vectors are affected in a similar way - , and the third one would be random errors associated with e.g. the evaluation process. The first two error sources would affect the appearance of spatial structures in the flow field only in a minor way. This means, we can demonstrate the relative accuracy of our system by successfully resolving spatial structures in flow fields with very small velocity fluctuations.

3. Aerodynamic applications

3.1. GRID TURBULENCE

Laminar and turbulent flow fields have been studied in order to obtain information on the accuracy attainable with the PIV technique. The

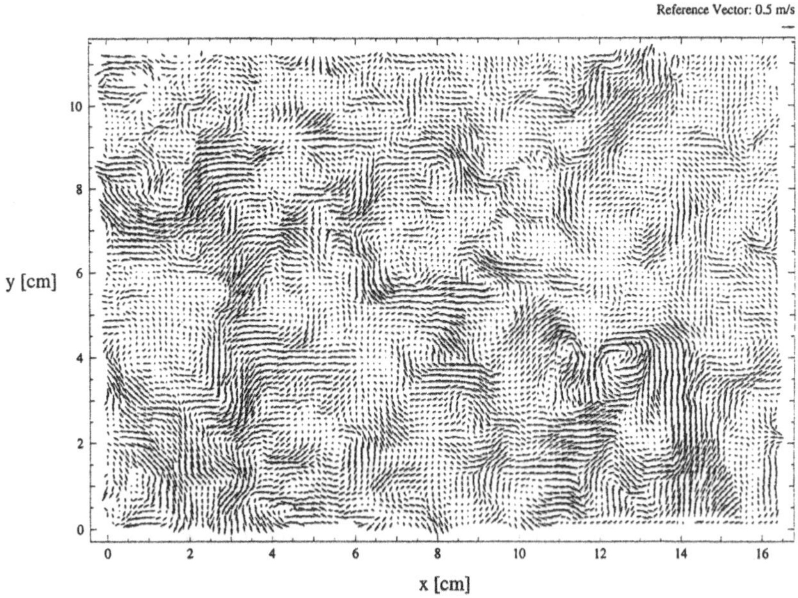

Figure 2. Instantaneous flow field of the fluctuating velocity component $(u-\overline{U},v)$ behind a grid at $\overline{U} = 10$ m/s with Tu ≈ 1.25%.

experiments were carried out in the DLR low turbulence wind tunnel (TUG), which is of an Eiffel type. Screens in the settling chamber and a high contraction ratio of 15:1 lead to a low turbulence level in the test section (cross section 0.3×1.5 m^2). The experiments were performed on the center line of the wind tunnel. Different grids can be installed near the end of the nozzle of the wind tunnel in order to generate turbulence. The basic turbulence level in the test section of the TUG of Tu $\approx$ 0.06 % (measured by means of a hot wire) can be increased by this modification. The flow was seeded in the settling chamber upstream of the screens used to reduce the turbulence of the wind tunnel flow.

As an example the instantaneous flow field behind a grid with a mesh size of 28 mm and a rod diameter of 8 mm is presented in Fig. 2. The PIV recordings were taken with a photographic camera and the high speed rotating mirror. The mean flow velocity in the wind tunnel was $\overline{U}$ = 10 m/s. The degree of turbulence was Tu $\approx$ 1.25 %, measured with a hot wire located downstream of the observation area. The distance between grid and observation area was $\approx$ 1.3 m (i.e. > 160 rod diameters, slowly decaying grid turbulence). The observation area was 16×12 cm^2 and is subdivided into 115×78 interrogation spots. In order to enhance the visual impression of the spatial structures in the flow field, a reference velocity U_{ref} is subtracted from the u-component of each local vector. In this case U_{ref} was set equal to the mean flow velocity $\overline{U}$. Thus, the instantaneous field of the fluctuating velocity component $(u-\overline{U}, v)$ is obtained. The results presented in [6] showed that it is possible to resolve spatial structures in a flow field with a degree of turbulence down to Tu $\approx$ 0.5 % by means of the PIV technique.

3.2. BOUNDARY LAYER INSTABILITIES

The transitional process in a boundary layer is determined by a mechanism of generation and interaction of various instabilities. Small oscillations may cause primary instability (two-dimensional waves, the so-called Tollmien-Schlichting (TS) waves). The growth of such TS waves leads to a streamwise periodic modulation of the basic flow, which gets sensitive to three-dimensional, spanwise periodic disturbances. These disturbances are amplified and lead to a three-dimensional distortion of the TS waves and farther downstream to the generation of three-dimensional Λ-vortices.

The extension of the knowledge about this mechanism enables the prediction and control of transition as required for applications in fluid mechanical engineering. In order to study the behavior of instabilities, quantitative data of velocity fields with known initial conditions have been acquired in a flat plate boundary layer, also in the TUG wind tunnel. In order to get reproducible and constant conditions for the development of the instabilities

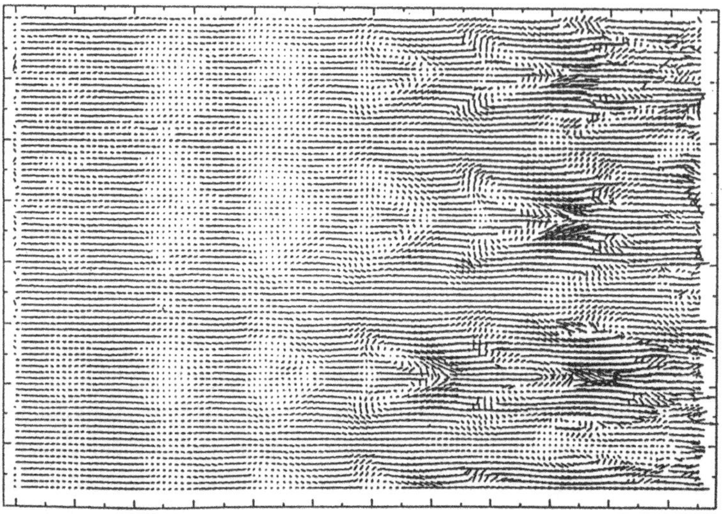

Figure 3. Field of instantaneous velocity fluctuations (u-U_{ref}, v) of b.l. instabilities above a flat plate, input signal is a 2D- and 3D-Wave.

it is necessary to know the initial amplitude of the velocity fluctuations at the beginning of the observation area. In our experiment [7] this is achieved by introducing controlled disturbances by means of a device for acoustic excitation which consists of a single, spanwise slot for the controlled input of two-dimensional disturbances and 40 separate slots (positioned spanwise as well) for the input of controlled three-dimensional disturbances. The velocity at the outer edge of the boundary layer was about U = 11 m/s. The average free stream turbulence level was Tu = 0.065 %. The light sheet (thickness 0.5 mm in the observation area) was oriented parallel to the plate. Its height above the plate could be varied but was usually 0.5 mm in the experiment. The observation area was 18 × 12 cm^2. The photographic recording technique was employed.

By applying different input signals to the acoustic excitation it was possible to excite different transition types. To mention here is the 'Fundamental Type', the 'Subharmonic Type ' and the 'Oblique Type'. Fig. 3 presents the field of the instantaneous velocity fluctuations (u-U_{ref}, v) obtained by exciting the 'Fundamental Type'. The direction of the flow is from left to right. The reference velocity U_{ref} (calculated by averaging over all velocity vectors in the recording) has been subtracted from all velocity vectors in order to show the fluctuating components of the velocity vector field. From left to right the beginning of the distortion of the TS waves and the development of the Λ-vortices can be clearly seen. It has to be realized that the PIV recording represents a cut through the three-dimensional structures in the boundary layer. The structures at the right side of the velocity field can be identified as the 'footprints' of the Λ-vortices. For the 'Fundamental Type' the Λ-vortices exhibit in an aligned pattern. The spanwise wavelength of these Λ-vortices (here 23.5 mm) matches with the wavelength of the controlled input of the 3D-waves. For the first time, the PIV technique allowed us the recording of a complete instantaneous velocity field in a plane of the boundary layer within microseconds.

3.3. TURBULENT BOUNDARY LAYER

The fully developed turbulent boundary layer on a plate has also been investigated in the same wind tunnel (TUG). The thickness of the boundary layer was $\delta \approx 7$ cm. The light sheet was oriented parallel to the main flow direction and perpendicular to the wall. The observation area was 0.07×0.1 m^2 at $\overline{U} = 10$ m/s. Fig. 4 shows the instantaneous flow field as captured by means of a photographic camera. It is remarkable how close to the wall velocity data could be recovered (appr. 1 mm). The main problems in this experiment were the scatter of the laser light from the surface of the wall and the strong velocity gradients within the interrogation spots close to the wall. During the same investigation several hundred PIV recordings were taken by means of the cross correlation CCD camera. In this case velocity vectors could be obtained down

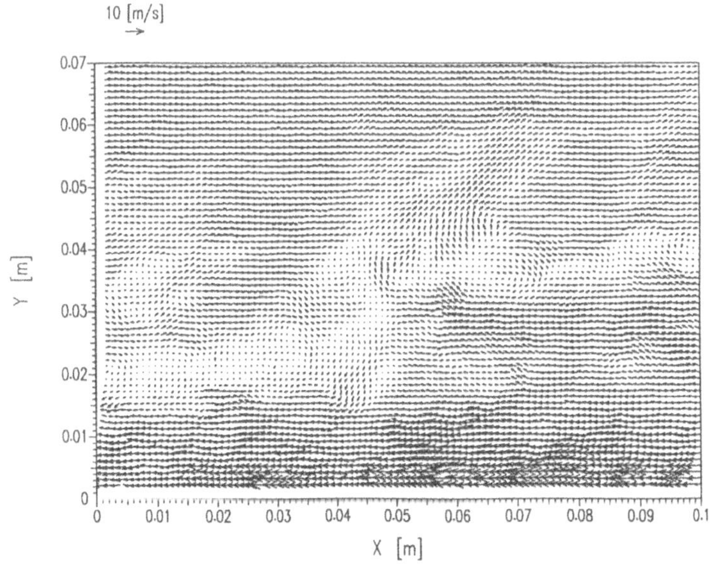

Figure 4. Field of instantaneous velocity fluctuations in a fully turbulent boundary layer, ($u - U_{ref}$, v).

to a distance of 0.15 mm from the wall at an observation area of $1.0 \times 1.0\ cm^2$. In a first part of the evaluation the boundary layer profile and the RMS components of the velocity fluctuations were calculated as average over all PIV recordings. These averaged quantities agree very well with the results from theory and pointwise velocity measurements as carried out by means of a hot wire. Work to analyze the spatial structures in the flow field, which can only be done with PIV data, is in progress.

3.4. VORTEX SHEDDING FROM A CYLINDER

The shedding of vortices from a cylinder has been subject of a number of investigations in the past. The regular, periodically alternating two-dimensional arrangement of vortices in a Kármán vortex street is well known. Recently, the three-dimensional structure of the wake behind the cylinder became of interest and was studied by means of flow visualization and hot-wire anemometry. Discontinuities in the relation between Strouhal number and Reynolds number were observed for Reynolds numbers $40 < Re < 160$ depending on the boundary conditions. Finally, different modes with oblique shedding of the vortices at different angles between the center line of the vortex and the axis of the cylinder have been identified. In order to study the spatial structure of the flow field in more detail, investigations by means of photographic PIV have

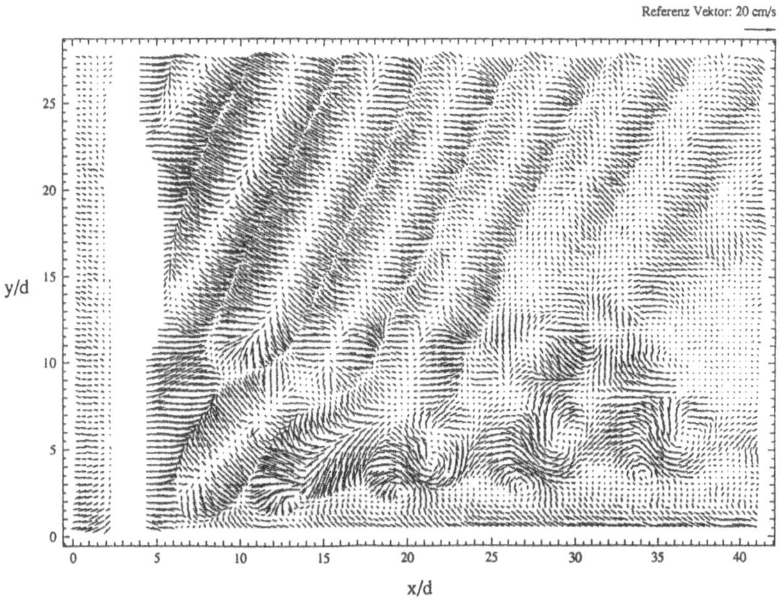

Figure 5. Field of instantaneous velocity fluctuations in the wake of a cylinder at $Re = 130$, $(u - U_{ref}, v)$.

been carried out in the low speed wind tunnel of MPI Göttingen at a mean flow velocity of $\overline{U}$ = 0.8 m/s [8]. As an example Fig. 5 shows details of the instantaneous flow field of vortices shedding from a cylinder of a diameter of 2.5 mm at Re = 130. The cylinder (left side, no vectors) is arranged vertically at 1.6 diameters behind the plane of the light sheet and attached to the wall of the wind tunnel (lower horizontal line of plot). Fig. 5 presents a cut through one side of the vortex street. It must be taken into account at the interpretation of the spatial structures shown in Fig. 5 that again the reference velocity U_{ref} obtained by averaging over all data has been subtracted. The end cell of the vortex, which forms the connection between the boundary layer and that vortex representing 'mode n= 3', can be clearly detected. The different angles at which the vortex sheds from the cylinder causes the branch like structure of the instantaneous velocity field. Details about the Kármán ratio, circulation and dissipation are also given in [8].

3.4. ROTOR

Helicopter noise has been subject of many research projects. Blade Vortex Interaction (BVI) was identified as a major source of impulsive noise. The measurement of the instantaneous velocity field close to the rotor blade tips is a demanding task for non-intrusive measurement methods as LDV and PIV [9]. PIV moreover allows to capture whole flow velocity fields within a few microseconds and is the only method to investigate the instantaneous vortical structures within the unsteady flow field on a rotor blade. Our experiments were carried out at the rotor test facility in the open test section of the low speed wind tunnel of the Department of Aerospace Engineering of RWTH Aachen. The helicopter rotor model had a radius of 0.5 m and consisted of four NACA 0015 blades (chord length = 54 mm) with rectangular tips. The rotor was fully articulated and had flapping hinges. The measurements were performed at a rotation of 25 cycles/s at a free stream flow velocity of U = 15.7 m/s. The Reynolds number based on the main chord length and the blade tip velocity was Re = 278000. Fig. 6 presents the unsteady flow field on a rotor blade at an azimuthal angle of $\Psi \approx 90°$ captured by means of the cross correlation video camera at time t_0. The observation area is 65 × 65 mm^2. The tip of the rotor blade is located at x = 0 cm and y = 0 cm. The blade moves in the z-plane and is oriented parallel to the x - axis. Three different vortices can be seen: the strong vortex to the left was generated by the rotor blade, which just passed the observation area, whereas the second, weaker vortex, which is located above the first vortex, was generated by the advancing blade at a time $t = t_0 - T_{period}/4$. At the right hand side of the plot (in the plane of the rotor) a part (left half) of a much weaker third vortex can be detected, which was generated at $t = t_0 - 2 \times T_{period}/4$, and which was just sliced by the rotor blade. Fig. 6b presents a plot of the instantaneous vorticity, which shows the vortical structures in some more detail.

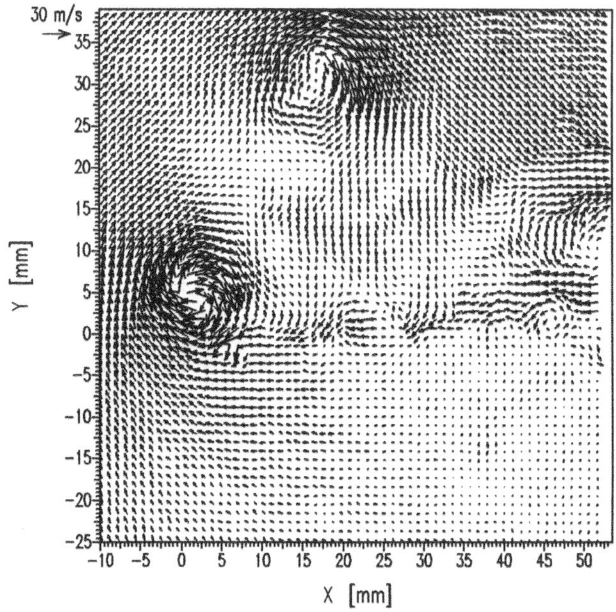

Figure 6a. Field of instantaneous velocity fluctuations immediately after the passage of a rotor blade.

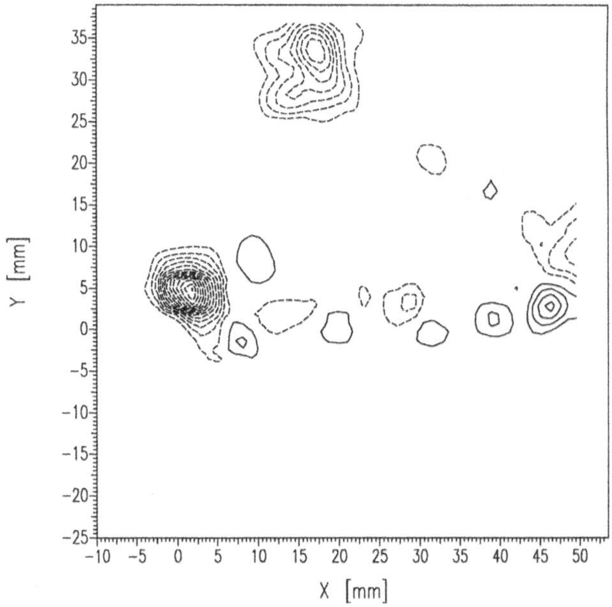

Figure 6b. Field of instantaneous vorticity immediately after the passage of a rotor blade.

3.5. CASCADE BLADE WITH COOLING AIR EJECTION AT MA = 1.27

Special problems appear with the application of PIV at high flow velocities: limited optical access in the high speed wind tunnel, problems of focusing the images of the tracer particles due to vibrations, velocity lag of tracer particles across shocks, etc. [11]. Nevertheless, the instantaneous flow field above a model of a cascade blade has been successfully investigated at transonic flow velocities by means of the PIV technique [10].

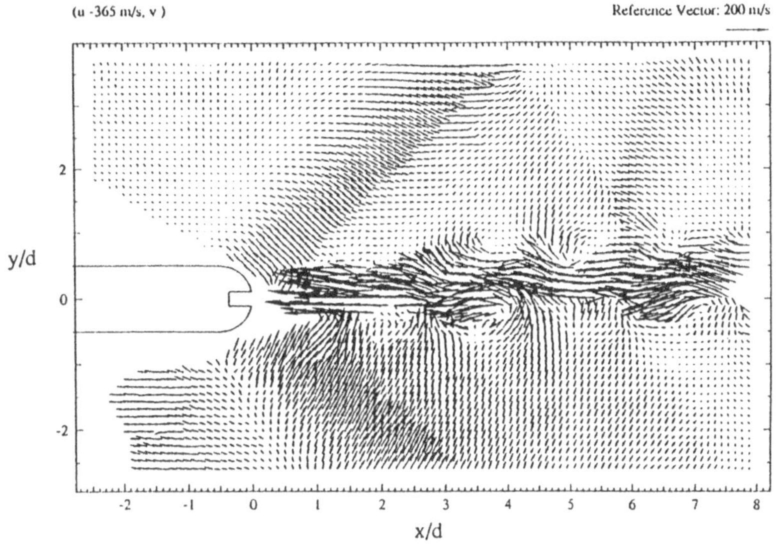

Figure 7a. Flow velocity field at Ma = 1.27 and 1.4% mass flow.

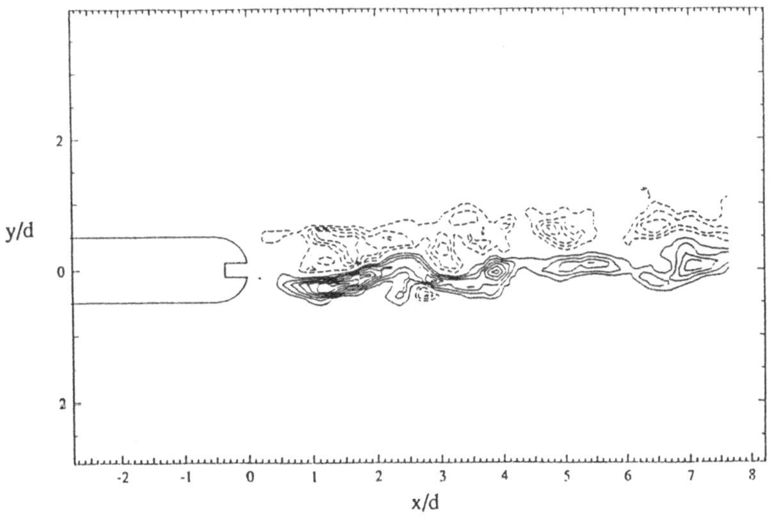

Figure 7b. Vorticity field at Ma = 1.27 and 1.4% mass flow.

The experiments were carried out in the DLR high-speed blow-down wind tunnel (HKG). Transonic flow velocities are obtained by sucking air from an atmospheric intake into a large vacuum tank. A quick-acting valve, located downstream of the test section, is rapidly opened to start the flow. Ambient air, which is dried before entering the test chamber, flows for max. 20 seconds through a test section with 725 mm spanwise extension. Grids in the settling chamber and a high contraction ratio of 10:1 lead to a low turbulence level in the test section.

The aim of this investigation was to study the effect of the ejection of cooling air on the wake behind a model of a cascade blade. Due to a specially adapted wind tunnel wall above and below the model and an adjustable tailboard above the model, the flow field of a real turbine blade could be simulated in a realistic manner. The PIV recordings were taken with the photographical PIV recording system utilizing the high speed rotating mirror for image shifting at a time delay between the two laser pulses of 3 μs. Fig. 7a presents the instantaneous flow velocity field at the trailing edge of the plate (thickness 2cm) for a cooling mass flow rate of 1.4% at a free stream Mach number Ma = 1.27. Expansion waves and terminating shocks can be easily seen. No data were obtained in the area above the model as the laser light was blocked off by the model. Data drop-out was also found in the area directly downstream of the model. The reason is mainly that the size of the interrogation area could not be further decreased at evaluation. This would have been necessary in order to satisfactorily resolve the strong velocity gradients close to the trailing edge of the model. Without ejection of cooling air, the wake behind the plate can be characterized as a vortex street. With ejection of air this is no longer true: two separate thin shear layers can be detected in the presentation of the instantaneous vorticity (Fig. 7b).

4. Conclusion

It has been demonstrated that investigations of the instantaneous flow field for different aerodynamic problems in different wind tunnels are possible by means of the PIV technique - in low speed flows as well as in transonic flows. It appears to us, that all major problems due to particles, optics, electronics and software have been solved in principle. However, there are still some refinements of the PIV technique necessary in order to be able to handle the special operational conditions in large wind tunnels. The results, presented in this paper, already show, that the PIV technique is a useful tool for aerodynamic investigations in the field of grid turbulence, boundary layer instabilities, rotors, and for the investigation of transonic flows in the wake of cascade blades. The quality of the instantaneous flow velocity data as measured by means of PIV is already adequate for comparison with the results of numerical calculations. A first comparison of the results obtained by experimental and numerical investigations of dynamic stall on a pitching airfoil has been already carried out [12].

The instantaneous velocity vector maps presented in this paper can be obtained in full detail via Internet [13].

5. References

[1] Adrian, R. J.: "Particle-Imaging Techniques for Experimental Fluid Mechanics", Annual Rev. Fluid Mechanics, Vol. 23, pp. 261 - 304, 1991.

[2] Willert, C.: "The fully digital evaluation of photographic PIV recordings", in press, Applied Scientific Research, 1996.

[3] Raffel, M., Leitl, B., Kompenhans, J.: "Data Validation for Particle Image Velocimetry", in 'Laser Techniques and Applications in Fluid Mechanics', Eds. R.J. Adrian et al, Springer-Verlag, pp. 210 - 226, 1993.

[4] Adrian, R. J.: "Image shifting technique to resolve directional ambiguity in double-pulsed velocimetry", Appl. Optics 25, pp. 3855 - 3858, 1986.

[5] Raffel, M., Kompenhans, J.: "Theoretical and experimental aspects of image shifting by means of a rotating mirror system for particle image velocimetry", Meas. Sci. Technol., vol. 6, pp. 795 - 808, 1995.

[6] Vogt, A.: "Optische Auswertung von Particle-Image-Velocimetry-Messungen im Nachlauf eines quer angeströmten Kreiszylinders", DLR Forschungsbericht, DLR-FB 93-49, 1993.

[7] Fischer, M., Wiegel, M., Kompenhans, J., Raffel, M., Vogt, A.: "Neue Wege in der experimentellen Transitionsforschung", DLR-Nachrichten, Heft 79, pp. 2-6, August 1995.

[8] Schröder, A.: "Untersuchung der Struktur des laminaren Zylindernachlaufs mit Hilfe der Particle Image Velocimetry", Diplomarbeit, Universität Göttingen, 1996.

[9] Raffel, M., Seelhorst, U., Bütefisch, K.A.: "Measurement of vortical structures on a helicopter rotor model in a wind tunnel by LDV and PIV", to be published, 8th Int. Symposium on Appl. of Laser Techniques to Fluid Mechanics, Lisbon, 1996, paper 14.3.

[10] Raffel, M., Kost, F., Willert, C., Kompenhans, J.: "Experimental aspects of PIV measurements of transonic flow fields and their benefit for the optimization of gas turbines", to be published, 8th Int. Symposium on Appl. of Laser Techniques to Fluid Mechanics, Lisbon, 1996, paper 28.1.

[11] Kompenhans, J., Raffel, M.: "Application of PIV technique to transonic flows in a blow-down wind tunnel", Proc. SPIE 2005, pp. 425 - 436, 1993.

[12] Wernert, P., Geissler, W., Raffel, M., Kompenhans, J.: "Experimental and numerical investigation of dynamic stall on a pitching airfoil", in press, AIAA Journal, vol. 34, May 1996.

[13] WorldWideWeb, http://www.es.go.dlr.de:8081/MP_PIV.

THREE-DIMENSIONAL PARTICLE VELOCIMETRY

KLAUS D. HINSCH and HEIKO HINRICHS
FB 8 - Physik, Carl von Ossietzky Universität Oldenburg
Postfach 2503, D-26111 Oldenburg, Germany

1. Introduction

In recent years, the problems tackled by experimentalists and theoreticians in fluid dynamics have grown considerably. Sophisticated flow facilities like large size or high speed wind tunnels are the basis for ambitious experiments, and powerful computers allow modelling of very complex flow fields. Interest has turned to nonstationary flows, their spatial structures and their development in time. An ultimate aim is to measure and understand the evolution of turbulence.

The metrological tools for the experimental study of flow fields have also seen an impressive improvement. Much early information about flows has been accumulated by mostly qualitative visualization techniques (Merzkirch, 1987). Smoke or dye line patterns have produced impressive milestones in the history of flow analysis (VanDyke, 1982). Attention then turned to the development of anemometers that provide quantitative velocity records of high accuracy at a selected location in space. Thermal probes like hot wire anemometers (HWA) have become a routine tool in velocimetry. Their spatial resolution, however, is limited by the finite dimensions of the probe. Furthermore, in delicate flows, the anemometer may introduce disturbances. Optical probes allow noncontact remote measurements within a very small sample volume. Thus, the favourite choice in single point measurements is the laser Doppler anemometer (LDA). It is noninterfering, but it relies on the presence of small particles in the flow. The technique assumes, of course, that the particle motion faithfully resembles the fluid motion. In gaseous flows this requires very small particles and sets limits to the largest velocity gradients acceptable.

Fluid velocity, the characteristic parameter of the flow, is a three-dimensional quantity. A complete specification calls for its three components u, v, and w in the direction of x, y, or z, respectively. Single point anemometers like HWA or LDA have been modified for 3-D operation. Yet, such a measurement provides only data from a single location in space. Many flow configurations of interest, however, are of three-dimensional nature. Especially the structural character of a flow, a topic of current interest, is only revealed by its spatial features. When the flows are very voluminous or measuring time is expensive, the effort or the cost to scan the flow with a single point

Th. Dracos (ed.), Three-Dimensional Velocity and Vorticity Measuring and Image Analysis Techniques, 129–152.

probe are enormous and may be even prohibitive. Even more, when the flows are non-stationary there is no time to move a single point probe through the flow field.

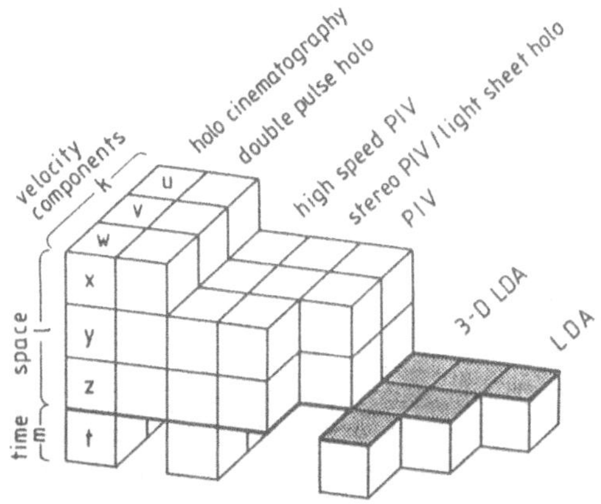

Figure 1. Techniques for optical flow velocimetry and the dimensions of data space.

In view of these problems, much recent effort has been dedicated to the development of flow velocimetry in several dimensions. We may classify techniques for flow velocimetry in terms of the dimensional size of the data that are covered (Figure 1). A technique can be labeled a (k,l,m)-method where k = 1,2,3 indicates the number of velocity components measured, l = 0,1,2,3 gives the number of dimensions in space, and m = 0,1 indicates whether a measurement yields velocity at only a single instant in time or a complete time record. Conventional one-dimensional LDA, for example, is a (1,0,1)-method, particle image velocimetry (PIV) in its usual form a (2,2,0)-method. The utmost in flow velocimetry is a (3,3,1)-method where the velocity vector is measured throughout a complete volume as a function of time. Holographic cinematography can yield such data.

Several innovative techniques are presently tackling the problem of volumic flow velocimetry. Examples are tomography, particle image velocimetry, global Doppler or laser induced fluorescence (LIF), to name a few. The present review is restricted to methods that obtain the velocity information from the motion of small tracer particles in the flow. We describe efforts to explore the dimensions of fluid dynamics beyond the established 2-D technique of particle image velocimetry, where both the in-plane components of the velocity are investigated within a thin sheet placed in the flow. The main emphasis of this paper is on optical concepts while the ever growing impact of computer based solutions (digital particle velocimetry) is left to other contributions in this volume.

2. Light sheet techniques

2.1. CLASSICAL PIV: A 2-D SLICE FROM THE FLOW

To begin with, let us recall briefly the great step forward in whole-field techniques that has become known as PIV (Adrian, 1991; Hinsch, 1993b; Grant, 1994). It can be a good starting point for extensions into more dimensions, because profitable use can be made of existing powerful PIV interrogation schemes. Furthermore, the thorough understanding of the imaging of tiny particles is prerequisite for all methods to follow.

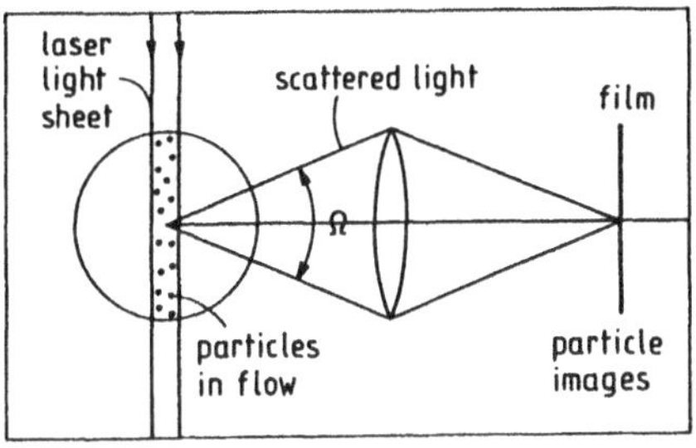

Figure 2. Optical setup for PIV (Particle Image Velocimetry)

A sheet from a powerful light source, usually a laser, is directed along a plane in the fluid (Figure 2). It illuminates tiny tracer particles of suitable properties within the flow. The particles are imaged at right angles to the sheet. When the light source is pulsed repeatedly, a series of particle images are recorded. Since the repetition period of the light pulses is known, magnitude and direction of the displacement of the particle images yield the velocity. Often, the recording is done photographically to produce high resolution transparencies for further evaluation. With increasing size of CCD-arrays and efficient computer power, direct video recording and digital processing of the images become popular.

PIV realizations can be classified by the number of pulses and frames that are employed (Adrian, 1991). In the most simple and commonly used version two pulsed images are superimposed on a single photographic frame. The evaluation thus calls for the determination of the separation of the pairs of particles. Usually this quantity is averaged over finite interrogation regions that must be much smaller than the size of the relevant flow structures in the transversal x- and y-dimensions. In other schemes the exposures are recorded seperately which eliminates the 180°ambiguity and offers better signal-to-noise ratio.

PIV has found successful applications in flows where the light sheet can be oriented parallel to a prevailing flow direction. This guarantees a sufficiently small out-of-plane velocity component and ensures that most particles stay in the light sheet between exposures. Unpaired particles contribute to noise and reduce the efficiency of the method. Usually a light sheet depth of 1 to 2 mm is appropriate which is made to fit the focal depth of the recording camera system. Thus a PIV study usually ends up with values of u and v at the interrogation locations (x,y,z_0) where z_0 is given with the accuracy of the light sheet depth.

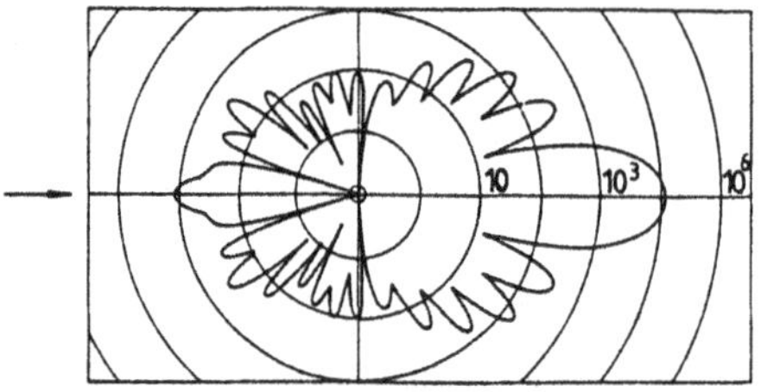

Figure 3. Scattering of 0.633-mm wavelength laser light by a 3-mm diameter water droplet in air (with permission from Invent GmbH, Erlangen).

It should be mentioned that PIV makes poor use of the light scattered by the particles. As is demonstrated by the example in Figure 3, the tiny particles suited for flow tracer applications scatter light mostly into the forward direction. At 90° observation the light intensity is down to about 0.1 percent. Thus, powerful laser illumination (ruby, Nd:YAG or copper vapor lasers for pulsed applications, Ar-Ion lasers for scanned beam light sheets) is required. Contrary to the forward direction, however, where the light is concentrated into a narrow lobe, the mean scattered light does not change very much within a large angular region around the direction of observation. This is a feature important for particle imaging.

Particle velocimetry calls for small and high quality particle images. Basically, the resolution in optical imaging is limited by diffraction and is determined by the angular spread of the light that contributes to image formation (angle Ω in Figure 2). Omitting nonessential numerical factors, we obtain for the transversal resolution length the familiar Rayleigh criterion (λ being the wavelength of light)

$$d_t \approx \lambda/\Omega \tag{1}$$

and for the longitudinal resolution length (Born and Wolf, 1980)

$$d_l \approx \lambda/\Omega^2 \tag{2}$$

When a particle is smaller in size than the resolution length it will not be resolved in the image, when it is larger its geometrical image has to be convolved with a point spread function whose width is given by the resolution length.

Due to the wide scattering around 90°, the angle Ω in PIV is determined by the aperture of the imaging system. Large aperture optics is therefore required for small diffraction limited particle images. At a magnification M=1 an f/2.8 lens yields some 6λ transversal and 30λ longitudinal resolution. Due to depth of focus considerations (particles throughout the complete light sheet depth must be recorded properly) the lens is usually stopped down to a compromise value. This is to make sure that particle images are almost identical irrespective of depth position.

More than a decade of intensive PIV development has produced many different processing routines. These range from digital particle tracking to optical or digital correlation analysis. In the latter, the cumbersome identification and pairing of particles is avoided by processing an ensemble of particles within the interrogation area. Details depend on the particle density, single or multiple frame techniques and preferences of the researcher concerning computerized or optical processing. In any case, there exist powerful hard- and software that are also available when 3-D problems can be decomposed into 2-D subtasks.

For later use in stereo extensions of PIV let us recall the typical steps in the partly optical correlation processing of a PIV double exposure record. The photographic recording (or a negative contact copy thereof) is interrogated with a narrow focused laser beam to produce a diffraction pattern known as Young's fringes system. The fringe pattern is usually read into a computer that performs a two-dimensional Fourier transformation of the pattern, ending up with the autocorrelation function of the input negative. This consists of a central peak accompanied by two side peaks at a distance equal to the wanted particle separation. Identification of the correlation peaks and determination of their position thus concludes the velocity measurement. In double-frame recording the according crosscorrelation between both the images is calculated. The final result of a flow analysis with this method usually is a field of small velocity vectors plotted at the interrogation locations.

The limits of particle image velocimetry in the form presented are obvious: a two-dimensional cut from the flow volume is extracted, only the in-plane components of the velocity are determined, and this at a single instant in time that is set by the pulsing of the laser. It is, of course, assumed that the pulse separation is much smaller than the time scales in the flow. Current progress in the generation of powerful laser pulses with repetition rates tailored to specific needs allows PIV to conquer the time domain. Thus, the design of high-speed PIV setups is possible (Lin and Rockwell, 1994).

A straighforward extension of light sheet PIV to volumic flows is to move the sheet rapidly in depth. This requires proper beam shaping optics (Prenel *et al.*, 1989). Brücker and Althaus (1992) and Brücker (1993) performed extensive studies of vortex breakdown and the evolution of vorticity (Brücker, 1995a) by scanning the depth with the aid of an oscillating mirror in the light path. Guezennec *et al.* (1994) used a series of small mirrors arranged helically on the circumference of a circular rotating drum to displace the laser beam before it passes the cylindrical expanding lens. The scanning time and sheet separation, of course, must be adjusted to the temporal and spatial scales of the flow. A video camera or a high speed movie camera have been employed in the recording of the particle configurations. The many advantages of ordinary PIV can thus be preserved while the sampling in depth adds volume information. The large size data provide for vorticity calculations and even allow to calculate the out-of-plane component of the velocity vector from mass conservation (Robinson and Rockwell, 1993).

We mentioned that the light sheet is directed in such a way that the main flow direction falls within the plane selected. Otherwise, the depth component of the flow would produce too many unpaired particle images. Within the limits given by light sheet depth and depth of focus, however, a displacement in z-direction is tolerated and its measurement opens a new dimension in the conquest of space. Modifications of the approved PIV setup have been proposed for this purpose. The time for a complete crossing of the depth of the light sheet, for example, has been measured by counting stroboscopically produced light streaks along the particle trajectory (Halloin and Jottrand, 1994). In a similar manner, Herpfer and Jeng (1995) combined a double pulsed Nd:YAG laser with an 800 μs pulse from a cw argon-ion laser to produce an ordinary PIV double image superimposed onto a streak. The spacing of the particle image pair allows to extract the in-plane component of the displacement, the supplementary knowledge of the length of the streak yields the out-of-plane component. In color coding, the flow is probed with parallel light sheets of different color and the color change along the trajectory yields the motion in depth (Cenedese and Paglialunga, 1988). In another configuration the light sheet is taylored for a well defined intensity profile in depth so that the brightness of a particle image reveals its depth coordinate (Dinkelacker *et al.*, 1992).

The size of an ordinary out-of-focus image may also be utilized to indicate a particle's depth location (Stolz and Köhler, 1994). Even better results are obtained when the image blur is combined with the image brightness. It was shown that the so-called depth ratio, i.e., the ratio of brightness to blur, is linearily related to depth position (Funes-Gallanzi *et al.*, 1995). More generally, size and shape of the point spread function can be utilized for depth determination when the particles are imaged with a coded aperture. In a three hole aperture, for example, each particle is imaged as a triangular set of three points. The size of the triangle indicates the distance from the plane of best focus and the orientation allows to decide whether the particle is closer or further away (Willert and Gharib, 1992). All of these cases, however, require specific evaluation strategies.

It should be mentioned that these techniques all require that the particle stays within the sheet for both exposures. Since displacements are linked to the velocity by the light pulse separation, the velocity in depth must be much smaller than the sidewise velocity. This requirement, however, can be largely relaxed by a bias translation of the image between exposures as it is usually employed in PIV image shifting for removal of the directional ambiguity. In-plane and out-of-plane displacement can be adjusted independently to suit the flow conditions best.

The out-of-plane velocity can also be determined from a correlation of particle records, because the number of unpaired particle images increases at higher velocities. Unfortunately, there are several other decorrelation mechanisms, above all the transverse escape of particles from the interrogation window. Thus, any decorrelation analysis needs a reference which can be obtained when a second light sheet is introduced parallel to the original one (Raffel *et al.*, 1995). Three images of the particle configuration are obtained at equal time separations, two in the ordinary manner with the first light sheet, the third one with the second sheet. The peak heights of the cross correlation between subsequent images are used for analysis. The basic principle is easily explained for the special case of two directly adjacent sheets. In the extreme case when the flow is strictly parallel to the light sheets, the complete package of particles stays within the sheet. Thus there will be good correlation between image one and two and no correlation between image two and three. For a pure out-of-plane flow, the result will be vice versa. In the general case, the depth component can be calculated from the cross-correlation heights taking into account the exact profile of the light intensity. In an earlier and related approach (Kimura and Kohno, 1991) two parallel sheets were placed a given distance apart and a time sequence of particle images was recorded. A certain particle configuration in the upstream sheet will show up in the downstream sheet after a time set by the depth component of the flow velocity. Thus, all the displacement components can be derived from a spatio-temporal correlation.

2.2. STEREOSCOPIC PIV

A promising concept that retains the possibility to apply PIV processing routines is stereo-recording (Dreesen, F. et al. ,1992; Gauthier and Riethmuller, 1988; Westerweel and Nieuwstadt, 1991; Grant *et al.*, 1991; Prasad and Adrian, 1993). Here, an object is viewed from two directions to extract its three-dimensionality. The underlying principle is easily summarized with the aid of Figure 4. Let us consider three particles, numbered 1 to 3, distributed in space. We are registering the relative separations between particles as seen from two different directions. These separations are identical for all particles that lie within planes normal to the bisector of the viewing directions. One such plane is the reference plane in the illustration. The separations differ, however, when the particle combination includes a partner from a position behind (particle 2) or in front (particle 3) of the reference plane. For the pair of particles 1 and 2, for example, D_2 is smaller than D_1 , because the second particle is behind the reference plane. It would be vice versa had we combined particle 1 and particle 3. From a combination of position information in two stereo photographs it is therefore possible to extract the depth information of the particle field. Obviously, the sensitivity for measuring the depth component increases with the angle between the viewing directions. The stereo cameras may also view from different sides of the sheet.

A stereo version of particle photography in PIV faces basic problems. There are two configurations to view the light sheet at an oblique angle and yet maintain focused particle images (Gauthier and Riethmuller, 1988):

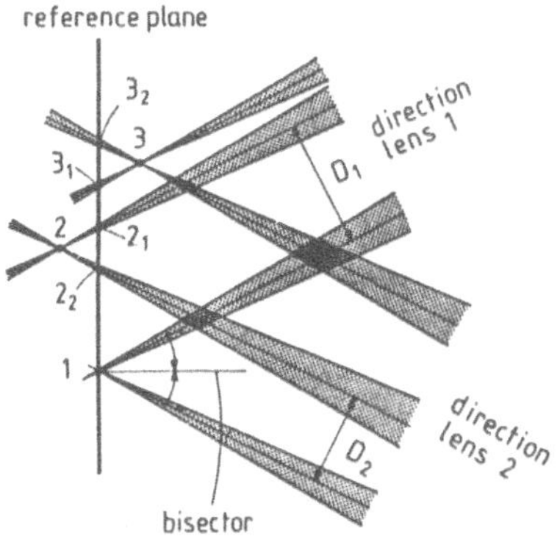

Figure 4. Schematic of stereo photography of particles and the determination of the depth coordinate by back-projection.

Angular displacement method. The imaging systems point at the particle field such that their optical axes form the required stereo angles with the light sheet (Figure 5a). The oblique orientation of the object plane results in a tilted image plane so that a normal camera will face focusing problems. Often the angle is therefore kept so small that the resulting defocusing is acceptable. For in focus imaging, even with large angles, camera lens and image plane must be tilted to intersect in a common line with the object plane (*Scheimpflug* condition). This, however, results in a geometrical distortion of the image with space variant magnification.

Translation method. The imaging systems are aligned at right angles to the light sheet and view the interesting section of the particle field far off-axis (Figure 5b). In this case the images are displaced off the usual imaging region and the backplanes of the cameras must be translated. While this translation has been avoided with an appropriate arrangement of mirrors by Reese *et al.* (1995) or Arroyo and Greated (1991) as in Fig. 5c, or by bending the optical path with prisms (Adrian *et al.*, 1993) or diffractive elements, the image is often subject to large aberrations due to the off-axis conditions. Usually this is critical when high quality particle images are required. Only by a costly method like phase conjugate holography (Adrian *et al.*, 1993) it is possible to compensate completely for such aberrations.

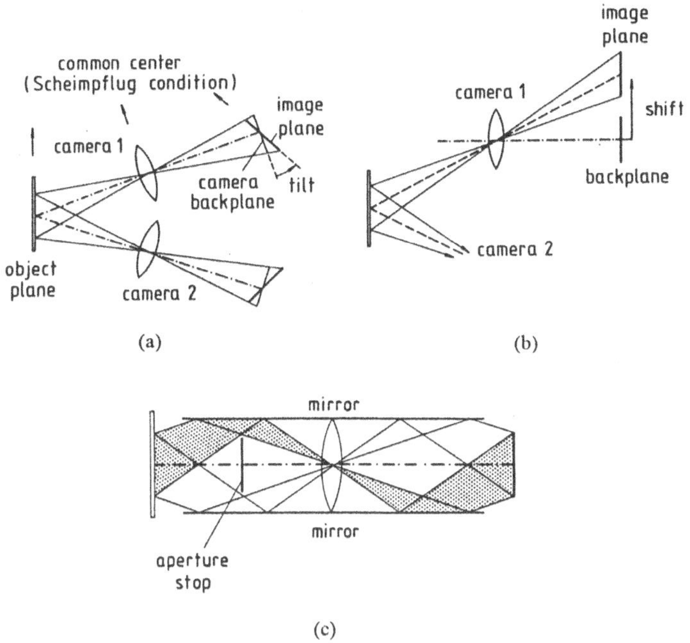

Figure 5. Stereoscopic recording of particles in a light sheet.
(a) Angular displacement method. Oblique axis viewing requires *Scheimpflug* condition.
(b) Translation method. Normal axis viewing requires backplane shifting.
(c) Introduction of mirrors for compact translation method.

In still a third approach that has been termed in-line stereoscopy both cameras image at right angles as in the translation method, but with common camera axes (Grant *et al.*, 1994a; Grant *et al.*, 1995). The stereo effect is introduced by different object-to-camera distances which makes an off-axis particle appear under different angles. Obviously, the sensitivity of the method is position dependent. The effect is most pronounced at the edge of the viewing field and vanishes on the axis. It should be mentioned that just this effect, the perspective projection in imaging, introduces an error in ordinary PIV when the viewing field extends over a large angle (Sinha, 1988; Grant *et al.*, 1994b; Raffel and Kompenhans, 1994). Yet, the method is easy to apply and robust in its application. Practical setups have been realized by either placing the cameras on the same side at different distances behind a beamsplitter or by imaging the flow from opposite sides.

Numerous stereoscopic studies of three-dimensional flows have been reported. A thorough analysis of the translation method, photographic recording, PIV-like evaluation of the separate records, matching of the data for three-dimensional velocity vectors and an application to a rotating disk flow is given by Prasad and Adrian (1993). Prasad and Jensen (1995) give a detailed ray tracing analysis of the Scheimpflug setup for the study of flows in a water tunnel introducing a liquid prism component to reduce distortions. New domains open up like the estimation of three-dimensional vorticity fields (Sinha and Kuhlmann, 1992). Funes-Gallanzi *et al.* (1995) combine stereoscopic triangulation with a defocus criterion for depth estimation. Brücker (1995b) combines stereoscopic recording with a scanning light sheet in a 3-D study of flow development around a spherical cap. Many recent stereoscopic investigations make use of the increased power of video imaging and computer image processing (Dracos and Malik, 1992; Kasagi and Sata, 1992; Maas *et al.*, 1993; Malik *et al.*, 1993; Funes-Gallanzi *et al.*, 1994; Kent *et al.*, 1994). Often even more than two cameras are installed to elimate ambiguities or to increase the resolution and the depth of field. The measurement volume can then be extended beyond a mere sheet to a substantial depth. Sophisticated processing schemes including photogrammetric tools have been developed.

Let us treat in more detail the angular displacement method in the *Scheimpflug* configuration as it combines high sensitivity (large angle between the cameras) with the possibility to apply the powerful PIV interrogation routines. Furthermore, the treatment will elucidate some basic optical relations (Hinsch, 1993a). Camera systems are available that allow not only the seperate alignment of the image plane - the swinging back camera already mentioned by Gauthier and Riethmuller (1988) - but also the required adjustment of the objective lens. In the interrogation of the stereo images, however, the geometric distortions must be accounted for. The magnification of corresponding regions in the object field differs from left to right image. To be independent of system parameters, a calibration with a precision grid in object space is advisable. The interrogation is then carried out on points of the distorted non-Cartesian grid in the image spaces to render regular sampling in object space.

Distortions can be avoided when a telecentric imaging system is used in the camera. This may consist of two lenses of focal length f_1 and f_2, respectively, separated by a distance $s = f_1 + f_2$. When the object is placed, say, in front of lens No. 1 there results a magnification $M = f_2 / f_1$ that is independent of object position. In this case, however, the size of the viewing field is severely restricted.

We have seen that the performance in determining the depth coordinate of a particle is greatly improved by a stereo arrangement. Basically, this can be attributed to the utilization of scattered light from widely separated angles. In a single lens system the aperture sets a practical upper limit to the angle Ω that determines depth resolution. It can only be increased further by employing two seperate lenses. In the usual stereo configurations, however, this advantage has to be paid for by two separate images that have to be matched for evaluation. Under *Scheimpflug* conditions this is even more severe because of the varying magnification. A method to overcome this has been proposed by the author's lab, called back-projection stereo. Its principle is explained by returning to Figure 4. Assume that we have a means to back-project the recorded images into object space. The back-projected light is registered in a reference plane within the light sheet. Thus, for particle *1* (in the reference plane) both the back-projected images overlap in focus at the former particle position in the reference plane. However, for particle *2* or *3* (behind or in front of the plane) there results a pair of images each (2_1, 2_2 and 3_1, 3_2, respectively). The intrapair separation is a measure of the depth position, the left-to-right sequence of the images yields the distinction between the space behind or in front of the reference plane.

Let us apply this reasoning to double pulse particle image velocimetry. Let particle *2* and *3* represent the first and second exposure of a particle travelling with the flow. Instead of the ordinary two images we now get four back-projected images for a particle pair, and from the separations between 2_1 and 3_1 and 2_2 and 3_2 both the in-plane as well as the out-of-plane displacement can be derived. When, for example, the particle moves parallel to the reference plane these separations are the same. When one of the particles is caught in the reference plane, there are only three images; when both are in the reference plane we find the usual two images. In PIV evaluation of such records the additional particle images produce corresponding peaks in the autocorrelation function from which the wanted information can be obtained.

(a)

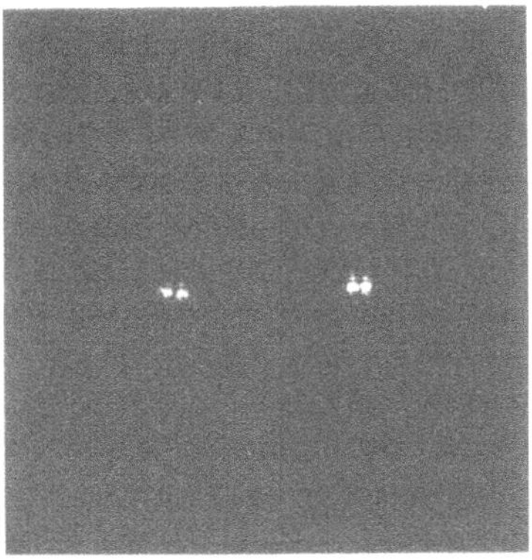

(b)

Figure 6. Interrogation of a transparency of a back projected stereo pair of 20-μm particles in plastic.
(a) A beat in the Young's fringes system indicates the out-of-plane displacement.
(b) Corresponding twin-peak side orders in the autocorrelation where the central peak has been blocked partly.

There are several possibilities for the necessary back-projection. We show an experiment with photographic recording of the particles in the image plane and back-imaging after production of a contact copy. Figure 6 demonstrates results obtained in a model environment of 20-μm particles embedded in plastic. The cameras were viewing a light sheet of 2 mm depth symmetrically at 20° each to the normal. The object was displaced by 1 mm in-plane and 0.5 mm out-of-plane. Due to the two predominant particle separations, the fringe system displays a beat modulation and the resulting autocorrelation output shows two side peaks. Unfortunately, the performance of this method is rather poor in real flows, because of photographic thresholding and repeated introduction of aberrations.

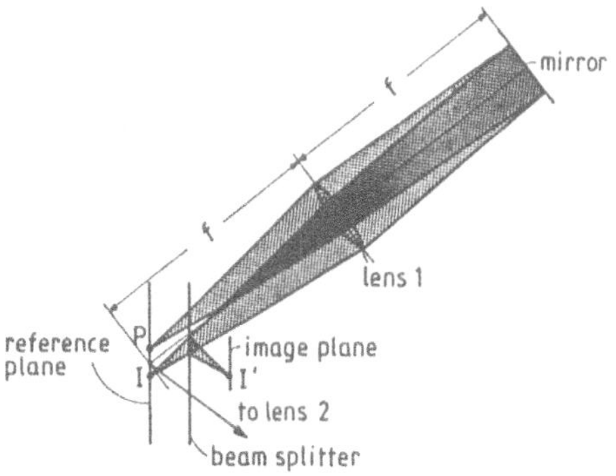

Figure 7. Back-projection of particle images by introduction of a mirror in the symmetry plane of a unit magnification telecentric optical system of 4f length.

A better result is expected from a direct stereo-photographic back-projection system. The optical setup for such an operation makes use of two unit magnification telecentric imaging optics as introduced above. Such a system is completely symmetric as it consists of two identical lenses separated by the double focal length. Therefore, half of the system can be omitted and a mirror can be introduced in the central plane of symmetry to reverse the imaging (Figure 7). Let the reference plane in the flow intersect the optical axis at a distance equal to the focal length. Then, any object point P in the reference plane is imaged back onto this plane (point I). Any point in front of (or behind) the reference plane is imaged symmetrically behind (or in front of) the plane. The back-projection configuration of Figure 4 is therefore realized with an inverted image of object space. Identical conditions hold for both the imaging directions. The introduction of a beam splitter parallel to the reference plane produces the image I' that can be collected on a photographic plate for further processing. For discrimination between the left and right images, either the reference plane can be placed outside the sheet or the images could be labelled by distinctively different apertures in the stereo branches. This technique has been used otherwise to discrimate between first and second exposure particle images (Lawson, 1994).

The back-projection principle has been treated in detail also because it forms the basis of the evaluation of particle image positions with high depth discrimination in the phase conjugate real image of a holographic particle recording (see section 3.2 and Adrian *et al.*, (1993)). In that case, the conventional mirrors in our system are replaced by a hologram that is operated as a phase conjugate mirror which produces a time reversed version of the original object light.

3. Holographic particle velocimetry

For true imaging of three-dimensional space, holography (Collier *et al.*, 1971) is the technique of choice. This very flexible method for recording with greatly relaxed depth restrictions has been rediscovered for flow velocimetry lately yielding techniques that have been summarized as holographic particle image velocimetry (HPIV) (Bryanston-

Cross *et al.*, 1992; Rood, 1993). Let us explain the principle in a setup for holographic recording of a particle field (Figure 8a). All particles of interest in the flow are illuminated with a properly expanded laser light wave. The complex light field scattered by the particles is stored by superposing a reference wave from the same laser source and recording the interference pattern on a photographic plate. Due to the oblique angle between reference wave and object light this is called off-axis holography. The coherence length of the laser must be adequate to guarantee interference over all path length differences encountered. The developed pattern (the hologram) is used to reconstruct an image of the object. For this purpose the hologram is illuminated with the original reference wave that reconstructs a virtual image of the particle field (Figure 8b). When looking or photographing through the hologram the particle configuration can be seen as it was during the recording. Different planes through the flow field may be selected for focused inspection. The maximum resolution angle Ω is now given by the size of the hologram.

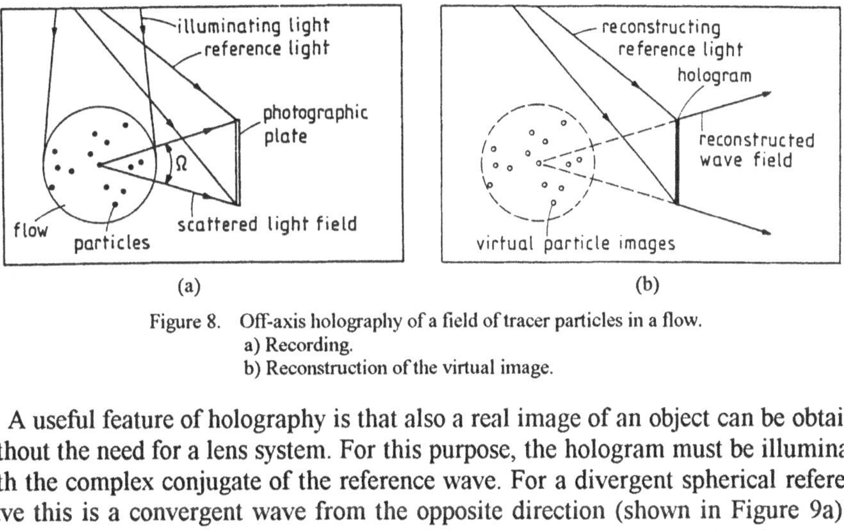

Figure 8. Off-axis holography of a field of tracer particles in a flow.
a) Recording.
b) Reconstruction of the virtual image.

A useful feature of holography is that also a real image of an object can be obtained without the need for a lens system. For this purpose, the hologram must be illuminated with the complex conjugate of the reference wave. For a divergent spherical reference wave this is a convergent wave from the opposite direction (shown in Figure 9a). In this case the hologram produces backward travelling object waves that form a real image of the light sheet. With a CCD-array or a photographic plate a plane anywhere through the field could be examined for particles. If aberrating media like windows or low quality lenses have distorted the object wave on its way to the holographic recording plate, these distortions can be compensated during the backward path of the conjugate wave provided the distorting media are still in the old position (Figure 9b). Then the particle images are formed by the original ideal wavefront.

For good particle image quality the reconstructing wave has to meet special requirements. The geometry of the recording wave has to be reproduced and the wavelength must be the same. In some lasers, this poses problems when a pulsed source is used for the recording and continuous illumination is wanted for viewing. The 10%

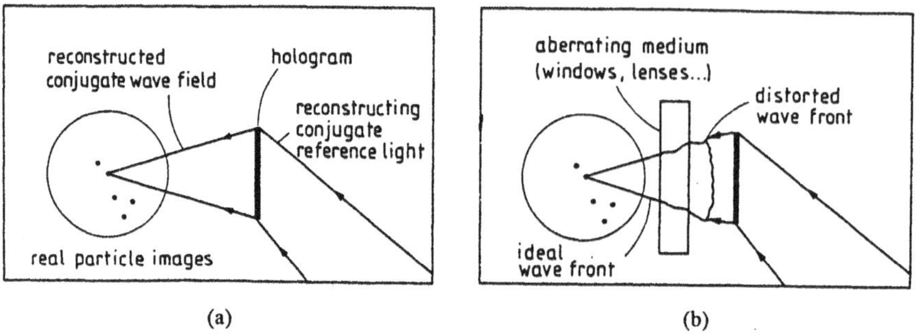

Figure 9. Generation of a real image of a tracer particle field by reconstruction with a phase conjugate wave.
(a) Ordinary reconstruction.
(b) Compensation of aberrations.

wavelength difference between ruby and He-Ne laser, for example, is detrimental to a faithful reconstruction. For high fidelity in ruby laser experiments of the author's group the images have been reconstructed with repeated pulses from the same laser. Pulsed Nd:YAG laser systems provide the advantage that CW lasers exist at the same wavelength. Another requirement is that distortions in the photographic material due to shrinkage, for example, must be avoided. In large distances between particles and hologram relay imaging of the particles onto a region close to the plate is of advantage because the area on the hologram that contributes to the particle image is reduced (Barnhart *et al.*, 1994).

The real power of holography lies in the almost unrestricted recording depth. But already holography of a light sheet can be utilized for various tasks because the depth position of the particles is preserved. When the stereoscopic techniques, for example, are applied to a holographic virtual image, a single camera system would suffice. The stereo images could be taken of the frozen-in particle distribution, one after the other. As mentioned earlier, the generation of real images could substitute the lens-and-mirror arrangement in back-projection stereoscopy. A hologram recorded in either arm and illuminated with the conjugate reference wave would reproduce the desired back-propagating waves (Adrian *et al.*, 1993; Barnhart *et al.*, 1994).

It has also been proposed to pass an interrogation section of the holographic real image of the particles in a light sheet through an aperture and use it as input for a 3-D autocorrelation evaluation analogous to 2-D Young's fringes production and processing (Coupland and Halliwell, 1992). The resulting autocorrelation peaks are located in 3-D space and their separation provides the 3-D displacement. In other efforts the small interrogation volume is selected from the deep reconstructed particle field by optical nonlinear processing and evaluated in a similar manner in the diffraction domain (Moraitis, 1993).

It should be mentioned that holographic particle imaging in the off-axis configuration has been used under still a different aspect. When operated with a separate reference beam for each of the pulses the first and second image can be distinguished thus removing the directional ambiguity that is inherent in simple double pulse recordings (Coupland *et al.*, 1987). A dual pulsed Nd:YAG laser system, for example, supplies suitable pulses that are aligned to travel identical object illumination paths but different reference paths (Hussain *et al.*, 1993, Adrian *et al.*, 1993, Barnhart *et al.*, 1994; Gray and Greated, 1994). Upon reconstruction, first and second exposure can be viewed separately. In this case, autocorrelation evaluation can be replaced by crosscorrelation processing that provides higher sensitivity and better signal-to-noise ratio.

3.1 OFF-AXIS HOLOGRAPHY OF DEEP PARTICLE FIELDS

The present state of off-axis holography for 3-D particle imaging in flow studies is illustrated by an arrangement that comprises many of the unique features holography offers (Adrian *et al.*, 1993). The recording setup is shown in Figure 10a. The illuminating light is produced by a Q-switched Nd:YAG laser that is pulsed twice at instants t_1 and t_2. Pulses are spatially seperated so that the first pulse forms, say, reference wave R_1 and the second pulse R_2. For the illuminating light both pulses are superposed. The light scattered by the particles is collected under two near-forward directions with relay optics that form particle images close to the holographic plate. A beam-bending prism is introduced in each imaging branch so that both object waves can be recorded on the same holographic plate. Thus, holograms of the particle fields are recorded at the instants of the laser pulses, each with a different reference wave.

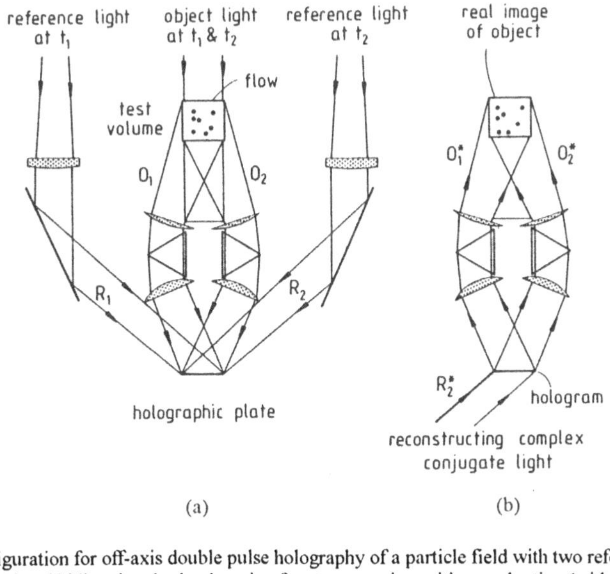

Figure 10. Configuration for off-axis double pulse holography of a particle field with two reference beams for ambiguity removal and bidirectional relay imaging for stereoscopic position evaluation (with permission after Adrian et al. (1993)).

(a) Recording.

(b) Phase conjugate reconstruction of real images for aberration compensation.

In the reconstruction process (Figure 10b) one complex conjugate reference wave is applied after the other to produce the backward travelling object wave at the times of first and second exposure. These are directed backwards through precisely the old imaging optics thus cancelling all prior aberrations during the recording. Both the temporal states of the particle field can thus be reconstructed consecutively by applying either reference wave. A light sensitive detector like a CCD target, for example, is placed somewhere in the real image. Now the situation is analogous to the back-projection stereo configuration that we have described earlier. Depending on the depth position of the particle with respect to the plane of the CCD we register identical images from both directions O_1* and O_2* (when the particle lies within this plane) or separated images that indicate the depth coordinate by the amount of separation and the left-right sequence. By moving the CCD on a precision traversing mount the whole volume is scanned.

The work shown gives a good example of how holography is suited to study arbitrarily complicated 3-D flows. In a series of planes in depth the 3-D velocity vector has been evaluated to yield very detailed information about the flow (Meinhart *et al.*, 1994). Alternatively to the plane-oriented evaluation, the particle field has been volume scanned and the velocity vectors computed by fully three-dimensional correlation (Gray and Greated, 1994). The large number of velocity vectors for evaluation require high computing power and effective processing algorithms. Furthermore, even the presentation of the data becomes difficult due to their multi-dimensional nature. Thus, it is obvious, that a demand exists also for holographic arrangements that require less effort though they may be restricted in their application.

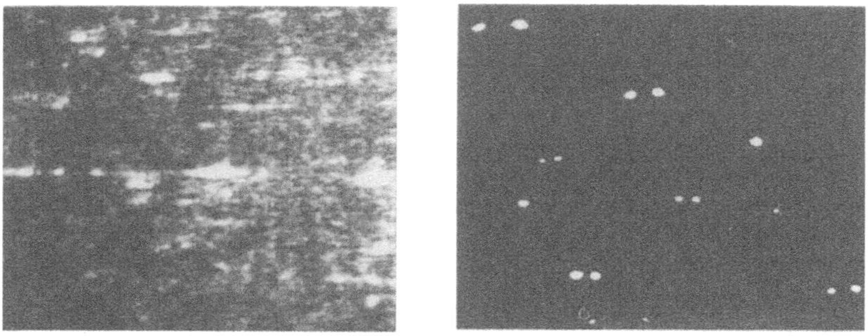

Figure 11. Reconstructed real particle images from a flow in a cylindrical glass pipe without (a) and with (b) substitute glass wall in the complex conjugate wave field (courtesy D.H. Barnhart).

The reconstruction of the phase conjugate object wave has still another useful property. When viewing the flow field through unavoidable complicated optical windows, it is possible to compensate many aberrations by positioning a substitute window in the reconstructed wave field and interrogating the real image at the former flow location (Barnhart *et al.*, 1995). An example from a cylindrical pipe flow is shown in Fig. 11, where a plane with reconstructed particle images is shown without and with the substitute glass pipe section.

3.2 IN-LINE PARTICLE HOLOGRAPHY

Off-axis holography is easily disturbed by vibrational noise, requires good coherence, and relies on an accurate reproduction of the reference wave for reconstruction. For small objects like the flow tracing particles, the so-called in-line arrangement is much simpler and provides better stability.

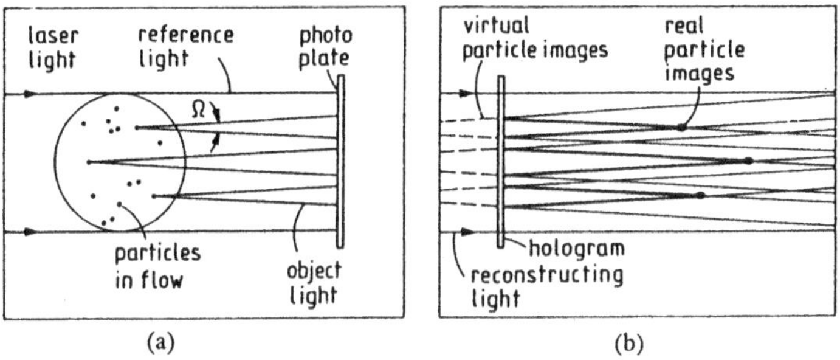

Figure 12. In-line holography of particle fields.
(a) Recording.
(b) Reconstruction.

Figure 12 illustrates a schematic of in-line particle holography. There is no extra reference wave, because the tiny particles leave enough of the illuminating plane wave light undisturbed so that it can serve this purpose. In-line holography relaxes many of the restrictions mentioned for off-axis holography. Its application in particle studies has a long history that has been summarized repeatedly (Thompson, 1989; Vikram, 1990). Since forward scattering is employed, the light is utilized much more efficiently. Due to dynamic limitations, however, the effective angular range Ω of the recorded light is usually smaller than in sidewise imaging. Because Ω determines the resolution, this can become a serious disadvantage. For reconstruction, the hologram is repositioned in the illuminating wave. Due to the on-axis geometry, both the virtual and real image are reconstructed simultaneously. Usually a microscopic analysis of the real particle images is carried out for evaluation. These images, however, are disturbed by the unused portion of the reconstructing wave and by noise from the virtual images (Meng *et al.*, 1993).

To understand the resolution limits in an in-line arrangement and provide for improvements, let us apply equations (1) and (2). Most registered light falls within the central lobe of the diffraction pattern of the particle which is given approximately by $\Omega = \lambda/d_p$. Thus the transverse and longitudinal resolution lengths are given by

$$d_t = d_p \tag{3}$$

$$d_l = d_p^2/\lambda \tag{4}$$

Clearly, in many practical cases where d_p is considerably larger than λ axial focusing according to Eq. (4) will be much lower than transverse focusing as in Eq. (3). It has been shown that the depth of field that can be covered with an in-line hologram is some 100times the longitudinal focus length (Bernal and Scherer, 1993). Thus, about 100 different planes can be distinguished. There are many applications of in-line particle holography that take the real images as they are and improve on resolution by sophisticated processing of the images (Stanton and Caulfield, 1984; Liburdy, 1987; Hobson, 1988; Dadi *et al.*, 1991). But the in-line setup can also be modified directly for improvements. Early solutions introduced an auxiliary illumination beam to enhance the scattered object light (Royer, 1977). Bernal and Scherer (1993) and Hussain *et al.* (1993) combined two traditional in-line setups at right angles so that the inaccurate depth position from, say, hologram 1 can be improved by an additional evaluation from hologram 2. In other approaches multibeam illumination was introduced to increase the aperture by recording several scattered fields, each of extent Ω, on the same hologram (Hussain *et al.*, 1993). Thus, highly effective forward scattering is combined with good axial focusing by a large aperture. Still in view of the poor longitudinal resolution a scheme for in-line recording, but off-axis viewing (IROV) has been proposed by Meng and Hussain (1995).

3.3 MULTIPLE LIGHT SHEET HOLOGRAPHY

The holographic study of a voluminous particle field provides three-dimensional information, but must be paid for by two principal disadvantages:

- The illuminating laser beam must be expanded to a larger cross section resulting in a loss of light.

- In the interrogation, particle selection is made by focusing into the reconstructed image volume (see the evaluation of virtual or real image in Figure 8b or Figure 9a) which results in an increase of the noise level by light from all those particles in out-of-focus regions that in ordinary PIV are dark.

While the former effect is obvious, the latter, i.e., the crowding of particle images in the investigated reference plane, must be studied as to its deteriorating influence on the rate of correct displacement vector identifications. It has been shown in autocorrelation simulations, for example, that the accepted success rate of 96% in a thin light sheet of 1 mm depth and with optimum particle density (Keane and Adrian, 1990) reduces to 86% or 44% in volumes of 10 mm or 20 mm depth, respectively (Hinrichs and Hinsch, 1994). For crosscorrelation conditions less severe effects are expected.

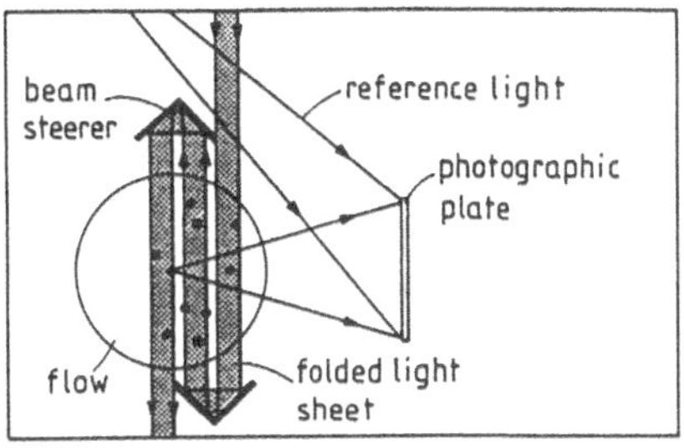

Figure 13. Holographic particle velocimetry with a folded light sheet.

An alternative approach has therefore been introduced in the holographic exploration of flows by recording a set of several light sheets that sample the flow field in depth (Hinsch *et al.*, 1990, Hinsch *et al.*, 1993). If the flow displays a predominant direction that determines the orientation of the light sheet, the velocity often changes only gradually in depth. If the spacing of the sheets is adjusted to the depth scale of the flow structures, much valuable information about the three-dimensionality of the flow can be obtained with a system of just a few light sheets. At the same time, favorable properties of ordinary PIV are preserved. First of all, optimum use is made of the available light by redirecting the sheet several times through the flow (Figure 13). Because the particles scatter only little, the quality of the light sheet stays sufficient for several passes. If needed, suitable beam shaping optics can improve the sheet dimensions after multiple passes through the measuring field. Second, each light sheet is available for any of the well established or developing evaluation techniques including three-dimensional aspects. Finally, a sophisticated balance of the limited coherence of the laser and the coherence requirements for holography ensures that each sheet can be reconstructed separately eliminating even most of the crosstalk from defocused particle images in other sheets.

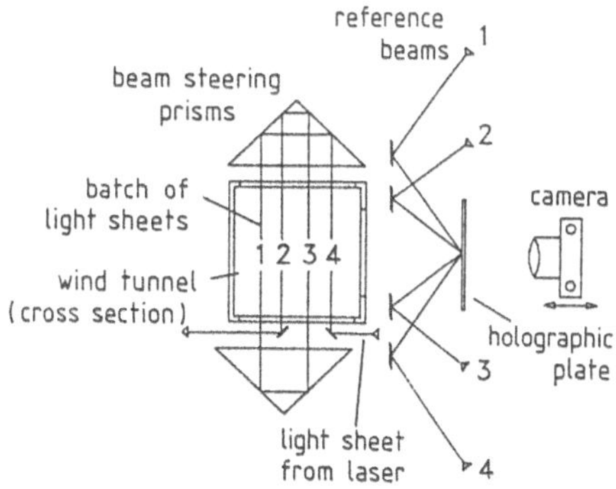

Figure 14. Multiple light sheet holography with controlled coherence in a windtunnel. Cross-sectional view of tunnel and optical beam configuration.

Figure 14 gives a schematic of a setup serving this purpose in a wind tunnel study. We are looking at the 0.25 x 0.25 m^2 cross section of the tunnel in the measuring region. Two prism beam steering devices produce four light sheets at 40 mm spacing. The scattered light is incident on a holographic plate together with four reference beams from different directions in space. All of these waves are present at the same time, but provisions have been made that light from each sheet interferes only with one reference beam each. For this purpose the limited coherence length of the laser light is used. Wavelength selective devices in the laser produce a coherence length that is just sufficient for the holographic recording of a single light sheet together with a reference beam that matches in optical path length. Optical detours in the reference beams make sure that this condition holds for the proper combinations of reference and object light. By reconstructing with a single reference beam at a time, particles in one sheet can be evaluated without disturbing light from any other plane in depth.

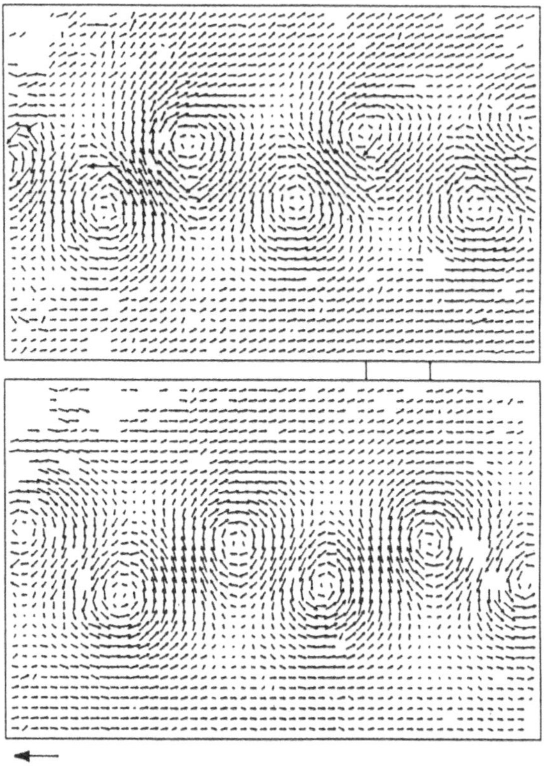

Figure 15. Double light sheet holography of oblique vortex shedding behind a cylinder. Velocity maps from two planes normal to the cylinder axis and spaced 40 mm. Mean velocity of 0.47 m/s has been subtracted. The arrow indicates 0.2 m/s.

As an example we present the flow field in two planes through the *von Kármán* vortex street behind a circular cylinder at Re = 190. Ideally, this is a two-dimensional structure. Often, however, it is observed that the vortices are not shed parallel to the cylinder axis which has been attributed to effects from the termination of the vortex axes at the ends. The slight three-dimensionality of this phenomenon is ideally suited to demonstrate the capability of our PIV method. Already two planes reveal important features of the flow structure. In our experiment, light originated from a 10-J Q-switched double pulse ruby laser (optical resonator length 0.7 m) that was modified for about 0.2-m coherence length. For this purpose one etalon in conjunction with the output mirror was removed. The original second etalon was exchanged for a model of specifically designed Q-factor to yield the desired coherence. Velocity vector plots of both light sheets that are spaced 40 mm in depth are shown in Figure 15. The cylinder is positioned 55 mm upstream to the right and halfway up the frame. The vortex street is evident in both the planes. A clearly visible spatial shift in the periodic structure is evidence of the oblique vortex shedding. It amounts to about 9 mm which yields an angle of about 13°.

It should be mentioned that multiple light sheet production simply by stripwise periodic blocking of the expanded laser beam has been proposed by Katz and Ni (1993) to reduce the adverse effects of blurred particle images. The reference presents several other holographic configurations for PIV.

4. Conclusions

Optical flow velocimetry is conquering more and more dimensions. We have seen that a hologram can store most of the spatial data about the particles that track a flow field. Current trends in digital holography (Schnars and Jüptner, 1994) suggest that before long powerful computers will be used for digital reconstruction of particle fields from computer holograms. Then the photographic recording step may become needless even in 3-D applications. Furthermore, the holographic techniques can be turned into cinematographic tools when the laser light source is operated at the appropriate repetition rate. In studies on cavitation bubble dynamics, for example, small holograms were recorded on rotating holographic plate material at 20 kHz rate with a dye laser pumped by a copper vapor laser (Lauterborn *et al.*, 1993). In other investigations, rotating drum cameras served a similar purpose (Ruff *et al.*, 1990; Eisfeld, 1992).

Our analysis has shown that powerful equipment is developed to reveal the spatial and temporal structures of complex flows in experimental investigations. The amount of experimental data, of course, grows tremendously, and the evaluation and visualization of multi-dimensional records present new challenges. Thus, each problem deserves special consideration to select the appropriate level in the dimensional hierarchy shown in Figure 1 and to choose the proper technique.

5. Acknowledgement

This work is a revised version of an earlier paper (Hinsch, 1995), parts of which have been reproduced with kind permission by IOP Publishing Limited, Bristol.

6. References

Adrian, R. J. (1991): Particle-imaging techniques for experimental fluid mechanics. Ann. Rev. Fluid Mech. **23**, 261-304.

Adrian, R. J., Meinhart, C. D., Barnhart, D. H., Papen, G. C. (1993): An HPIV system for turbulence research. In: *Holographic Particle Image Velocimetry* (ed. E. P. Rood), ASME FED **148**, 17-21.

Arroyo, M. P. and Greated, C. A. (1991): Stereoscopic particle image velocimetry. Meas. Sci. Technol. **2**, 1181-1186.

Barnhart, D. H., Adrian, R. J., Papen, G. C. (1994): Phase conjugate holographic system for high resolution particle image velocimetry. Appl. Opt. **33**, 7159-7170.

Barnhart, D. H., Adrian, R. J., Meinhart, C., Papen, G. C. (1995): Phase-conjugate holographic system for high resolution particle image velocimetry through thick-walled, curved windows. In: *Interferometry VII* (eds: Pryputniewicz, R. J., Brown, G. M., Jüptner, W. P.) Proc SPIE **2545**, 165-175.

Bernal, L. P. and Scherer, J. (1993): HPIV measurements in vortical flows. In: *Holographic Particle Image Velocimetry* (ed. E. P. Rood), ASME FED **148**, 43-50.

Born, M. and Wolf, E. (1980): *Principles of Optics* (Oxford: Pergamon Press) 441.

Brücker, C. and Althaus, W. (1992): Study of vortex breakdown by particle tracking velocimetry (PTV). Part 1: Bubble-type vortex breakdown. Exp. Fluids **13**, 339-349.

Brücker, C. (1993): Study of vortex breakdown by particle tracking velocimetry (PTV). Part 2: Spiral-type vortex breakdown. Exp. Fluids **14**, 133-139.

Brücker, C. (1995a): Digital-Particle-Image-Velocimetry (DPIV) in a scaning light-sheet: 3D starting flow around a short cylinder. Exp. Fluids **19**, 255-263.

Brücker, C. (1995b): 3-D DPIV using a scanning light-sheet and stereoscopy: study of flow development around a spherical cap. ASME-FED **229**, 497- 503.

Bryanston-Cross, P. J., Funes-Gallanzi, M., Quan, C., Judge, T. R. (1992): Holographic particle image velocimetry (HPIV). Opt. Laser Tech. **24**, 251-256.

Cenedese, A. and Paglialunga, A. (1988): A new technique for the determination of the third velocity component with PIV. Proc. 4th Int. Symp. Appl. Laser Anemometry to Fluid Mechanics (Lisbon), paper 3.14.

Collier, R. J., Burckhardt, Ch. B., Lin, L. H. (1971): *Optical Holography* (New York: Academic Press).

Coupland, J. M., Pickering, C. J. D., Halliwell, N. A. (1987): Particle image velocimetry: Theory of directional ambiguity removal using holographic image separation. Appl. Opt. **26**, 1576-1578.

Coupland, J. M. and Halliwell, N. A. (1992): Particle image velocimetry: Three-dimensional fluid velocity measurements using holographic recording and optical correlation. Appl. Opt. **31**, 1004-1008.

Dadi, M., Stanislas, M., Rodriguez, O., Dyment, A. (1991): A study by holographic velocimetry of the behavior of free small particles in a flow. Exp. in Fluids **10**, 285-294.

Dinkelacker, F., Schäfer, M., Ketterle, W., Wolfrum, J. (1992): Determination of the third velocity component with PTA using an intensity graded light sheet. Exp. Fluids **13**, 357-359.

Dracos, T. and Malik, N. A. (1992): 3D particle tracking velocimetry - its possibilities and limitations. In: *Flow Visualization VI* (ed. Y. Tanida and H. Miyashiro) (Berlin: Springer-Verlag), 785-791.

Dreesen, F., Hinrichs, H., Hinsch, K., Roshop, A. (1992): Holographic particle image velocimetry (HPIV) - an extension of PIV to a 3-D measuring method. In: *Flow Visualization VI* (ed. Y. Tanida and H. Miyashiro) (Berlin: Springer-Verlag), 822-826.

Eisfeld, F. (1992): High-speed photography, high-speed cinematography and high-speed holography as tools to investigate fast flows. In: *Flow Visualization VI* (ed. Y. Tanida and H. Miyashiro) (Berlin: Springer-Verlag), 419-425.

Funes-Gallanzi, M., Bryanston-Cross, P. J., Udrea, D. D. (1994): High accuracy techniques applied to the extraction of absolute position estimates in a 3DPIV System. Proc. Seventh Int. Symp. on Appl. of Laser Techniques to Fluid Mechanics (Lisbon), paper 35.7.

Funes-Gallanzi, M., Udrea, D. D., Bryanston-Cross, P. J. (1995): Advances of the Warwick real-time 3DPIV system for turbomachinery research. ASME-FED **229**, 465-472.

Gauthier, V. and Riethmuller, M. L. (1988): Application of PIDV to complex flows: measurement of the third component. In: VKI-LS 1988-06 *Particle Image Displacement Velocimetry*. Von Karman Institute for Fluid Dynamics, Rhode-Saint-Genèse, Belgium.

Grant, I. (1994): *Selected Papers on Particle Image Velocimetry* (Bellingham: SPIE Milestone Series MS 99).

Grant, I., Zhao, Y., Tan, Y., Stewart, J. N. (1991): Three component flow mapping: experiences in stereoscopic PIV and holographic velocimetry. In: *Fourth International Conference on Laser Anemometry, Advances and Applications*; American Society of Mechanical Engineers ASME, 365-371.

Grant, I., Fu, S., Pan, X., Wang, X., Aroussi, A. (1994a): An in-line stereoscopic, colour PIV system for 3-component velocity measurements with ambiguity resolution. Proc. Seventh Intl. Symp. on Appl. of Laser Techniques to Fluid Mechanics (Lisbon), paper 26.3.

Grant, I., Pan, X., Wang, X., Stewart, N. (1994b): Correction for viewing angle applied to PIV data obtained in aerodynamic blade vortex interaction studies. Exp. Fluids **18**, 95-99.

Grant, I., Fu, S., Pan, X., Wang, X. (1995): The application of an in-line, stereoscopic, PIV system to 3-component velocity measurements. Exp. Fluids **19**, 214-221.

Gray, C. and Greated, C. A. (1994): Three-dimensional correlation analysis for particle field hologram velocimetry. In: *Optical Methods and Data Processing in Heat and Fluid Flow* (London: City University, Inst. of Mech. Engineers), 145-149.

Guezennec, Y. G., Zhao, Y., Gieseke, T. J. (1994): High-speed 3-D scanning particle image velocimetry (3-D SPIV) technique. Proc. Seventh Int. Symp. on Appl. of Laser Techniques to Fluid Mechanics (Lisbon), paper 26.1.

Halloin, V. L. and Jottrand, R. (1994): A new technique for the experimental study of 3D velocity fields. Exp. Fluids **17**, 115-118.

Herpfer, D. C. and Jeng, S. M. (1995): Planar measurement of three component velocity by streaked-particle-imaging velocimetry. Appl. Opt. **34**, 2301-2304.

Hinrichs, H. and Hinsch, K. D. (1994): Multiple light sheet particle holography for 3-D flow velocimetry. Proc. Seventh Intl. Symp. on Appl. of Laser Techniques to Fluid Mechanics (Lisbon), paper 26.2.

Hinsch, K., Hinrichs, H., Kuhfahl, G., Meinlschmidt, P. (1990): Holography with controlled coherence for 3-D particle image velocimetry. In: *Proc. Holographics* 1990 (ed. T. Tschudi), 51-57.

Hinsch, K. D. (1993a): The many dimensions of optical flow diagnostics. In: *Laser Anemometry: Advances and Applications - Proc. of the 5th Intl. Conf.* Proc. SPIE **2052**, 63-78.

Hinsch, K. D. (1993b): Particle Image Velocimetry (PIV). In: *Speckle Metrology*, ed. R. S. Sirohi (New York: Marcel Dekker), 235-324.

Hinsch, K. D., Hinrichs, H., Roshop, A., Dreesen, F. (1993): Holographic and stereoscopic advances in 3-D PIV. In: *Holographic Particle Image Velocimetry* (ed. E. P. Rood), ASME FED **148**, 33-36.

Hinsch, K. D. (1995): Three-dimensional particle velocimetry. Meas. Sci. Technol. **6**, 742-753.

Hobson, P. R. (1988): Precision coordinate measurements using holographic recording. J. Physics E **21**, 139-145.

Hussain, F., Liu, D. D., Simmons, S., Meng, H. (1993): Holographic particle velocimetry: prospects and limitations. In: *Holographic Particle Image Velocimetry* (ed. E. P. Rood), ASME FED **148**, 1-11.

Kasagi, N. and Sata, Y. (1992): Recent developments in three-dimensional particle tracking velocimetry. In: *Flow Visualization VI* (ed. Y. Tanida and H. Miyashiro) (Berlin: Springer-Verlag), 832-837.

Katz, J. and Ni, H. (1993): Problems and solutions associated with utilization of holography for three dimensional velocity measurements. In: *Holographic Particle Image Velocimetry* (ed. E. P. Rood), ASME FED **148**, 51-56.

Keane, R. D. and Adrian, R. J. (1990): Optimization of particle image velocimeters. Part I: Double pulsed systems. Meas. Sci. Technol. **1**, 1202-1215.

Kent, J. C., Trigui, N., Choi, W. C., Guezennec, Y. G. (1994): Detailed measurement on in-cylinder fluid dynamics using water analog simulation and 3-D particle tracking velocimetry. In: *Optical Methods and Data Processing in Heat and Fluid Flows* (London: City University, Inst. of Mech. Engineers), 69-74.

Kimura, I. and Kohno; Y. (1991): Measurement of three-dimensional velocity vectors in a flow field based on spatio-temporal image correlation. *FLUOCOME '91* (American Society of Mechanical Enginers) 609-615.

Lauterborn, W., Judt, A., Schmitz, E. (1993): High-speed off-axis holographic cinematography with a copper-vapor-pumped dye laser. Opt. Lett. **18**, 4-6.

Lawson, N. J., Halliwell, N. A., Coupland, J. M. (1994): Particle image velocimetry: Image labeling by use of adaptive optics. Appl. Opt. **33**, 4241-4247.

Liburdy, J. A. (1987): Holocinematographic velocimetry: resolution limitations for flow measurement. Appl. Opt. **26**, 4250-4255.

Lin, J. C. and Rockwell, D. (1994): Cinematographic system for high-image-density particle image velocimetry. Exp. Fluids **17**, 110-118.

Maas, H. G., Gruen, A., Papantoniou, D. (1993): Particle tracking velocimetry in three-dimensional flows. Part 1: Photogrammetric determination of particle coordinates. Exp. Fluids **15**, 133-146.

Malik, N. A., Dracos, Th., Papantoniou, D. A. (1993): Particle tracking velocimetry in three-dimensional flows. Part II: Particle tracking. Exp. Fluids **15**, 279-294.

Meinhart, C. D., Barnhart, D. H., Adrian, R. J. (1994): An interrogation and vector validation system for holographic particle image fields. Proc. Seventh Intl. Symp. on Appl. of Laser Techniques to Fluid Mechanics (Lisbon), paper 1.4.

Meng, H., Anderson, W. L., Hussain, F., Liu, D. D. (1993): Intrinsic speckle noise in in-line particle holography. J. Opt. Soc. Am. **A 10**, 2046-2058.

Meng, H. and Hussain, F. (1995): In-line recording and off-axis viewing technique for holographic particle velocimetry. Appl. Opt. **34**, 1827-1840.

Merzkirch, W. (1987): *Flow Visualization* (Orlando: Academic Press).

Moraitis, C. S. (1993): Three-dimensional particle image velocimetry by optical processing of holographic records. Intl. Congr. on Instrumentation in Aerospace Simulation Facilities ICIASF'93 (IEEE Publ 93CH3199-7), paper 39.

Prasad, A. K. and Adrian, R. J. (1993): Stereoscopic particle image velocimetry applied to liquid flows. Exp. in Fluids **15**, 49-60.

Prasad, A. K. and Jensen, K. (1995): Scheimpflug stereocamera for particle image velocimetry in liquid flows. Appl. Opt. **34**, 7092-7099.

Prenel, J. P., Porcar, R., El Rhassouli, A. (1989): Three-dimensional flow analysis by means of sequential and volumic laser sheet illumination. Exp. Fluids **7**, 133-137.

Raffel, M. and Kompenhans, J. (1994): Error analysis for PIV recording utilizing image shifting. Proc. Seventh Intl. Symp. on Appl. of Laser Techniques to Fluid Mechanics (Lisbon), paper 35-5.

Raffel, M., Gharib, M., Ronneberger, O., Kompenhans, J. (1995): Feasibility study of three-dimensional PIV by correlating images of particles within parallel light sheet planes. Exp. Fluids **19**, 69-77.

Reese, J., Chen, R. C., Fan, L.-S. (1995): Three-dimensional particle image velocimetry for use in three-phase fluidization systems. Exp. Fluids **19**, 367-378.

Robinson, O. and Rockwell, D. (1993): Construction of three-dimensional images of flow structure via particle tracking techniques. Exp. in Fluids **14**, 257-270.

Rood, E. P., ed (1993): *Holographic Particle Velocimetry.* ASME FED **148.**

Royer, H. (1977): Holographic velocimetry of submicron particles. Opt. Commun. **20**, 73-75.

Ruff, G. A., Bernal, L. P., Faeth, G. M. (1990): High speed in-line holocinematography for dispersed-phase dynamics. Appl. Opt. **29**, 4544-4546.

Schnars, U. J. and Jüptner, W. P. O. (1994): Digital recording and reconstruction of holograms in hologram interferometry and shearography. Appl. Opt. **33**, 4373-4377.

Sinha, S. K. (1988): Improving the accuracy and resolution of particle image or laser speckle velocimetry. Exp. Fluids **6**, 67-68.

Sinha, S. K. and Kuhlman, P. S. (1992): Investigating the use of stereoscopic particle streak velocimetry for estimating the three-dimensional vorticity field. Exp. Fluids **12**, 377-384.

Stanton, A. C. and Caulfield, H. J. (1984): An approach for automated analysis of particle holograms. Opt. Eng. **23**, 577-582.

Stolz, W. and Köhler, J. (1994): In-plane determination of 3D-velocity vectors using particle tracking anemometry (PTA). Exp. Fluids **17**, 105-109.

Thompson, B. J. (1989): Holographic methods for particle size and velocity measurement - recent advances. In: *Holographic Optics II: Principles and Applications* (ed. G. M. Morris), Proc. SPIE **1136**, 308-326.

Van Dyke, M. (1982): *An Album of Fluid Motion* (Stanford: Parabolic).

Vikram, C. S. (1990): *Holographic Particle Diagnostics* (Bellingham: SPIE Milestone Series MS21).

Westerweel, J. and Nieuwstadt, F. T. M. (1991): Performance tests on 3-dimensional velocity measurements with a two-camera digital particle-image velocimeter. In: *Laser Anemometry Advances and Applications* (eds: Dybbs, A. and Ghorashi, B.; American Society of Mechanical Engineers), Vol. **1**, 349-355.

Willert, C. E. and Gharib, M. (1992): Three-dimensional particle imaging with a single camera. Exp. Fluids **12**, 353-358.

CHAPTER III

PARTICLE TRACKING VELOCIMETRY

PARTICLE TRACKING VELOCIMETRY (PTV)
Basic Concepts

TH. DRACOS
Prof. em. ETH Zürich
Untergasse 14
CH - 8126 Zumikon

1. Introduction

Particle Tracking Velocimetry (PTV) is one of the oldest measuring techniques in fluid mechanics. It is often called low particle number density Particle Image Velocimetry (PIV) (Adrian, 1991). The two methods have indeed in common that they determine the velocity from the displacement of particles in a moving fluid during a prescribed time interval. In contrast to PIV, in which the mean displacement of a small *group* of particles is sought for, PTV tracks the trajectories of *individual* particles in three-dimensional space. Particle identification and determination of its position in space is a very important step in PTV. This is only possible if the number density, i.e. the number of particles per unit volume, $\rho_N = N/V$, where N is the number of particles in an observation volume V, is not too high.

Both methods make use of the definition of velocity

$$u = lim_{\Delta t \to 0} \frac{\Delta s}{\Delta t}$$

In practice, the time interval is small but finite. It is prescribed in PIV by the pulse Illumination frequency and in PTV by the imaging frequency. The pulse illumination frequency can be made sufficiently high to allow measurements in high speed flows. If we disregard high speed cinematography, the imaging frequency of video cameras with sufficient spatial resolution is limited. Under these circumstances PTV can only be applied to low speed flows. This is a much more severe limitation of PTV versus PIV than the low particle number density used in PTV.

Particles do not have to be necessarily solid. Gas bubbles in liquids, soap bubbles filled with gas in another or the same gas, drops in gases or encapsulated fluid in another or the same fluid can also act as particles, provided their dimensions are small compared to the distance travelled. Particles can have positive $\rho_p < \rho_f$ or negative $\rho_p > \rho_f$ buoyancy or

Th. Dracos (ed.), Three-Dimensional Velocity and Vorticity Measuring and Image Analysis Techniques, 155–160.

be neutrally buoyant. Here ρ_p is the density of the particle and ρ_f the density of the fluid.

Tracking the movement of particles is as important for two-phase flows as for the determination of the velocity field in a moving fluid. The latter is in fact only possible when the particles faithfully follow the movement of the fluid particle they mark. The method presented in the following is accordingly not only a method for velocity field measurements but in many cases a method suitable for the study of two-phase flows as well.

The use of particles of different kinds can allow a simultaneous measurement of the velocity field and of the paths of particles which do not follow the flow. Examples in this context are the study of the movement of buoyant particles(e.g. air bubbles) in quiescent fluid and of the flow field possibly induced by their movement, or the settlement of particles in a turbulent flow field with zero mean velocity. These are characteristics that distinguish PT/PTV from PIV. The two techniques have different merits and, what is also important, limitations. They do therefore not compete but complement one another.

2. Basic Concepts of Particle Tracking

Particle streak velocity measurements have been used for a long time. As long as the extraction of velocity information from the images was dependent on manual analysis, the number density was very small. Digital image analysis techniques opened the way to an automated analysis and an increase of the number density (Dimotakis et al., 1981). When CCD cameras are used, errors in the determination of the streak length can become considerable. In addition, in three-dimensional flows the results may be biased towards small velocities (Adrian, 1991). It is therefore preferable to have instantaneous images of particles taken by multi-pulse illumination on a single photographic plate or in a sequence of photographic or video frames and try to establish paths of individual particles and therefrom determine velocities. Using one camera, a number of two-dimensional tracking procedures have been developed (Willis & Deardorff, 1974, Utami & Ueno, 1984, Kobayashi & Saga, 1988, Adamczyk & Rimai, 1988a , Hassan & Canaan, 1991, and references therein).

By adding a second camera for a stereoscopic view of the movement of the particle, the step to three-dimensional tracking was accomplished (Chiu & Rib, 1956). The addition of a second camera and later of a third and fourth poses a new problem: the establishment of correspondences between the images of individual particles taken simultaneously by two or more cameras. Accordingly, three-dimensional particle tracking always involves two steps, particle image matching and coordinate determination at a given time and particle path tracking in time. Basically, two approaches have been put forward to solve this problem.

The one starts by determining particle tracks in two-dimensional space from frame sequences of each camera used. Correspondences of all particles in a track from one camera with the particles of a track from one of the other cameras are then established. Once this is done, the three-dimensional track of the particle is determined too and the velocity of the particle at a point along the track can be computed. The first attempts in this direction were made by Chiu & Rib (1956) and by Adamczyk & Rimai (1988b). Later a fully automated and efficient procedure was developed by Kent et al., 1993, Guezennec et al., 1994, Trigui et al., 1995. A simplified flow chart of the procedure developed by these authors is shown in Fig.1. The two-dimensional particle tracking technique they used is schematically shown in Fig.2(a). It is based on a search tree and the minimisation of a penalty function for the different paths. The penalty function evaluation is essentially based on the deviation of the position of a particle from its predicted position. Experience has shown that the optimal number of particle positions

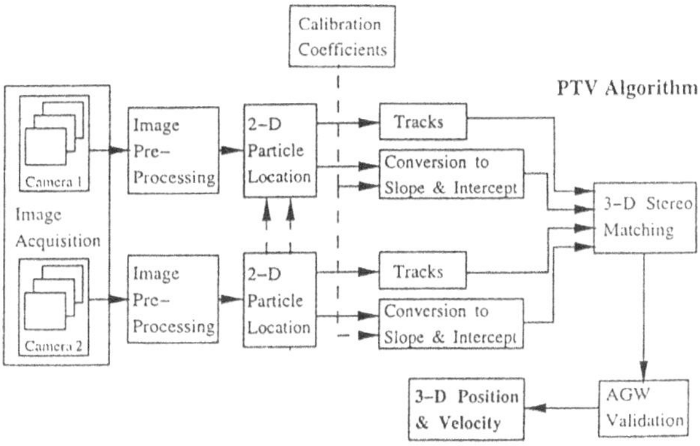

Figure 1 Flow chart of the tracking code adapted from Guezennec et al. (1994)

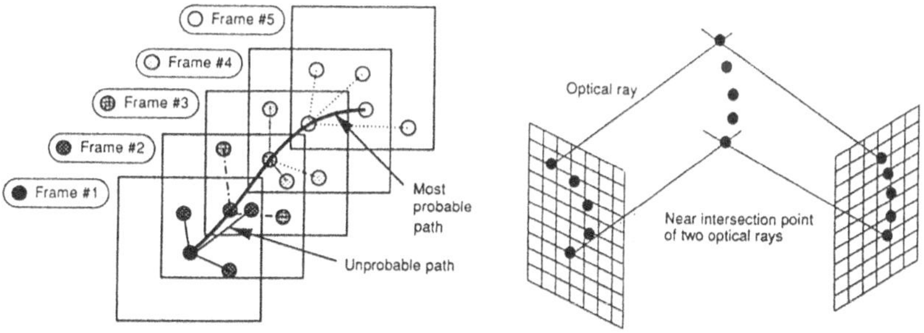

(a) (b)

Figure 2 (a) Principle of two-dimensional tracking and (b) link establishment between tracks adapted from Guezennec et al. (1994)

used for the determination of a two-dimensional track is five. The main difficulty in two-dimensional tracking arises from intersections of projections of particle trajectories on two-dimensional images. Of course any other two-dimensional tracking procedure can be employed instead. One such procedure is presented by Hassan & Canaan (1991) and was used to determine the movement of bubbles in a two-phase flow.

The stereo matching of two-dimensional tracks is shown in Fig.2(b). Using two commercially available CCD cameras with 512×512 pixel sensors the maximum number of velocity vectors determined at each time step is between 400 and 500.

The technique has been successfully applied to different flows with stationary cameras. In this case the radius of search areas in a next frame is usually smaller than the travel distance of the particles, and reversals in the path direction of a particle are not probable. This simplifies the tracking but limits the length of the tracks as the particles will soon leave the observation volume and are no longer trackable.

The importance of following the movement of the particles for the determination of long trajectories and the complicated form of these trajectories is illustrated in Fig.3(a),(b),(c). Fig.3(a) shows the projection of a three-dimensional trajectory of a particle on the x,y-plane detected by moving cameras. Flow direction is from bottom to top. The movement of the particle is followed over a period of 20s corresponding to 500 time steps, i.e. the trajectory contains 500 successive positions of the particle. During this time the carriage on which the cameras were mounted travelled 4.69m downstream the channel in which the observation was made. In the moving coordinate frame the trajectory is entangled. A small part of this trajectory extending over 2s and containing 50 particle positions is

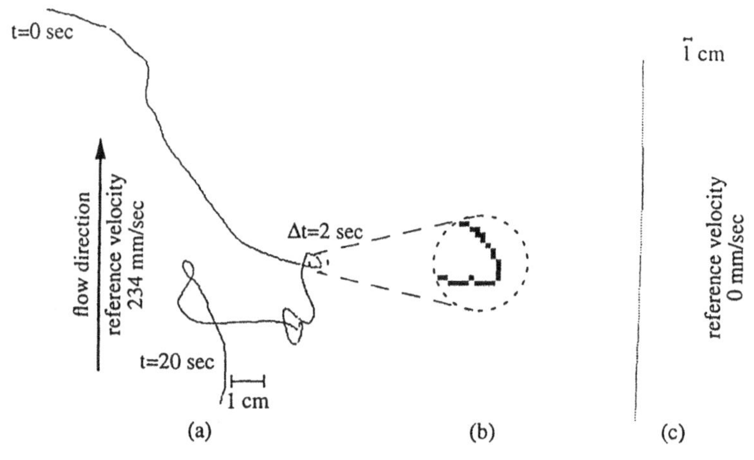

Figure 3 Track projections on the x,y-plane. (a) Track containing 500 successive positions taken by moving cameras. (b) Blow-up of a part of the trajectory containing 50 particle positions.(c) The same part of the trajectory shown from stationary cameras.

marked in Fig.3(a) and shown enlarged in Fig.3(b). Note that although it corresponds to 10% of the observation time, its length in x,y projection is only 3% of the length of the projection of the whole trajectory. This is because during that period of time the particle was moving in x-direction approximately with the reference velocity, i.e. the velocity of the carriage, and its lateral movement was directed mainly in z-direction. The part of the trajectory shown in Fig.3(b) is depicted in Fig.3(c) as it would be seen from a stationary camera assembly. The trajectory is nearly straight and most probably much easier to detect, as intersections of the projection of particle trajectories are less probable to occur. However, it is approximately 470mm long and can be tracked only if the observation volume is very large. The trajectory shown in Fig.3(a) is confined inside a rectangle 120mm long and 90mm wide, although it contains 10 times more particle positions. The observation volume needed can be four times smaller in x-direction than the one in the previous case. Projections of such trajectories on an image plane often intersect and their detection becomes cumbersome.

The first method will not be treated in detail. The reader is referred to the bibliography quoted.

The other approach starts with the determination of the coordinates of particles in three-dimensional space at a given time and consequent tracking of particle positions in time in three-dimensional space. The problem of track intersections does not arise in this case, as particle trajectories do not intersect in tree-dimensional space. This technique was simultaneously developed by Nishino et al. (1989), Kobayashi et al. (1989), Papantoniou & Maas (1990). When applied with moving cameras, it allows to determine long trajectories and will be described to some detail and critically assessed in the following.

References

Adamczyk, A.A., Rimai, L. 1988a. 2-Dimensional particle tracking velocimetry (PTV) Technique and image processing algorithms, *Exp. Fluids,* **6**, 373 - 380

Adamczyk, A.A., Rimai, L. 1988b. Reconstruction of a 3-dimensional flow field from orthogonal views of seed track video images, *Exp. Fluids,* **6**, 380 - 386

Adrian, R.J. 1991. Particle Imaging techniques for experimental fluid mechanics, *Ann. Rev. Fluid Mechanics,* **23**, 261 - 304

Chiu, W.C., Rib, L.N. 1956. The rate of dissipation of energy and the energy spectrum in a low-speed turbulent jet, *Trans. Am. Geophys. Union,* **37**, 13 - 26

Dimotakis, P.E., Debussy, F.D., Koochesfahani, M.H. 1981. Particle streak velocity field measurements in a two-dimensional mixing layer,

Guezennec, Y.G., Brodkey, R.S., Trigui, N., Kent, J.C. 1994. Algorithms for fully automated three-dimensional particle tracking velocimetry, *Exp. Fluids,* **17,** 209 - 219

Hassan, Y.A., Canaan, R.E. 1991. Full-field bubbly flow velocity measurements using a multi-frame particle tracking technique, *Exp. Fluids,* **12**, 49 - 60

Kent, J.C., Trigui, N., Chui, W.C., Guezennec, Y.G., Brodkey, R.S. 1993. Photogrammetric calibration for improved three-dimensional particle tracking velocimetry (3D-PTV), *SPIE Proceedings,* **2005,** 400 - 412

Kobayashi, T., Saga, T. 1988. A real-time velocity measurement algorithm for two-dimensional flow fields, *Proc. Int. Symp. on FLUCOME '88*

Kobayashi, T., Saga, T., Sekimoto, K. 1989. Velocity measurement of three-dimensional flow around rotating parallel discs by digital image processing, *ASME Publ. FED-85,* 29 - 36

Nishino, K., Kosagi, N., Hirata, M. 1989. Three-dimensional particle tracking velocimetry based on automated digital image processing, *J. Fluid Eng.* **111**, 384 - 391

Papantoniou, D., Maas, H.-G. 1990. Recent advances in 3-D particle tracking velocimetry, *Proc. Sth. Int. Symp. on Appl. of Laser Techs in Fluid Mech.*, Lisbon

Trigui, N., Guezennec, Y.G., Brodkey, R.S. 1995. Fully automated three-dimensional particle tracking velocimetry: Theory, implementation and validation, *Proc. Int. Workshop on PIV - Fukui'95*, 59 - 70

Utani, T., Ueno, T. 1984. Visualisation and picture processing of turbulent flow, *Exp. Fluids ,* **2,** 25 - 32

Willis, G.E., Deardorff, J.W. 1974. A laboratory model of the unstable planetary boundary layer, *J. Atmos. Sciences,* **31**, 1297 - 1307.

VIDEOGRAMMETRY: METHODOLOGY AND PTV/LIF APPLICATIONS

A. GRUEN
ETH Zurich
Institute of Geodesy and Photogrammetry
ETH Hoenggerberg
8093 Zurich, Switzerland

1. Introduction

This contribution is meant to introduce into the fundamentals of videogrammetry. Videogrammetry is a vision-based measurement technique using video components (but not exclusively) in order to perform non-contact, on-line, precise measurements. As such it is of crucial importance to both our PTV and LIF systems, which have been developed at ETH Zürich in recent years, applied to various kinds of flow measurements, and which are described in various contributions of this volume (compare chapters 3 and 4).

After a classification of the term „videogrammetry" in section 2 we describe the basic measurement concept as it is realized in our PTV system, which consists of applying more than two CCD cameras simultaneously. The underlying sensor model is that of the „bundle method", used in form of a least squares estimation procedure (section 3). Section 4 introduces into some of the system aspects, and addresses both functional and hardware issues.

If precise measurements and reliable results are required, a good system calibration is very critical. In section 5 we present several calibration techniques and focus on the most general and flexible method of self-calibration. Finally, section 6 describes the algorithms used for high precision image measurements.

For the reader it is important to note that this contribution is meant to provide for the technical background of some aspects of videogrammetry. For the implementation details concerning our PTV and LIF systems the other contributions of chapters 3 and 4 must be conducted.

2. Defining „videogrammetry"

"Photogrammetry" is the science and art of making precise and reliable measurements from images. While the term "photogrammetry" may evoke the notion of "photographic" images, the technique as such is by no means restricted to those but incorporates all sorts of non-photographic images as well. The term "digital" in „digital photogrammetry" emphasizes the use of digital images, in whatever form, be it from the X-ray, visible, near and far infrared range, or microwave portion of the electromagnetic spectrum. Even ultrasound and electron-microscopic images and the like could be concerned.

Over the years photogrammetry has focussed on some particular aspects of the overall data acquisition, reduction, analysis and representation process. Special expertise has been accumulated in the areas of geometric modelling of sensor imaging processes, observation network design and analysis, execution of precise and reliable measurements,

Th. Dracos (ed.), Three-Dimensional Velocity and Vorticity Measuring and Image Analysis Techniques, 161–190.

refinement of estimation models and statistical analysis of results including comprehensive error propagation through the whole data reduction process. This has been amended in recent years by appropriate image analysis, feature extraction and object reconstruction techniques, thus giving digital photogrammetry a relevant status as a discipline in its own right within the broad "vision community" (computer-, machine-, robot vision).

Figure 1 shows the role of machine vision (and its subcategories robot vision and videogrammetry) with respect to computer vision.

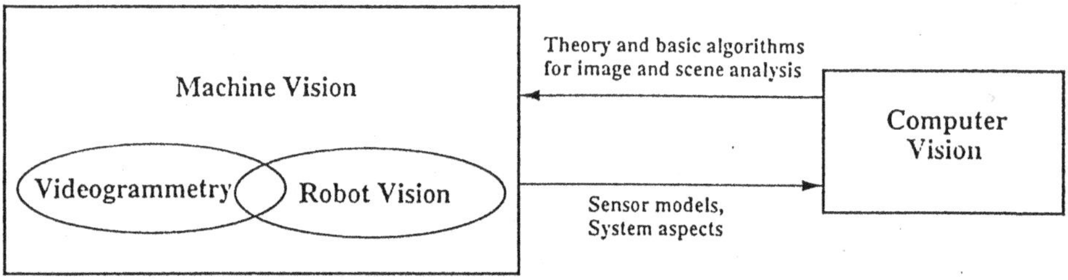

Figure 1. Classification of various vision disciplines

Whereas computer vision addresses theory and basic algorithmic questions of image and scene analysis, machine vision provides for sensor models and system aspects (including hardware issues). Robot vision is a subclass of machine vision in the sense that it uses the machine vision principles for strictly time-constrained solutions.

"Videogrammetry", as a subcategory of both photogrammetry and machine vision, uses video components such as CCD cameras, video recorders and frame grabbers for image acquisition.

The terms computer vision, machine vision, robot vision, digital close-range photogrammetry and videogrammetry are often being used as synonyms for very similar techniques and applications. All of them are based on the acquisition of digital imagery and due to the nature of digital images, the automation of data processing is an inherent system characteristic, as well as real-time and on-line processing aspects. While computer vision often deals with more qualitative aspects of image understanding, digital close-range photogrammetric systems and many machine vision systems aim at the derivation of metric information from digital images. Robot vision, as the name suggests, provides a robot with the necessary information about its environment in order to be able to either act on some object and/or to position itself within this environment for self-navigation purposes. It is the nature of robot vision that it requires strongly time-constrained solutions and thus poses strict requirements on algorithmic design and integration of processing hardware.

A system which is called „photogrammetric" or „videogrammetric" has to meet the following requirements:

(a) capability for self-diagnosis (quality control)
(b) potential for high precision and reliability (redundant sensor data), and
(c) task flexibility with respect to 3-D object reconstruction functions.

Especially the requirements (a) and (b) ask for the simultaneous use of more than two CCD-camera frames. The concept is known as multi-station approach in photogrammetry.

This clearly expresses the fact that modern photogrammetry operates beyond stereo vision. For further details on system aspects compare sections 3 and 4.

Throughout the past ten years videogrammetry has proved to be a very appropriate measurement technique to solve a great number of different problems. After a decade of intense research and development this technique has reached a degree of maturity such that it can be used in very diverse applications. Our group for instance has worked on projects such as:

- Surface reconstruction of human faces (medicine, doll manufacturing)
- Tooth measurement - "videodentography" (dentistry)
- Human movement studies (biomechanics)
- Determination of insect trajectories (biology, pharmacy)
- Architectural structures (architecture, monument preservation)
- Measurement of cars (industrial design, quality control and crash testing)
- Measurement of airplane surface patches and jet engine parts (material testing and industrial inspection)
- High-resolution 3-D particle tracking in turbulent flows (hydromechanics, space science)
- Chemical reactions in turbulent flow (chemistry, hydromechanics)
- Table tennis robot (robotics)
- Monitoring of crystal growth (material sciences)
- Self-orientation and egomotion determination of a vision-based robot (robotics).

Compared to other measurement techniques videogrammetry can be used advantageously in the case of

- 3-D and non-contact measurements
- Large number of objects (targets)
- Moving objects
- Good temporal resolution requirements
- Precise and reliable results requirements
- Fast recording and processing requirements (on-line solutions)

Obviously, since PTV applications ideally show all these characteristics, videogrammetry is an appropriate technique to provide for solutions there.

In addition, videogrammetry generates permanent records (stored images), which can be used for remeasurement, additional measurements, and verification at any time. Also, since videogrammetric hardware components are normally off-the-shelf equipment, the respective systems are inexpensive.

3. Basic measurement concept

In videogrammetry predominantly the principle of optical triangulation is applied, both in passive and active mode. „Passive" means that illumination (natural or artificial) is used such that it provides only for the marking or visualization of the object (e.g. tracer particles in PTV). In the „active" mode the artificial illumination (also called structured lighting) is introduced into the measurement process as a geometrical constraint. As an example, consider our LIF system, where a laser sheet is shifted incrementally and thus defines the location of volume image elements („voxels"), compare chapter 4 of this volume.

3.1. THE MULTI-IMAGE MEASUREMENT MODE

In videogrammetry, characteristically, the object to be measured is recorded (imaged) by a certain number of CCD frames, taken from different locations. A single frame is not sufficient information to reconstruct a 3-D object in general. Two frames (a „stereomodel") provide, mathematically, for the required solution. Although the stereomodel measurement concept is widely used in applications, it has some serious drawbacks: (a) it does not provide for sufficient redundancy in observations for the detection and location of blunders in measurements, (b) in case of a more complex object not all parts might be visible from only two camera positions, (c) some parts of the object, while visible, may generate a bad measurement geometry, thus leading to inaccurate results, and (d) large objects cannot be covered with only two images at a given high accuracy level.

Figure 2 shows a photogrammetric (or videogrammetric) network arrangement, also called a „block". The object points P_i (i = 1,...,n) are imaged onto the CCD frames F_j (j = 1,...,m) according to the laws of perspective projection. Any one frame may include a large number of points, up to a few thousands, depending on the type of CCD sensor used. Any one object point may be imaged onto an arbitrary number of CCD frames (usually a maximum of about nine are used). The triangles denote „control points". These are points with known object space coordinates which relate the image measurements to an object space („world") coordinate system.

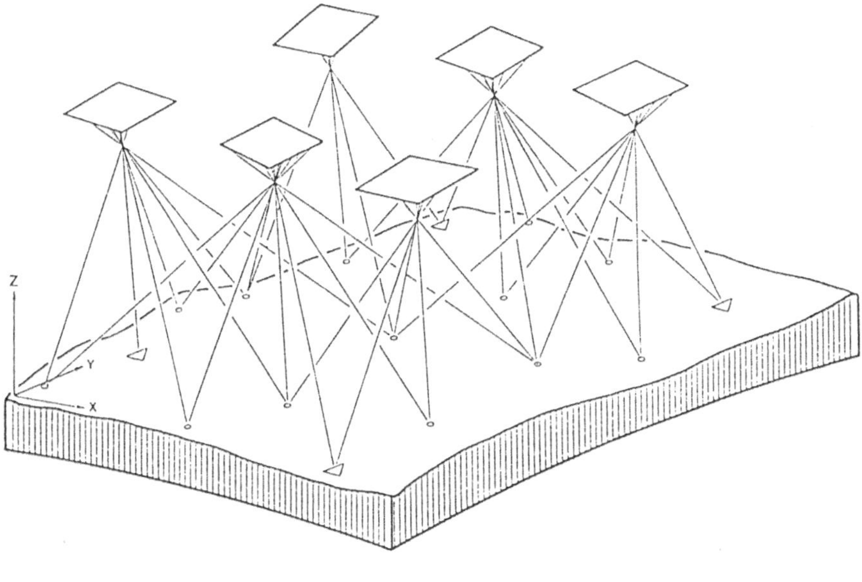

Figure 2. Videogrammetric block

We recommend to use more than two frames, thus generating a geometrically strong and statistically reliable network. Figure 3 shows such a typical measurement configuration. The object (a 3-D point cloud as in PTV) is imaged onto four CCD frames.

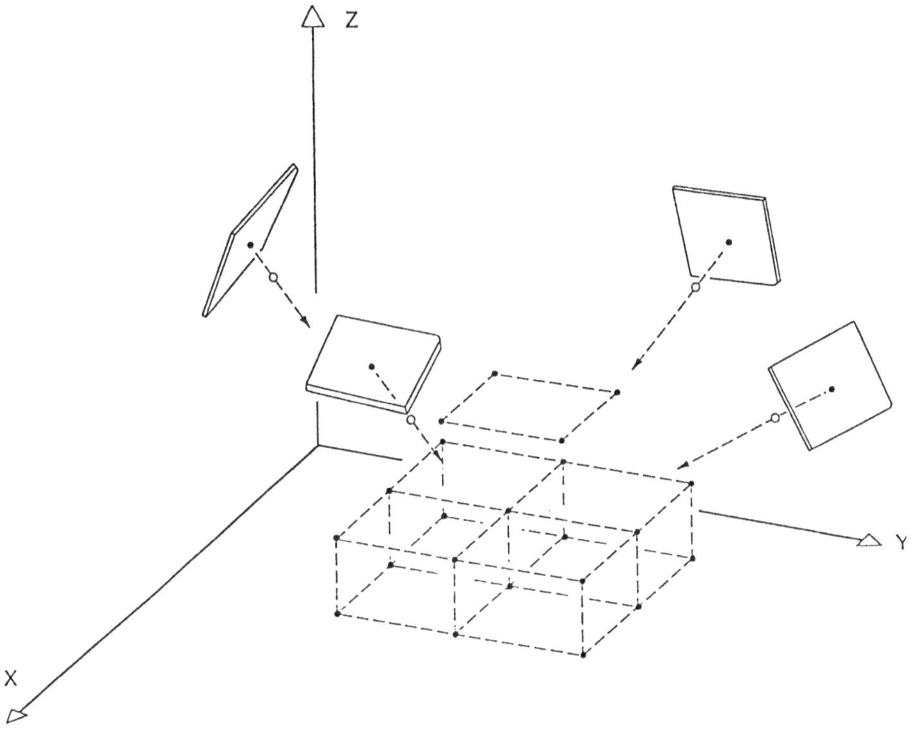

Figure 3. Recording of 3-D point cloud with multi-station configuration

If the point cloud is static it is sufficient to use just one CCD camera and to move it from one position to the others. In case of a moving point cloud four cameras would be used for simultaneous imaging.

3.2. THE BUNDLE METHOD

3.2.1. *The mathematical model*

The „bundle method“ is the underlying mathematical model (also called a „sensor model“) of photographic or CCD single frame cameras. It is based on the laws of perspective projection. In Figure 4 the object point P is imaged by a straight line through the perspective center O into the image point P'.

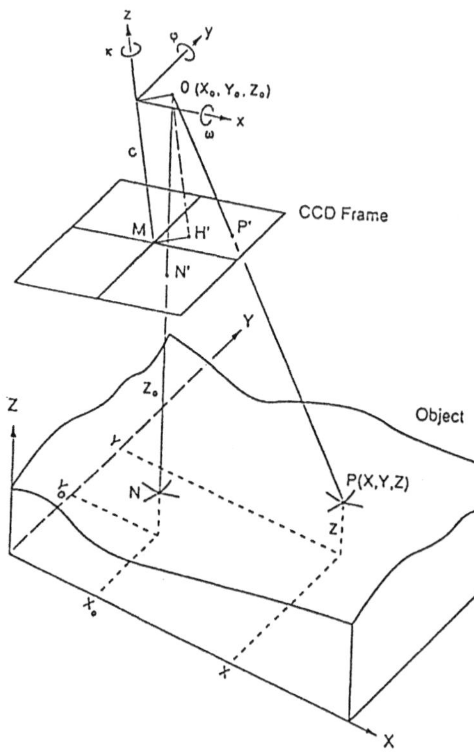

Figure 4. Sensor model of bundle method

This straight line condition is called „collinearity condition". It can be formulated as

$$\begin{bmatrix} X \\ Y \\ Z \end{bmatrix}_i = \lambda_{ij} R_j \begin{bmatrix} x_i - x_o \\ y_i - y_o \\ 0 - c \end{bmatrix}_j + \begin{bmatrix} X_o \\ Y_o \\ Z_o \end{bmatrix}_j \qquad 1)$$

with

X_i, Y_i, Z_i . . . Object space coordinates of object point (P_i)
X_{oj}, Y_{oj}, Z_{oj} . . Object space coordinates of perspective center (O_j)
x_{ij}, y_{ij}, 0 . . . . Measured image coordinates of point (P'_{ij})
x_{oj}, y_{oj} . . . . . . Image space coordinates of principal point (H'_j)
c_j . . . . Camera constant of CCD-frame (j)
R_j . . . . Rotation matrix (orthogonal) between image and object space coordinate systems
λ_{ij} . . . . Scale factor for imaging ray (ij)

$i = 1, \ldots, n$ (n = number of object points)
$j = 1, \ldots, m$ (m = number of CCD frames)

In (1) the interior orientation of a CCD-frame (j) is defined by the parameters x_{oj}, y_{oj}, c_j while the parameters X_{oj}, Y_{oj}, Z_{oj}, φ_j, ω_j, κ_j define the exterior orientation. Here φ_j, ω_j, κ_j are the three rotation angles which build up the rotation matrix R_j. The three component equations in (1) are reduced to two by cancelling out the scale factor λ_{ij}, and then rearranged according to

$$x_{ij} = -c_j f_{ij}^{x} + x_{oj} = -c_j \frac{r_{11j}(X_i - X_{oj}) + r_{21j}(Y_i - Y_{oj}) + r_{31j}(Z_i - Z_{oj})}{r_{13j}(X_i - X_{oj}) + r_{23j}(Y_i - Y_{oj}) + r_{33j}(Z_i - Z_{oj})} + x_{oj}$$

$$y_{ij} = -c_j f_{ij}^{y} + y_{oj} = -c_j \frac{r_{12j}(X_i - X_{oj}) + r_{22j}(Y_i - Y_{oj}) + r_{32j}(Z_i - Z_{oj})}{r_{13j}(X_i - X_{oj}) + r_{23j}(Y_i - Y_{oj}) + r_{33j}(Z_i - Z_{oj})} + y_{oj} \qquad (2)$$

$r_{11j} \ldots r_{33j}$ are the elements of R_j.

Since (2) provides for a general sensor model it accommodates easily some specific cases. Depending on the parameters which are considered either known a priori or treated as unknowns these equations may result in the following cases, which are also listed in Table 1 (the image coordinates x_{ij}, y_{ij} are always regarded as observed quantities):

(a) General bundle method: All parameters on the right hand side of (2) are unknown (interior orientation, exterior orientation, object point coordinates).
(b) Bundle method for „metric camera" systems: x_{oj}, y_{oj}, c_j (interior orientation) are given, all others unknown.
(c) Spatial resection:
  (ca) Interior orientation and object point coordinates (X_i, Y_i, Z_i) are given, the exterior orientation has to be determined.
  (cb) Only object point coordinates are given, the interior and exterior orientation have to be determined.
(d) Spatial intersection: The interior and exterior orientation are given, the object point coordinates have to be determined. This includes the stereo as well as the multiframe approaches.

Any combination of these procedures is possible within the general bundle concept. Also, incomplete parameter sets of exterior/interior orientation and object points can be handled.

Table 1. The various modes of sensor model application

Procedure		Given Parameters	Unknown Parameters
General bundle		-	$(X,Y,Z)_i$; IO_j; EO_j
Metric camera bundle		IO_j	$(X,Y,Z)_i$; EO_j
Spatial resection	(a)	IO_j; $(X,Y,Z)_i$	EO_j
	(b)	$(X,Y,Z)_i$	IO_j; EO_j
Spatial Intersection (Stereo or multiframe)		IO_j; EO_j	$(X,Y,Z)_i$

IO ... Interior orientation, EO ... Exterior orientation

A standard CCD camera is a „non metric" camera in the sense that its interior orientation is not given a priori. The camera constant c and the parameters of the principal point x_o, y_o have to be determined by the user. This procedure is called „camera calibration". In our

PTV system calibration is performed by using a calibration frame with given object point coordinates (compare section 5 and Maas, chapter 3 of this volume) and applying case (cb) of Table 1.

3.2.2. *Least squares estimation*

Equations (2) are considered observation equations functionally relating the observations x_{ij}, y_{ij} to the parameters of the right hand side according to

$$l = f(x) \quad . \tag{3}$$

For the estimation of x the Gauss-Markov model of least squares estimation is used. After linearization of (3) and the introduction of a true error vector e we obtain

$$l - e = Ax \quad . \tag{4}$$

The design matrix A is a n x u matrix (n ...... number of observations, u ...... number of unknown parameters), with $n \geq u$ and, in general, rank (A) = u.
With the assumed expectation E(e) = 0 and the dispersion operatur D we get

$$\begin{aligned} E(l) &= Ax \ , \\ D(l) &= C_{ll} = \sigma_0^2 P^{-1}, \\ D(e) &= C_{ee} = C_{ll} \ . \end{aligned} \tag{5}$$

P is the „weight coefficient" matrix of l and C stands for covariance operator, σ_0^2 is the variance factor.

The estimation of x and σ_0^2 is usually (but not exclusively) attempted as unbiased, minimum variance estimation, performed by means of least squares, and results in

$$\begin{aligned} \hat{x} &= (A^T P A)^{-1} A^T P l \ , \\ v &= A\hat{x} - l \ , \\ \hat{\sigma}_0^2 &= \frac{v^T P v}{r} \ , \quad r = n - u \ . \end{aligned} \tag{6}$$

v ... Residuals of least squares.

The structure of A is determined by the procedure applied (compare Table 1), it reflects the overall network design and thus also the geometrical and numerical stability of the arrangement.

For $A^T PA$ to be uniquely invertible, as required in (6), the network needs an external „datum", that is, the seven parameters of a spatial similarity transformation of the network need to be fixed. This is usually achieved either by introducing control points with at least seven fixed coordinate values, or by fixing seven appropriate elements of the exterior orientations of two frames.

The precision of the parameter vector x is controlled by its covariance matrix
$C_{xx} = \sigma_0^2 \, (A^T PA)^{-1}$.

3.3. PRECISION AND RELIABILITY

Within the least squares estimation context, the terms „accuracy", „precision" and „reliability" are well defined. Accuracy, or „model quality" (this refers to our estimation model) consists of the parts precision and reliability. They are defined as follows:

Precision: Statistical quality of the estimated system parameters x (e.g. object point coordinates), if the underlying estimation model assumptions (functional, stochastic) are true.

Measures: Covariancematrix K_{xx}, standard deviations σ_x

Reliability: Quality of the estimation model with respect to the detection of model errors (systematic, stochastic, blunders).
Measures: Internal, external reliability.
Error detection techniques: Data snooping, robust estimation

There are well developed theories available for both precision and reliability. For details we refer to Gruen (1980, 1986), where photogrammetric applications of these concepts are specifically demonstrated.
The main parameters influencing a system's precision are

- network configuration (geometrical arrangement of cameras/images)
- number and location of control points
- type and quality of camera components (lens, sensor, electronics, etc.)
- quality of camera calibration
- quality of data transfer between camera and computer
- quality of definition of object points
- measurement device/algorithm for image coordinates

This results in a fairly complex parameter space and all components have to be tuned in an optimum way if high precision results are to be expected. There is plenty of videogrammetric literature available on this topic, compare e.g. Beyer (1992).

Figure 5 demonstrates the impact of imaging geometry on precision. Shown are error ellipsoids on measured points of an antenna dish and their variation with changing imaging geometry. It can clearly be seen how a strong convergence of the camera axes can improve precision as opposed to a parallel camera axes set-up.

The main parameters influencing a system's internal reliability are:

- number of imaging rays per object point
- number and distribution of image points per CCD frame
- number and distribution of control points
- measurement precision of image points

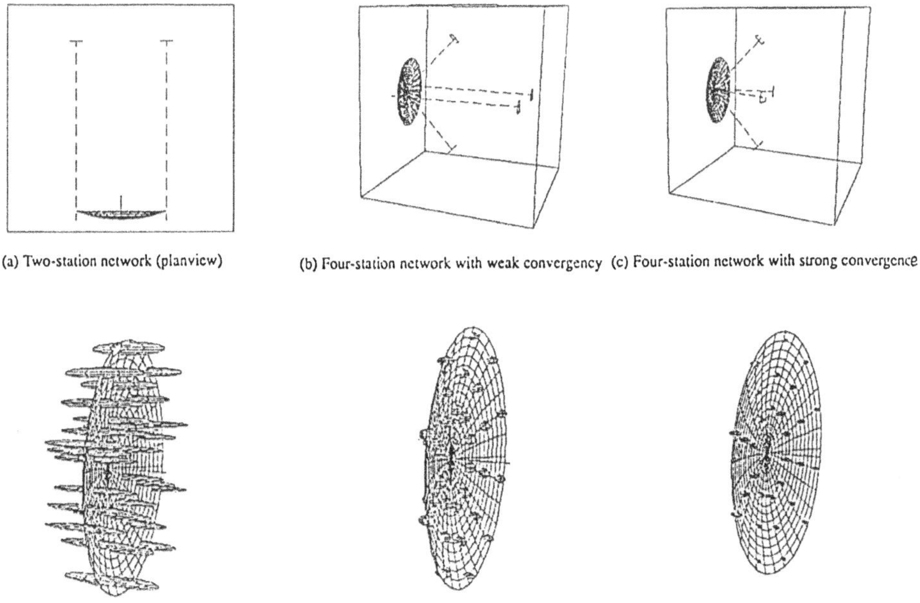

Figure 5. Impact of imaging geometry on precision

The first two parameters are well taken care of in our PTV system by (a) looking with four CCD cameras at all particles, and (b) by imaging a very large number of particles with more or less even distribution. In situ control points are not required, because we determine the orientation and calibration of our CCD cameras in a separate step through reference frame calibration (compare Maas, chapter 3 of this volume).

4. System aspects

Generally, a videogrammetric system consists of the data reduction processes and hardware components as depicted in Figure 6.

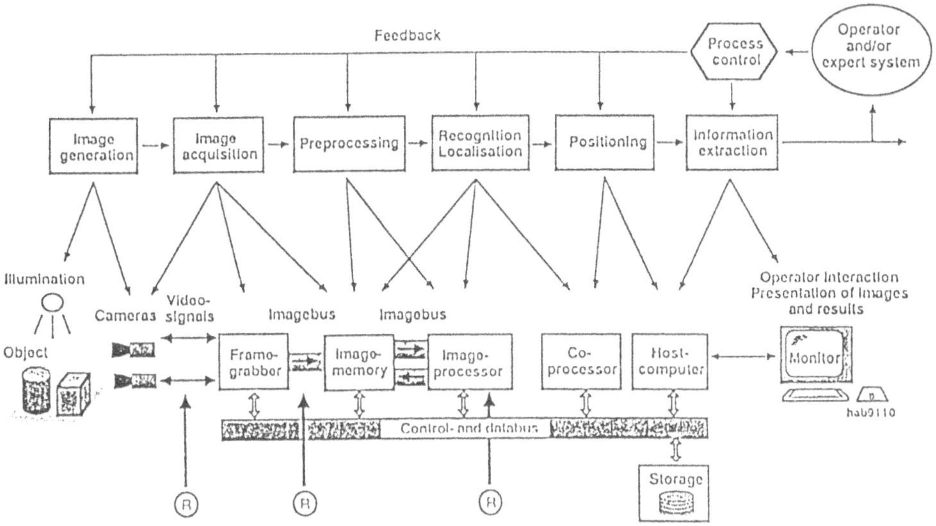

Figure 6. Functional modules and associated hardware components of a videogrammetric system ®. . . Added hardware for very fast (robotic) processing

Image generation involves the light source and the object reflecting the light (laser sheet and particles in PTV, laser sheet and fluorescent material in LIF). The image is formed as an optical image in the focal plane of the camera. A videogrammetric system should be capable of accommodating any number of cameras and not be restricted to two or even just one. The number of cameras to be employed jointly or even simultaneously is dictated by the application requirements.

With increasing number of cameras the **image acquisition** process becomes more and more demanding. Image acquisition transfers the optical image into a two-dimensional discrete mathematical function, stored as a matrix of grey values. The hardware involved includes sensors (semi-conductor type), amplifiers, filters, frame grabbers with analog to digital converters, and timing devices. In many systems the use of television standards plays a role because the images are often transferred from the camera to the frame grabber via analog video signal. The European video norm CCIR specifies an image frequency of 25 Hz. Since each B/W image consists of 256 Kbyte pixel data a two camera system would require already a bandwith of 6.25 Mbyte for data transfer. Equally demanding are the storage requirements. Therefore, in our PTV system, which operates with four cameras simultaneously, we store the analog video images intermediately on four videorecorders. From there they are digitized and transferred to a Sun Workstation for further processing (compare Virant, Dracos, chapter 3 of this volume). Video cameras and the related transfer norms were not built for measurement but rather for entertainment purposes. Therefore a careful analysis and calibration of image capturing and transfer devices with respect to geometric and radiometric characteristics and distortions is mandatory for obtaining high accuracy results.

The aim of **preprocessing** is to improve the radiometric and geometric quality of the image, and to perform data reduction. Image quality improvement includes noise removal, contrast enhancement, blurr removal and grey level calibration. In our PTV system a typical operation is highpass filtering to adjust for the Gaussian intensity distribution of the laser light profile (Virant, Dracos, chapter 3 of this volume). In our LIF system preprocessing is used for destriping the images (Deusch, chapter 4 of this volume). We did some experiments with JPEG compression on PTV image data, but with only little success (the high image frequencies caused by small particle images were severely destroyed).

The **recognition and localization** phase involves the detection of features that are relevant to the reconstruction of the object. Features of interest can be point-shaped targets, edges, corners, texture elements, etc. In case of PTV imaging the particles have to be detected in image space, for each image separately, and their pixel coordinates to be determined with high accuracy. This is achieved in our system with a center of gravity algorithm. A special technique has been developed for the separation of overlapping particles (Maas, chapter 3 of this volume).

In applications leading to more complex images, techniques for image understanding have to be activated for feature detection. This usually proves to be the most difficult part towards full system automation.

The **positioning** phase comprises the derivation of 3-D metric information in object space from features recognized and located in image space. Here the camera model has to be utilized. This is usually done in form of one of the variants of the bundle method (compare section 3.2.1).

The derivation of 3-D object space coordinates requires the establishment of correspondences between homologous image features in different images. Our PTV system applies a multi-image search technique based on the utilization of epipolar constraints, derived from the camera model (Maas, chapter 3 of this volume).

As a result of this phase we obtain a set of 3-D coordinates for the particles. This point cloud is generated once for every video cycle (25 times per second). The covariance matrix for these coordinates is available at any time, such that statistically sound precision measures like standard deviations for the particles' positions can be derived. In a subsequent step a 3-D tracking algorithm is activated, which connects identical particles and results in 3-D trajectories. This is an example for the **information extraction** task. This phase uses parameters and precision estimates from the previous positioning module in order to derive 3-D object models and related information, whatever is required by the task at hand.

In our LIF system the 3-D voxel space is subdivided into a regular grid of image cuboids. Each cuboid is tracked by an algorithm called „least squares matching“ (compare Maas, chapter 4 of this volume) through the image sequences. Thus recognition/localization and positioning tasks are integrated here into a simultaneous approach.

While some of the hardware components of Figure 6 will be closer addressed in chapter 4.1., Figure 7 shows a brief summary of the process components (hardware, software, algorithms) of a videogrammetric 3-D PTV system.

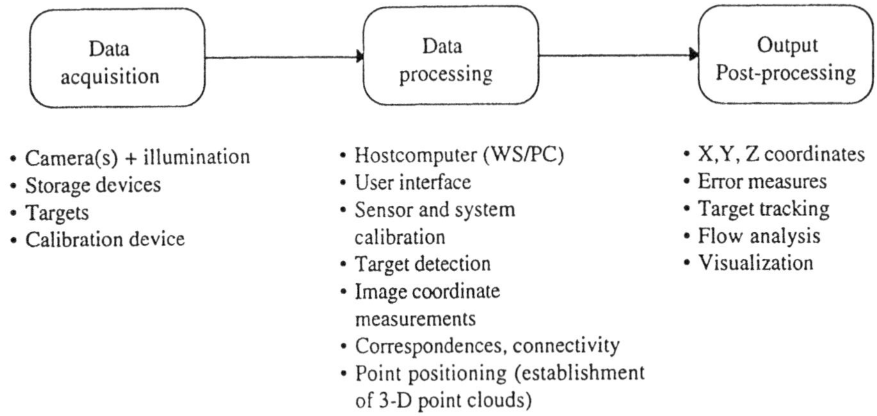

Figure 7. Process components of a videogrammetric 3-D PTV system

We distinguish data acquisition, data processing, and output/post-processing phases. More details about the individual components can also be found in Maas, chapter 3 and Virant, Dracos, chapter 4 of this volume.

4.1. CCD CAMERAS AND IMAGE DATA TRANSFER

Image acquisition is a key element in any videogrammetric system. It consists of electrooptical scanning, image data transfer, A/D convertion and image storage (compare Figure 8).

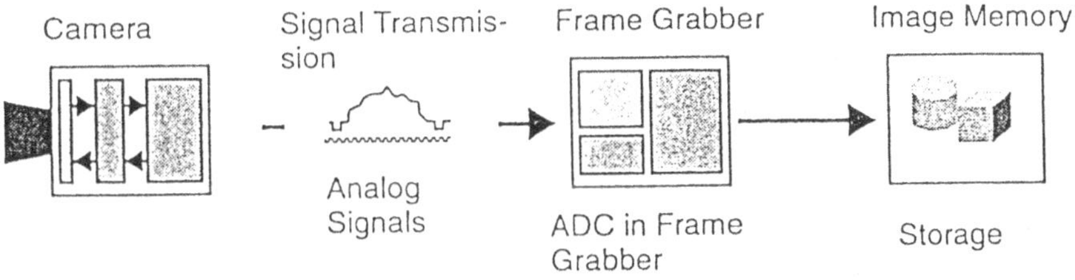

Figure 8. Image acquisition with solid state cameras

In the following we will address some of these components more in detail.

Since the early 1960s self-scanning solid state chips based on silicon have been used. Nowadays solid state imagers are almost synonymous with CCD sensors. A brief survey of the development of modern solid state sensors is given in Real (1986). More detailled technical descriptions can be found in Beynon, Lamb (1980) and Weimer, Cope (1983). Solid state devices can be used for imaging and storage. As photodetectors most solid state imaging sensors are based on silicon design, having a MOS (Metal-Oxide-Semiconductor) structure. Solid state photodetectors come as single element opto-mechanical scanner, linear array and area (matrix) array sensors.

Another classification scheme is based on the type of readout for the charge, and distinguishes

- CCD (Charge Coupled Device)
- CID (Charge Injection Device)
- Photodiodes

Photodiodes and CID arrays have the advantage of being XY-addressable, meaning that each individual pixel or pixelgroup can be read out individually (assuming the camera allows for that). CCDs belong to the class of charge transfer devices (CTD). A mixture is the Charge Priming Transfer (CPT), which has charge transfer for an XY-photodiode array. For further details on these matters see Weimer, Cope (1983).

CCDs come as linear array and area array imagers. Linear arrays are preferred in remote sensing to allow for a large image format, large S/N ratio, and a limited data transfer rate. They are also favoured in the scanning of graphical products. Area arrays are standard in machine vision and of much interest for videogrammetry because they allow to apply the conventional well established mathematical model of collinearity. The principle of charge generation and transfer is explained in Beynon, Lamb (1980). Dähler (1986) gives an introductory and critical account of this imaging technology. CCD area arrays are either interline or frame transfer devices. Figure 9 shows the functional principle of both designs.

The interline transfer device is organized into pairs of columns, alternating a photosensitive (imaging) column with an opaque vertical shift register. During the integration period charge is accumulated in the imaging column and subsequently transferred to the opaque column. Then the charge is shifted vertically through this column into a horizontal shift register. From there the signal is read out line by line during the next integration period.

In the frame transfer device the sensor area is divided into two zones of equal size. One zone is photosensitive, the other serves as the (opaque) storage area. Both zones consist of vertical columns of shift registers. Through integration this charge is transferred in parallel to the storage zone. The readout of the stored charges through a horizontal shift register is performed during the subsequent integration period.

Frame transfer makes the most efficient use of the imaging area. The contiguous pixel arrangement allows for superior resolution, i.e. smaller pixel spacing, and thus less MTF degradation and less Moiré, improved sensitivity through larger effective pixel (photosensitive area) size and a smaller chip area.

During transfer time the „exposure“ (integration) continues. This produces „smear“. With interline transfer the transfer times are smaller, thus the smear is reduced. In frame transfer cameras the relation integration to transfer time is about 19 :1.

CCD matrix array cameras come in a great variety of different types. We distinguish „industrial cameras“, still video cameras, micro- und macroscanning cameras, camcorders, etc. They vary in terms of number of pixels, chip size, pixel pitch (resolution), output norm, type of synchronization, internal or external storage devices, camera control elements, etc. Some of them have special features as miniaturization (e.g. endoscope cameras), cooling (for very long exposure times), intensification (low light/ultrashort shutter times applications), high speed (>100 frames per second, compare our LIF system), and infrared imagers. However, most of them do have in common some essential functional elements, as they are depicted in Figure 10.

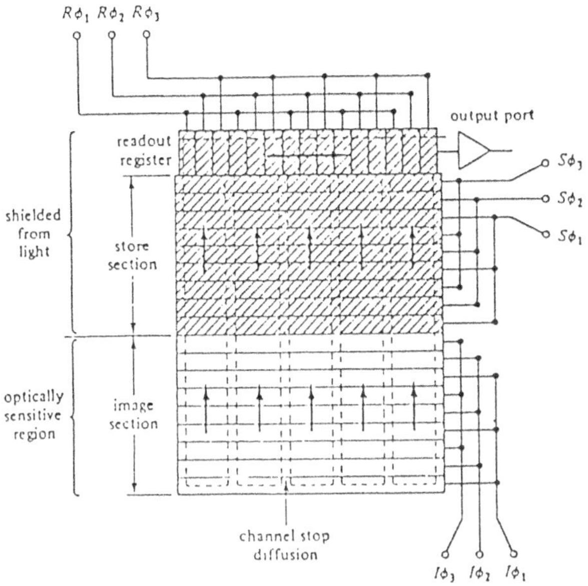

(a) Frame transfer

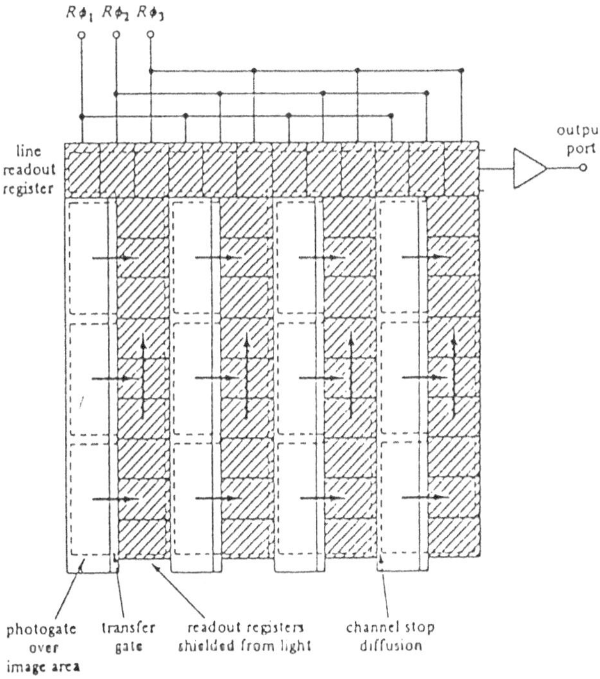

(b) Interline transfer

Figure 9. Interline and frame transfer in CCD technology

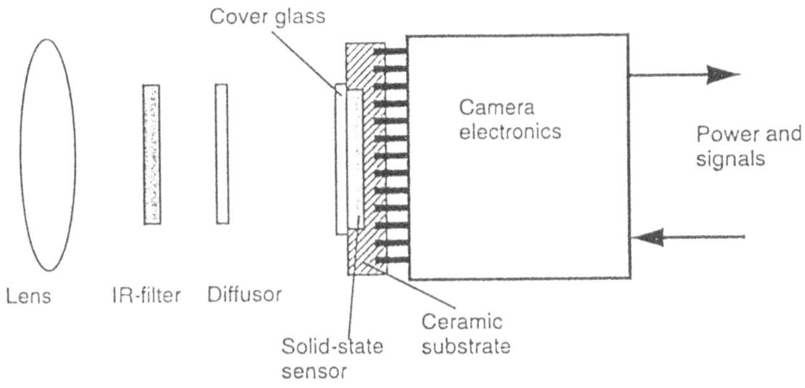

Figure 10. Functional elements of a solid state camera

The lens is usually a CCTV-lens (Close Circuit Television) with a C-, CS- or even a F-mount (like the Megaplus). Typical sensor sizes are of 1/2 and 2/3 inch diameter, or 4.8 x 6.4 and 6.6 x 8.8 mm^2 format. The IR-filter is used to cut off wavelenghts above 700 nm to avoid optical scattering in the solid state layer (large wavelenghts tend to penetrate deeply into the substrate). The diffusor, usually a birefringent quartz plate, is found in most cameras with interline transfer sensors. It serves as an optical low-pass filter to suppress aliasing.

The camera electronics serves different purposes, details can be found in Beyer (1992). At least it has to provide for images of maximum quality at the camera output.

The readout of a CCD sensor is defined by the video norm, which is used for image data transfer between camera and frame grabber. We distinguish the four most important norms RS-170, CCIR (both B/W) and NTSC, PAL (both color). Table 2 gives the basic data for RS-170 (US norm) and CCIR (European norm).

TABLE 2. Data for the RS-170 and CCIR video standards

	RS 170	CCIR
Frame Rate/Field Rate	30/60 Hz	25/50 Hz
Interlacing	2:1	2:1
Number of lines	525	625
Number of active lines	480	576
Field time	16 2/3 ms	20 ms
Time per line	43.49 µs	64 µs
Line frequency	15750 Hz	15625 Hz
Active line period	-52.5 µs	52 µs
Nominal video bandwidth	4.5 Mhz	5.5. Mhz
Resolution (Nyquist)	472	572
Aspect Ratio	4:3	4:3

To provide for an image display free of flicker both norms use the principle of interlacing. This means that the full video **frame** is acquired and displayed in form of two half images (**fields**) which are produced by the odd and even line numbers.
Figure 11 shows how the odd and even fields are combined into a frame.

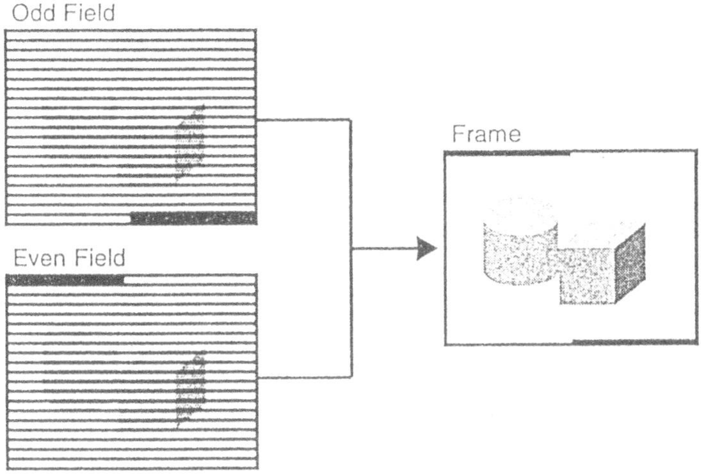

Figure 11. Interlaced video with fields generating a frame

Together with image data the video signal also includes synchronization information - vertical synchronization (VSYNC) for the start of the even or odd field, and horizontal synchronization (HSYNC), containing a pulse for the start of each video line.

„Frame grabbing“, as a part of the image acquisition process, refers essentially to the A/D conversion and storage of a complete image frame. For the storage of a sequence of images at video rates the memory must be fast and large (one 512x512x8bit image occupies 256 Kbytes). This demonstrates the need for very fast image preprocessing and data reduction. Standard single circuit frame grabber boards usually come with an A/D converter, one or more frame memories, arithmetic processors, control and clocking circuits and three D/A-converters for red blue, green (RGB) outputs to a color monitor from look-up tables. On-board memory can run as high as 32 Mbytes, with ever increasing local processing capabilities. Often the frame grabber is an integral part of a set of boards, providing for much more than just simple „grabbing“ functions. In a recent development, frame grabbers channel the image data directly into the host computer memory at video rate (available for PCs only).

In case of video signal-based data transfer the frame grabber has to derive the sampling clock from the horizontal synchronization via phase-locked-loop (PLL). This kind of synchronization leads to a number of unwanted effects, as for instance

- shear of image due to disturbance of PLL in vertical retrace
- line-jitter (individual shift of each video line), typically more than 0.1 pixel (compare Figure 12)
- individual scaling of each video line

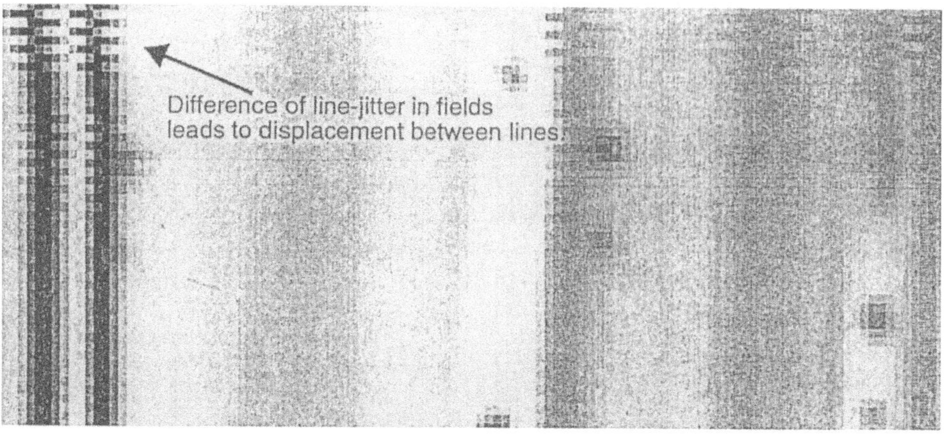

Figure 12. Line-jitter, shown for an image grabbed from a video recorder (from Beyer (1992))

A signal not exhibiting these problems is the **pixel-clock**. It is a signal with a frequency identical to the sensor element clock. This mode provides for pixel-synchronous frame grabbing, that is, every sensor pixel information is mapped one-to-one into the image memory.

An even better image data transfer mode is provided by digital transmission. This results in better noise performance, but requires the A/D-conversion already within the camera.

For videogrammetric applications it is crucial to fully understand the nature of a pixel. From its generation at the sensor to its use in the computer each pixel undergoes a number of operations, geometrical and radiometric in nature, which may lead to distortions both in geometry and radiometry. Figure 13 shows the operations and resulting distortions which act on a pixel on its way from sensor to computer.

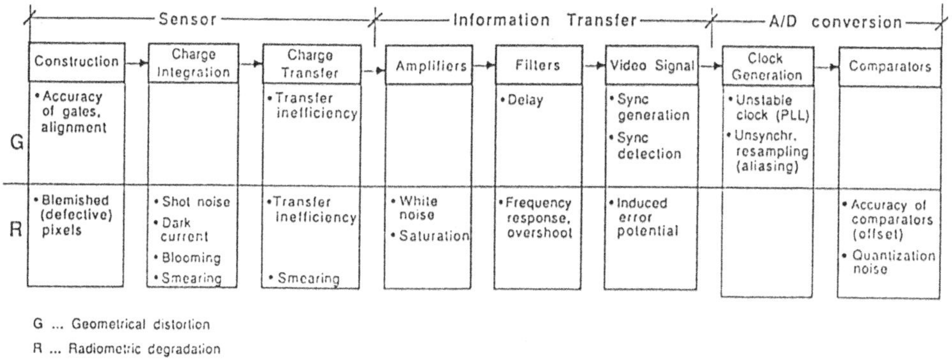

Figure 13. A pixel on its way from sensor to computer

When analyzing error sources and their effect on a pixel it is important to keep the duality between pixel radiometry and information location in mind. A change in pixel grey value leads to a geometrical distortion of an image feature (although the physical location of the pixel itself is not altered) and vice versa. Therefore the G- and R-distortions may have similar effects on the final results.

Again, most of these distortions can be avoided by evading the video standard. But, since even the original (sensor) pixel definition (pixel size and pixel pitch) is sometimes hard to get at, a good calibration approach is urgently needed for geometrical data processing.

5. Calibration and self-calibration

Traditionally, in photogrammetry, the term „calibration" refers to the determination of the parameters of interior orientation x_0, y_0, c of a particular camera or photograph (compare section 3.2). In the modern notation of videogrammetry „calibration" is the procedure that allows beyond that the determination of all deviations of the bundle of imaging rays from a truly perspective model. These deviations are regarded as systematic errors. In any CCD-camera/frame grabber system we may expect the following major systematic errors:

- Lens distortion (radial and tangential)
- Unflatness of imaging plane
- Warm-up effects
- Unequal and/or unknown pixel spacing in x and y
- PLL-synchronization effects (image shear, line-jitter, etc.)
- Other electronically caused errors

Over the years many calibration techniques have been developed. For videogrammetric close-range applications the most relevant are

(a) Laboratory calibration
A collection of techniques using special equipment and various different methods to determine the interior orientation and the systematic errors of a CCD camera. This specialized equipment (goniometers, etc.) is expensive and the procedures are mostly very timeconsuming.

(b) Reference frame calibration
Also known as „testrange calibration". A reference object, consisting of a certain number of precisely measured points, is imaged. From the mathematical relations between reference frame coordinates X, Y, Z and measured pixel coordinates x, y the required parameters can be derived. The testrange may be represented by (preferably) 3-D, 2-D or even 1-D object.
The so-called „plumbline technique", which is used in some places for the calibration of radial and tangential distortion, belongs to the last category (Fraser (1992)).

(c) Pointcloud calibration
A 3-D cloud of well signalized points is used for calibration. The advantage of this technique is that the points need not to be measured beforehand. This makes it fast and economic. This is counterbalanced by the disadvantage that in order to determine all necessary parameters at least four images from different locations have to be taken (Gruen, Beyer (1992), compare also Figure 3)).

(d) Self-calibration
Originally a technique which was meant to determine the calibration parameters from project data, simultaneously with the other project parameters (object points and exterior orientation parameters).

Self-calibration can be considered the most universal and flexible approach, while the other methods can be regarded as special cases therefrom. For example, although we apply in our PTV system the method of reference frame calibration, it is based on self-calibration software (compare Maas, chapter 3, of this volume).

Therefore, we will briefly describe here the basics of self-calibration. Taking the bundle model as the idealized sensor model we can express systematic errors in image space by expanding equations (2) by additional functions with „additional parameters“. This leads to

$$x_{ij} = -c_j f_{ij}^{x} + x_{oj} + \Delta x_{ij} \quad , \qquad y_{ij} = -c_j f_{ij}^{y} + y_{oj} + \Delta y_{ij} \tag{7}$$

The terms Δx_{ij} , Δy_{ij} can be understood as corrections to the image coordinates x_{ij} . y_{ij} in order to reduce the physical reality of the sensor geometry to the perspective model. Obviously the principal point coordinates x_{oj} , y_{oj} can be considered part of this correction term. If a parameter for the camera constant is added, the full interior orientation is included in the additional parameter set.

The formulation of the additional parameter function is very crucial for a successful self-calibration. The actual physically caused deformations, which are a priori unknown in structure and magnitude, should be modelled as closely and as completely as possible. On the other hand, the chosen parameter set must be safely determinable in a given network arrangement.

In CCD camera-based systems the following functions have proved to be effective:

$$\begin{aligned} \Delta x &= -\Delta x_o + \frac{\bar{x}}{c}\Delta c + \bar{x}s_x + \bar{y}a + \bar{x}r^2k_1 + \bar{x}r^4k_2 + \bar{x}r^6k_3 + \left(r^2 + 2\bar{x}^2\right)p_1 + 2\bar{x}\bar{y}p_2 , \\ \Delta y &= -\Delta y_o + \frac{\bar{y}}{c}\Delta c + \;0\; + \bar{x}a + \bar{y}r^2k_1 + \bar{y}r^4k_2 + \bar{y}r^6k_3 + 2\bar{x}\bar{y}p_1 + \left(r^2 + 2\bar{y}^2\right)p_2 , \end{aligned} \tag{8}$$

with $\bar{x} = x - x_o$, $\bar{y} = y - y_o$, $r^2 = \bar{x}^2 + \bar{y}^2$

The indices $_{ij}$ are left out here for the sake of simplicity.

This is called a „physical model“, because all its components can directly be attributed to physical error sources. The individual parameters represent:

Δx_o , Δy_o , Δc . . . Change in interior orientation elements
s_x . . . . Scale factor in x („affinity“)
a . . . . Shear factor (jointly in x, y)
k_1 , k_2 , k_3 . . . . . First three parameters of radial symmetric lens distortion (k_3 is a priori disregarded if normal- and wide-angle lenses are used)
p_1 , p_2 . . . . First two parameters of decentering distortion

The location of the principal point is not specified for most CCD-cameras, varies from camera to camera and depends on the configuration of the frame grabber. The scale factor in x is required to model the imprecise specification of the sensor element spacing and additional imprecisions introduced with PLL line-synchronization. In the latter case the pixel spacing in x must be computed from the sensor element spacing, the sensor clock frequency and the sampling frequency with:

$$psx = ssx \frac{f\text{sensor}}{f\text{sampling}} \tag{9}$$

with psx .... Pixel spacing in x
ssx .... Sensor element spacing in x
$f$sensor ..... Sensor clock frequency
$f$sampling ... Sampling frequency of frame grabber

The shear factor (a) must be included to compensate for the geometric deformation which can be induced by PLL line-synchronization. The use of additional parameters leads to an extended bundle model

$$l - e = Ax + A_3 z \ . \tag{10}$$

z, A_3 .... Vector of additional parameters and associated design matrix

In a general bundle concept all unknown parameters are treated as stochastic variables. This permits to consider a priori information about these parameters to be included and includes both extreme cases where parameters are either excluded from the model or may be treated as free unknowns. If we split the vector x into its components x_p (for object point coordinates) and t (for exterior orientation parameters) we obtain the following estimation model:

$$\begin{aligned} -e_B &= A_1 x_p + A_2 t + A_3 z & -l_B \quad ; \quad & P_B \\ -e_p &= I x_p & -l_p \quad ; \quad & P_p \\ -e_t &= \qquad\qquad I t & -l_t \quad ; \quad & P_t \\ -e_z &= \qquad\qquad\qquad\quad I z & -l_z \quad ; \quad & P_z \end{aligned}$$

e_B , e_p , e_t , e_z .... Vectors of true errors of image coordinates, object point coordina tes, exterior orientation elements, additional parameters
l_B , l_p , l_t , l_z ..... Vectors of observations of image coordinates (minus constant term from Taylor expansion), object point coordinates, exterior orien tation elements, additional parameters
P_B , P_p , P_t , P_z ... Associated weight coefficient matrices
x_p , t , z .... Parameter vectors of object point coordinates, exterior orientation elements, additional parameters
A_1 , A_2 , A_3 ..... Associated design matrices
I .... Identity matrix

If a reference frame with wellknown object point coordinates is used, these control point coordinate observations receive infinite weights (diag (P_p) $\rightarrow \infty$), which freezes the estimated parameter vector $\hat{x}_p$ to the introduced value l_p , such that the system then effectively only contains the parameters for exterior orientation and the calibration parameters. For more details on self-calibration see Gruen, Beyer (1992).

The need for a good calibration of CCD camera-based systems is demonstrated by Figure 14, which shows the deformation of an image grid of a CCD frame by systematic errors. The amount of deformation attains almost 100 µm in the corners of the frame, which corresponds roughly to ten times the pixel size. This is a dramatic distortion considering the fact that good image targets can be measured with an accuracy of 1/50 of a pixel.

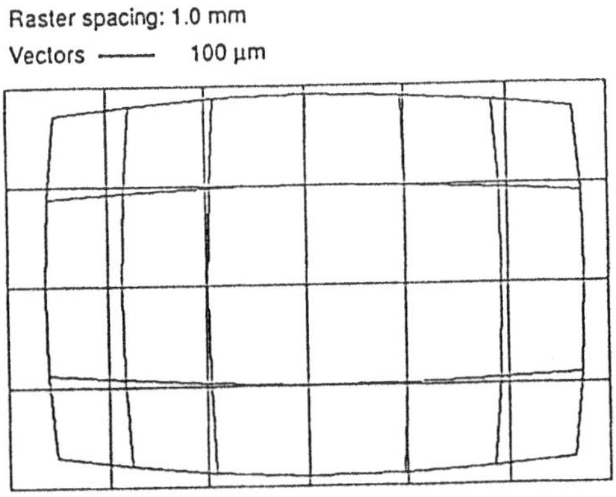

Figure 14. Deformation of a CCD image grid by systematic errors; recoverd by self-calibration

6. High precision image measurement

As mentioned in section 3.3., one of the parameters responsible for a system's overall precision and accuracy is the quality of the measurement algorithm for image targets.

In our PTV system distinctive blob shaped targets (particles) are imaged (Figure 15, center). They are, however, inferior to those circular retroreflective targets which are often used in industrial quality control applications (Figure 15, left). Figure 15, right shows, in constrast, a 2-D image patch with complex image structure from our LIF system.

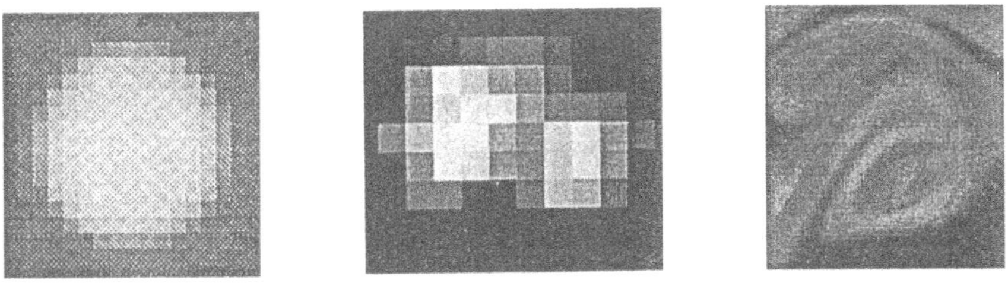

Figure 15. Various target images
Left: Retroreflective circular object
Center: Particles in our PTV system
Right: 2-D image patch from our LIF system

In videogrammetry, the method of Least Squares (LS) Image and Template Matching is a standard technique for high accuracy image measurements. It has been shown that with image targets of the type of Figure 15, left a system accuracy of 0.02 pixels can be achieved by using LS-Template Matching. The PTV targets are of lesser quality and of very small size (3x3 pixels on the average), so that we can only expect an accuracy of about 0.1 pixels. For that it is sufficient to use a less refined, but computationally faster approach, e.g. the „center of gravity" algorithm.

6.1. „CENTER OF GRAVITY" ALGORITHM

Here the center coordinates x_c , y_c of a blob are determined by

$$x_c = \frac{\sum g_i x_i}{\sum g_i} \quad , \qquad x_c = \frac{\sum g_i y_i}{\sum g_i} \quad , \tag{12}$$

with x_i , y_i the pixel coordinates of the pixels of the image patch containing the blob, and g_i the associated grey values. This can be interpreted as a general arithmetic mean with the grey values as the weighting parameters.

6.2. 1-D LEAST SQUARES MATCHING

Since LS-Matching, in a 3-D variant, is a key element for cuboid tracking in our LIF system, we will give here in the following a brief introduction to LS-Matching. Assume we have a one-dimensional image profile (a videoline for instance) and we shift this profile by a small amount. We can determine the shift in the following manner.

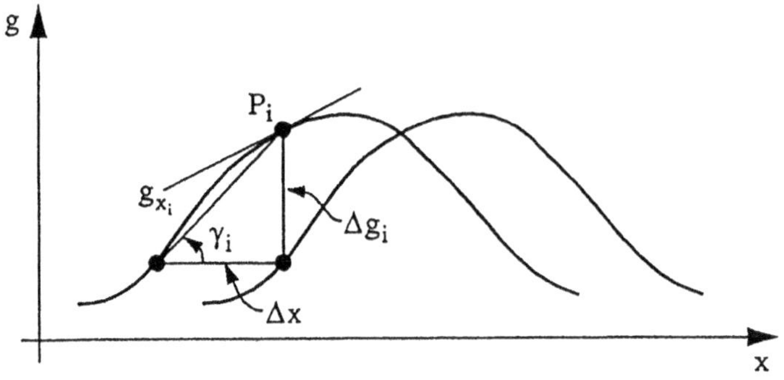

Figure 16. Shift Δx of a grey value profile
Δg_i ... grey value difference between original and shifted profile
g_{xi} ... first derivative of grey value function in P_i

According to Figure 16 we obtain the relation

$$\tan\gamma_i\, \Delta x = \Delta g_i \quad . \tag{13}$$

Since γ_i is not available we replace $\tan\gamma_i$ by g_{xi} , the first derivative of the grey value function in P_i and obtain so

$$g_{xi}\Delta x = \Delta g_i \quad . \tag{14}$$

This one equation could be used to determine the shift Δx (in an iterative manner, because (14) is an approximation of the correct relation (13)). In fact we use all pixels P_i (i=1, ..., n) to set up n equations of type (14). The parameter Δx is then obtained via least squares estimation. In this very simple case of just one parameter the solution is given directly by

$$\Delta\hat{x} = \frac{\sum g_{x_i} \Delta g_i}{\sum g_{xi}^2} \quad , \tag{15}$$

with the standard deviation

$$\sigma_{\Delta x} = \hat{\sigma}_0 \left(\frac{1}{\sum g_{xi}^2} \right)^{1/2} \quad . \tag{16}$$

σ_0 is estimated via

$$\hat{\sigma}_0^2 = \frac{\sum v_i^2}{n-1} \quad , \tag{17}$$

with the least squares residuals

$$v_i = g_{x_i} \Delta\hat{x} - \Delta g_i \quad . \tag{18}$$

These simple relations show us some of the properties of LS-Matching:

(a) With this technique we can
track a feature in image space
derive the position of a feature in the image by computing grey value differences between an image or template patch and this feature patch.

(b) The precision of the position depends on
the amount of image noise (through σ_0)
the size of the first derivatives (steep gradient images lead to precise location determination and vice versa, compare (16))

(c) We note that with this technique we use as observations the differences of grey values between two images. The first derivatives g_{xi} of one of these images are only used as the coefficients of the shift parameter in (14). Small variations of these coefficients have only little influence on the solution, so that the image noise does not constitute a serious problem. Image noise is expressed by $\hat{\sigma}_0$ and the residuals v_i .

6.3. 2-D LEAST SQUARES MATCHING

This basic concept of Least Squares Matching can be extended to 2-D image patches. In videogrammetry this algorithm is usually applied to establish correspondences between conjugate image features for the purpose of object reconstruction.

In LIF applications we can use it to track flow patterns through a sequence of images. Figure 17 shows two images out of a LIF sequence. A small patch in the left

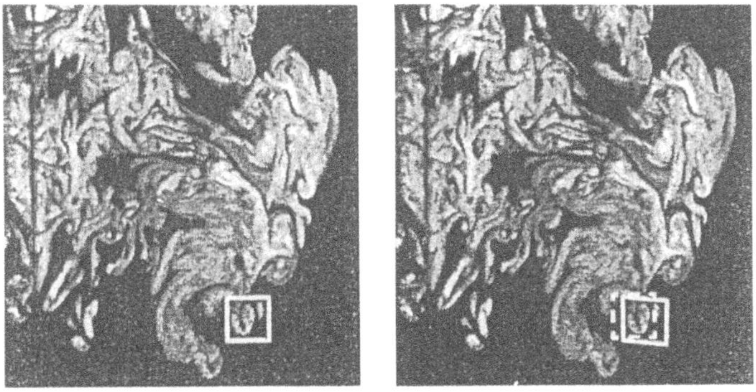

Figure 17. Tracking of a 2-D image patch in a LIF sequence by LS-Matching

image is selected and the corresponding patch in the right image is located by LS-Matching. In order to allow for a optimal fit of both patches a six parameter affine patch deformation is introduced.

Besides the two shift parameters Δx , Δy also four reshaping parameters (two scales and two shears) are applied.

In the following we will briefly introduce the principle of 2-D LS-Matching. A straightforward extension to the tracking of 3-D image cuboids, as it is used in our LIF system, can be found in Maas, chapter 4 of this volume.

Assume two image regions are given as discrete two-dimensional functions $f(x,y)$, $g(x,y)$, which can be defined as conjugate regions of a stereopair in the „left" and the „right" image respectively. $f(x,y)$ is interpreted in the following as the „template", $g(x,y)$ as the „picture". Matching is (ideally) established if

$$f(x,y) = g(x,y) \quad . \tag{19}$$

Because of random effects (noise) in both images (or in one set only if the template is assumed to be free of noise), equation (19) is not consistent. Therefore a noise vector e(x,y) is added, resulting in

$$f(x,y) - e(x,y) = g(x,y) \quad . \tag{20}$$

The location of the function values g(x,y) must be determined in order to provide for the match-point. This is achieved by minimizing a goal function which measures the distances between the grey values in template and picture. The goal function to be minimized in this approach is the L_2-norm of the residuals of least squares estimation.

In the least squares context equation (20) can be considered as a nonlinear observation equation, which models the vector of observations $f(x,y)$ with a function g(x,y), whose location in the „right" image needs to be estimated. The location is described by shift parameters Δx, Δy, which are counted with respect to an initial position of g(x,y), the approximation of the conjugate picture region $g^o(x,y)$.

In order to account for a variety of systematic image deformations and to obtain a better match, image shaping parameters and radiometric corrections are introduced besides the shift parameters.

If the grey values are given over a grid, the image shaping is achieved by resampling of $g^o(x,y)$ over the transformed grid points. The geometrical transformation is modelled by a bivariate polynomial

$$\begin{aligned} x &= t_y^T \bar{A} t_x \quad , \\ y &= t_y^T \bar{B} t_x \quad , \end{aligned} \tag{21}$$

with

$$\begin{aligned} t_x^T &= \left\{1 \; x_o \; x_o^2 \ldots x_o^{m-1}\right\} \quad , \\ t_y^T &= \left\{1 \; y_o \; y_o^2 \ldots y_o^{m-1}\right\} \quad , \end{aligned} \tag{22}$$

and the parameter matrices $\bar{A}$, $\bar{B}$

$$\bar{A} = \begin{bmatrix} a_{11} & a_{12} & \ldots & a_{1m} \\ \cdot & & & \cdot \\ \cdot & & & \cdot \\ \cdot & & & \cdot \\ a_{m1} & a_{m2} & \ldots & a_{mm} \end{bmatrix} \quad , \quad \bar{B} = \begin{bmatrix} b_{11} & b_{12} & \ldots & b_{1m} \\ \cdot & & & \cdot \\ \cdot & & & \cdot \\ \cdot & & & \cdot \\ b_{m1} & b_{m2} & \ldots & b_{mm} \end{bmatrix} \tag{23}$$

x_o , y_o are the grid locations of the data points of $g^o(x,y)$.

The transformation parameters a_{11} , . . ., a_{mm}, b_{11} , . . . , b_{mm} need to be estimated from (20).

In order to be able to operate with the conventional least squares approach the function g(x,y) in (20) must be linearized.

This yields

$$f(x,y) - e(x,y) = g^o(x,y) + \frac{\partial g^o(x,y)}{\partial x} dx + \frac{\partial g^o(x,y)}{\partial x} dy \quad . \tag{24}$$

$$dx = \frac{\partial x}{\partial p_i} dp_i \quad , \qquad p_i = \text{ith transformation parameter in (21)}$$
$$dy = \frac{\partial x}{\partial p_i} dp_i \quad . \tag{25}$$

As explained earlier, we specify $\bar{A}$, $\bar{B}$ to

$$\bar{A} = \begin{bmatrix} a_{11} & a_{12} \\ a_{21} & 0 \end{bmatrix} \quad ,$$
$$\bar{B} = \begin{bmatrix} b_{11} & b_{12} \\ b_{21} & 0 \end{bmatrix} \quad , \tag{26a}$$

and obtain thus the transformation

$$x = a_{11} + a_{12}x_o + a_{21}y_o \quad ,$$
$$y = b_{11} + b_{12}x_o + b_{21}y_o \quad . \tag{26b}$$

which allows for a fully affine image shaping of the picture function.
Equations (26b) also include the shift parameters Δx , Δy, which are denoted here by a_{11} , b_{11} .
Differentiation of (26b) gives

$$dx = da_{11} + x_o da_{12} + y_o da_{21} \quad ,$$
$$dy = db_{11} + x_o db_{12} + y_o db_{21} \quad . \tag{27}$$

Using for the first derivatives of the picture function the simplified notations

$$g_x = \frac{\partial g^o(x,y)}{\partial x} \quad , \quad g_y = \frac{\partial g^o(x,y)}{\partial y}$$

and adding a radiometric shift parameter r_s and a scaling parameter r_t to the system equation (24) results with (27) in

$$f(x,y) - e(x,y) = g^o(x,y) + g_x da_{11} + g_x x_o da_{12} + g_x y_o da_{21} + g_y db_{11} + g_y x_o db_{12} + g_y y_o db_{21} +$$
$$+ r_s + g^o(x,y) r_t \quad . \tag{28}$$

Combining the parameters in (28) in the parameter vector x

$$x^T = (da_{11}, da_{12}, da_{21}, db_{11}, db_{12}, db_{21}, r_s, r_t) \quad ,$$

their coefficients in the design matrix A, and the vector difference $f(x,y) - g^o(x,y)$ in l, the observation equations are obtained in classical notation (with $e = e(x,y)$) as

$$l - e = Ax \quad . \tag{29a}$$

With the statistical expectation operator E and the assumptions

$$E(e) = 0\,,\ E(ee^T) = \sigma_o^2\, P^{-1} \quad . \tag{29b}$$

the system (29a) is a Gauss-Markov estimation model.
The least squares estimation in model (29a) leads to the unbiased minimum variance estimators

$$\hat{x} = (A^T PA)^{-1} A^T Pl \quad , \quad \text{solution vector} \tag{30a}$$

$$\hat{\sigma}_o^2 = \frac{1}{r}\, v^T Pv \quad , \quad \text{variance factor} \tag{30b}$$

$$v = A\hat{x} - l \quad , \quad \text{residual vector} \tag{30c}$$

$r = n - u$
r . . . redundancy
u . . . number of transformation parameters
n . . . number of observations

The residuals v_i can be interpreted as the differences in grey values between the estimated picture patch (region surrounding the matchpoint) and the template patch.

$$v_i = \hat{g}(x,y)_i - f(x,y)_i \quad , \quad i = 1, \ldots, n \tag{31}$$

^ . . . stands for least squares estimator

Hence $\hat{\sigma}_o$ is an *a posteriori* estimator for the difference of the local template noise and picture noise.

Because of the nonlinearity of the original setup (20), the final solution is obtained iteratively. With the first approximations

$$a_{11}^o = b_{11}^o = a_{21}^o = b_{12}^o = 0 \quad , \quad a_{12}^o = b_{21}^o = 1 \tag{32}$$

(r_s and r_t are linear *a priori*)

results the set of coordinates

$$x_i = x_{oi} \ , \ y_i = y_{oi} \quad , \quad i = 1, \ldots, n$$

for the first iteration step (n = number of grid points (grey values) in template or picture).

After the solution vector (30a) is obtained the transformation (26b) is applied and $g^o(x,y)$ is resampled over the new set of coordinates, and the design matrix A is re-evaluated. The iteration stops if each element of the alteration vector $\hat{x}$ in (30a) goes below a certain limit ($c_1, \ldots, c_6$)

$$|da_{11}| < c_1 \quad , \quad |db_{11}| < c_2 \quad ,$$
$$|da_{12}| < c_3 \quad , \quad |db_{12}| < c_4 \quad , \qquad (33)$$
$$|da_{21}| < c_5 \quad , \quad |db_{21}| < c_6 \quad .$$

The functional model consists essentially of two parts. One set of parameters is designed to model geometrical distortions („shaping function"), the other set corrects the radiometric anomalies. Two different considerations of diametrical kind lead to the selection of a particular parameter set. The estimation model should accommodate enough parameters in order to be able to model the distortions as comprehensively as possible. This, however, induces the danger of over-parameterization. Therefore the quest for many modelling parameters is balanced by the need to restrict the type and number of parameters to those which are safely determinable. Nondeterminable parameters do have an impairing effect on the estimation model and deteriorate the quality of the match. Hence it is of crucial importance for least squares matching to include an appropriate test procedure for parameters and to be capable of excluding those individual parameters which are nondeterminable. The designation „adaptive" indicates such a capability. This is a „self-tuning" approach, the tuning being performed simultaneously with the parameter estimation. The nonlinearity of the original estimation model requires iterations of the linearized model anyway, so that the tuning process can be advantageously incorporated into the sequence of the iteration steps. The iteration starts with a full parameter set. These parameters might even be considered as stochastic variables. The associated weights must then relate to the size of the distortions to be expected. Even small weights keep the system from becoming singular. In the course of the iterations the nondeterminable parameters are excluded. For details on a possible rejection strategy see Gruen (1984).

Analysis of results and computational aspects
The analysis of results is a key element in this matching procedure, because it is expected that this approach allows for a better assessment of the quality of results than the conventional correlation methods. For the computation of precision measures compare section 3.2.2.

For application in our LIF system, this 2-D Least Squares Matching technique has been extended to 3-D, such that volume elements (cuboids), consisting of an array of volume image elements (voxels), can be tracked through 3-D space. 3-D LS-Matching allows for three shifts and a fully 3-D affine reshaping of the cuboids, thus applying a twelve parameter affine transformation. Some of these transformation parameters give direct information on crucial flow parameters (Deusch, chapter 4 of this volume). In LIF applications 3-D LS-Matching must be amended with geometric and radiometric constraints in order to provide for stable solutions in certain cases of weak image signal content (Maas, chapter 4 of this volume).

7. References

Beyer, H.A. (1992) Geometric and Radiometric Analysis of a CCD-Camera Based Photogrammetric Close-Range System, *Dissertation ETH Nr. 9701*, Zürich, May 1992.

Beynon, J.D.E. & Lamb, D.R. (1980) Charge coupled devices and their applications, *McGraw-Hill, London.*

Dähler, J. (1986) Charge Coupled Devices (CCD) (in German), *Institute of Geodesy and Photogrammetry*, Report No. 107, March 1986.

Fraser, C.S. (1992) Photogrammetric camera component calibration - a review of analytical techniques, *Invited Paper, „Workshop on Camera Calibration and Orientation in Computer Vision", XVIIth ISPRS Congress, Washington, USA, August 2.*

Gruen, A. (1980) Precision and reliability aspects in close-range photogrammetry, *Invited paper to Commission V at the XIVth Congress of the ISP, Hamburg,* 13-17 July 1980. *The Photogrammetric Journal of Finland,* Vol. 8, No. 2, 1981, pp. 117-132.

Gruen, A. (1984) Processing of amateur photographs, *Presented Paper to Commission V at the XVth Congress of the ISPRS, Rio de Janeiro, June 1984. The Photogrammetric Record,* 11 (65), April 1985, pp. 567-579.

Gruen, A. (1986) Photogrammetric point positioning by bundle adjustment (in German), *Institute of Geodesy and Photogrammetry, Mitteilungen Nr. 40, 87 pages.*

Gruen, A., Beyer, H.A. (1992) System calibration through self-calibration, *Invited Paper, „Workshop on Camera Calibration and Orientation in Computer Vision", XVIIth ISPRS Congress, Washington, USA, August 2.*

Real R.R. (1986) Components for video-based photogrammetry of dynamic processes. *Int. Arch. of Photogrammetry and Remote Sensing,* Vol. 26, Part 5, pp. 363-373.

Weimer, P.K. & Cope, F.D. (1983) Image sensors for television and related applications, *In: Advances in Image Pickup and Display, 6, Academic Press, New York.*

CONTRIBUTIONS OF DIGITAL PHOTOGRAMMETRY TO 3-D PTV

H.-G. Maas
Institute of Geodesy and Photogrammetry,
Swiss Federal Institute of Technology, ETH Hoenggerberg,
CH-8093 Zurich, Switzerland

1. Introduction

In the following, a 3-D PTV data processing chain from stereoscopic (or rather: multi-camera) digital image sequences to 3-D particle coordinates will be shown. Herein, we assume the availability of digital image sequences imaging each flow marker only once per image; see (Virant/Dracos, chapter 3) for the experimental realization and the acquisition of multiple digital image sequences. As shown in Figure 1, the 3-D PTV data processing chain from digital image sequences to 3-D particle coordinates contains digital image processing, digital photogrammetry and object tracking.

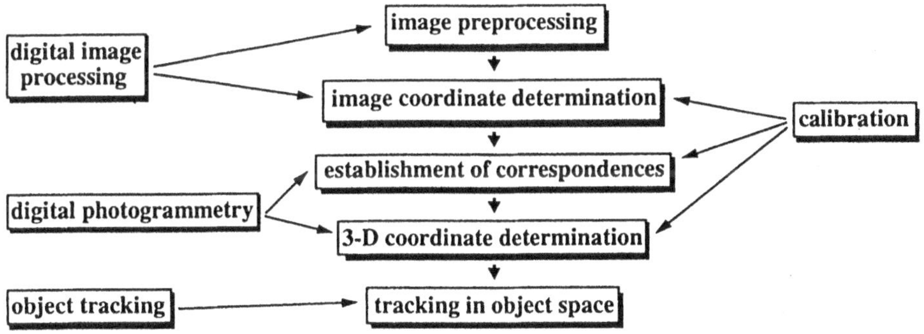

Figure 1: Flow scheme of 3-D PTV

This data processing chain contains two matching steps:

- The establishment of stereoscopic correspondences (i.e. the establishment between particle images acquired from different viewing directions at one time instant) as a necessary prerequisite for 3-D coordinate determination.
- The establishment of correspondences on the time axis (tracking).

Th. Dracos (ed.), Three-Dimensional Velocity and Vorticity Measuring and Image Analysis Techniques,
191–207.

As already mentioned by (Dracos, introduction to chapter 3), there are two options for the data processing chain:

- If tracking is performed firstly in image space, the establishment of stereoscopic correspondences becomes easier, as it can be supported by relational information provided by the 2-D trajectories.
- If 3-D coordinate determination is performed firstly, tracking becomes more reliable, as it can be based on 3-D information. This procedure leads to less ambiguities in tracking, as trajectories cannot intersect in 3-D space, while unfavourable trajectory intersections will occur quite frequently in 2-D image space.

The first approach has for example been followed in (Netzsch/Jähne, 1993). In the 3-D PTV implementation described here, the latter approach will be followed, as it promises a higher yield of trajectories, if the establishment of stereoscopic correspondences is supported by the use of multiple cameras rather than only two. For the future, a combination of both methods is envisaged.

To achieve homogeneous accuracy in all three coordinate directions and to optimize the common field of view of the cameras, a stereoscopic observation system has to be arranged convergent as shown in Figure 2.

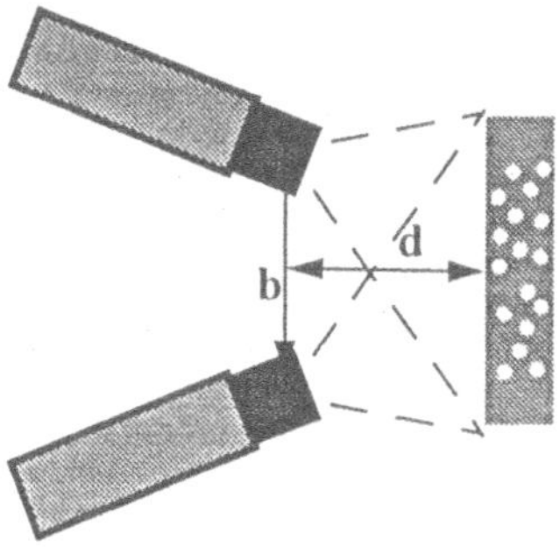

Figure 2: Convergent camera setup

A reasonable base-to-distance ratio will be of order b:d = 1:1; depth-of-focus considerations, however, may necessitate the use of a smaller convergence angle, especially in combination with very small observation volumes. As it will usually not be possible to fine-adjust the cameras with respect to pre-defined values, the relative orientation between the cameras and the absolute orientation of the observation system with respect to the observation volume is not known a priori. Moreover, the standard video components which are often used in PTV systems, show strong systematic distortions. For these reasons, the observation system has to be calibrated. To avoid expensive and inflexible laboratory equipment, photogrammetric methods can be used for the system calibration.

In the following, the mathematical model of 3-D coordinate determination and system calibration will be shown. After that, algorithmic aspects of image processing and image matching will be discussed, and some results will be given, indicating the potential and limitations of the methods.

2. Mathematical Model

The fundamental mathematical theorem of photogrammetric 3-D particle coordinate determination is the collinearity condition (Gruen, chapter 3), which states that object point, camera projective centre and image point lie on a straight line. This mathematical formulation applies to a pinhole camera model not considering any distortions mainly introduced by the use of lenses, and has to be extended to meet the physical realities. One has to consider the fact that each ray from a particle to the sensor passes the three optical media water, glass and air with different refractive indices, which causes a twice broken beam to be handled in the collinearity equation. Furthermore, the image coordinates have to be corrected for effects of lens distortion and of the A/D conversion of the image data.

2.1. Additional parameters for systematic image deformation

The effects caused by lens distortion (e.g. barrel or pillow form distortion) and A/D conversion are usually compensated by a set of additional parameters, which have to be determined in the calibration together with the parameters of exterior and interior orientation of the cameras. Lens distortion is compensated by a set of 5 additional parameters modeling radial and decentering distortion (Gruen, chapter 3). For the compensation of effects of image digitization and storage, especially of the often a priori unknown difference between the pixel rate of the CCD cameras and the clock rate of the framegrabber and the low frequency part of linejitter, an additional affine transformation has to be performed (e.g. El-Hakim, 1986):

$$\bar{\bar{x}}_i' = a_0 + a_1\bar{x}_i' + a_2\bar{y}_i' \qquad \bar{\bar{y}}_i' = b_0 + b_1\bar{x}_i' + b_2\bar{y}_i' \tag{1}$$

Due to linear dependencies on the parameters of the collinearity condition only two of the six parameters of the affine transformation can be introduced as unknowns: a scale in horizontal image coordinate direction and a shear (a_1, a_2).

In total each of the cameras of a 3-D PTV system is described by 16 parameters: 6 for exterior orientation, 3 for interior orientation, 5 for lens distortion and 2 for influences of the A/D conversion. These parameters have to be determined in a system calibration.

2.2. Multimedia geometry

The effect of a ray being twice broken due to different refractive indices on the optical path through water, glass and air leads to a multimedia geometry model. Knowing the camera calibration data, the interfaces between the different optical media and the refractive indices, this effect can be modeled strictly.

2.2.1. Development of a multimedia module

If the glass wall separating water and air can be considered a plane parallel plate, a simple module based on Snell's Law can be developed, which can be implemented into the collinearity condition equation. This module computes a radial shift of each object point relative to the nadir point of the respective camera (Figure 3), which can be used as a correction term in the collinearity condition. If the X-Y plane of the coordinate system is chosen parallel with the plane interface glass/water (resp. air/glass), some simplifications are possible. The procedure is shown in Figure 3: If the point $P(X, Y, Z)$ in object space is shifted to $\bar{P}(\bar{X}, \bar{Y}, Z)$, the collinearity condition can be applied for P using the object coordinates of the shifted point $\bar{P}$. Only a radial shift by ΔR ($\Delta R > 0$ if $n_2 > n_1$ and $n_3 > n_1$) parallel to the X-Y plane has to be computed for each point relative to the nadir point of each camera. Thus, rays from different cameras C_j to an object point P are calculated with different points $\bar{P}_j$, with the broken beams still intersecting in P.

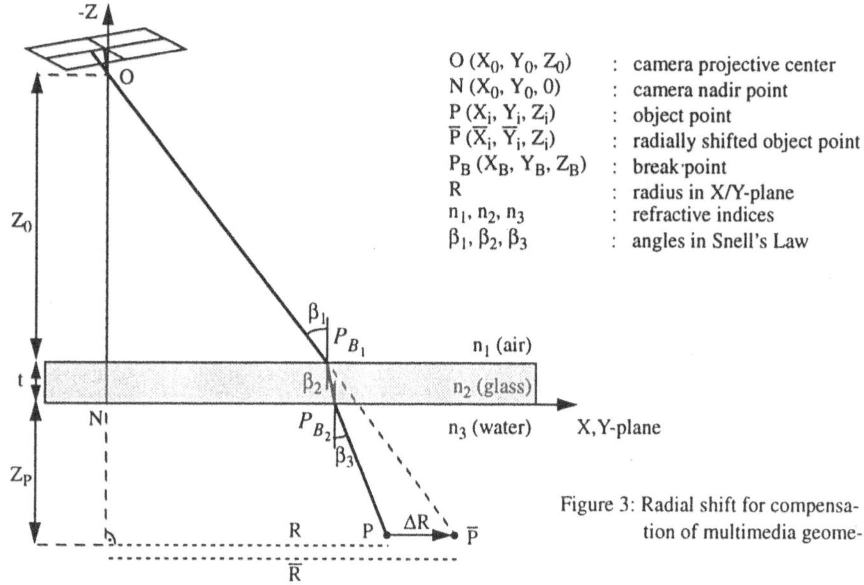

Figure 3: Radial shift for compensation of multimedia geome-

From Figure 3 can be derived:

$$Z_0 \cdot \tan\beta_1 + t \cdot \tan\beta_2 + Z_p \cdot \tan\beta_3 = R \quad , \qquad \bar{R} = (Z_0 + t + Z_p) \cdot \tan\beta_1 \,. \tag{2}$$

With Snell's Law

$$n_1 \cdot \sin\beta_1 = n_2 \cdot \sin\beta_2 = n_3 \cdot \sin\beta_3 \tag{3}$$

the system describing the multimedia geometry is complete. Equations (2), (3) can only be solved iteratively due to the trigonometric functions. If P is chosen as a first approximation for $\bar{P}$ and

$$\bar{R}_{(0)} = \sqrt{(X_P - X_0)^2 + (Y_P - Y_0)^2} \,, \tag{4}$$

the angle of incidence in the medium n_1 in the first iteration becomes

$$\beta_1 = \operatorname{atan}\left(\frac{\bar{R}_{(0)}}{Z_0 + t + Z_p}\right). \tag{5}$$

From Snell's Law one gets the angles of incidence and refraction in the other two media:

$$\beta_2 = \operatorname{asin}\left(\frac{n_1}{n_2} \cdot \sin\beta_1\right), \qquad \beta_3 = \operatorname{asin}\left(\frac{n_1}{n_3} \cdot \sin\beta_1\right) \tag{6}$$

and the correction for $R_{(0)}$:

$$dR = R - (Z_0 \cdot \tan\beta_1 + t \cdot \tan\beta_2 + Z_p \cdot \tan\beta_3)\,. \tag{7}$$

The equations (4) - (7) are used iteratively

$$\bar{R}_{(1)} = \bar{R}_{(0)} + dR \qquad \Rightarrow \beta_1, \beta_2, \beta_3 \qquad \Rightarrow dR \qquad \text{etc.}$$

until $dR < \varepsilon$ (e.g. with $\varepsilon = 0.001$ mm).

Transforming back from polar coordinates to the cartesian system, one obtains the coordinates of the radially shifted point $\bar{P}$

$$\bar{X}_P = X_0 + (X_P - X_0) \cdot \frac{\bar{R}}{R}\,, \qquad \bar{Y}_p = Y_0 + (Y_P - Y_0) \cdot \frac{\bar{R}}{R}\,, \qquad \bar{Z}_P = Z_P\,. \tag{8}$$

The nadir point N is herein defined as the orthogonal projection of the camera projective centre O onto the X-Y plane of the chosen local coordinate system. The collinearity condition can then be used with the radially shifted point $\bar{P}$ ($\bar{X}$, $\bar{Y}$, Z) instead of P (X,Y, Z).

The convergence of the iterative scheme equations (4) - (7) can be accelerated considerably by introducing an overcompensation-factor ($ocf \approx 1.1 \ldots 1.8$) for the calculation of ΔR. The choice of *ocf*, however, depends on the refractive indices, the ratio of the light pathlengths in water and air and the incidence angle itself, so that a constant *ocf* can only represent an average state. A consequent extension of this idea is the implementation of the radial shift into a lookup-table (input: Depth in water Z_i and radial distance R from nadir point, output: Radial displacement factor $\bar{R}$/R). The computational effort needed for the initialization of these lookup-tables (one for each camera) will be compensated by the reduced processing time for the particle image sequences, even if they are short.

A more general treatment of the multimedia problem has been published by Kotowski (1988). Kotowski allows for an arbitrary number of optical media with different refractive indices separated by plane interfaces, which do not have to be parallel.

2.2.2. Direct solution for the spatial intersection

If short computation times are required and the least squares adjustment consumes too much computation time, a direct ray tracing and intersection solution can be developed for the spatial intersection following Okamato and Hoehle (1972) under the premises of plane parallel interfaces as made before. This solution shown in the following, however, can only be implemented as a spatial intersection and does not represent a functional model suited for a least squares adjustment.

From the image coordinates x', y' of a point corrected for the influence of lens distortion and digitization, one can calculate the direction cosine in air

$$\begin{bmatrix} a_1 \\ b_1 \\ c_1 \end{bmatrix} = A \cdot \begin{bmatrix} x'/s' \\ y'/s' \\ -c/s' \end{bmatrix} \qquad \text{with} \qquad s' = \sqrt{x'^2 + y'^2 + c^2} \tag{9}$$

and determine the intersection of the ray with the air/glass interface (Figure 3)

$$\begin{bmatrix} X_{B_1} \\ Y_{B_1} \\ Z_{B_1} \end{bmatrix} = \frac{Z_0 - t}{c_1} \cdot \begin{bmatrix} a_1 \\ b_1 \\ c_1 \end{bmatrix} + \begin{bmatrix} X_0 \\ Y_0 \\ Z_0 \end{bmatrix}, \tag{10}$$

where the beam is broken.

The new direction cosine in glass becomes

$$\begin{bmatrix} a_2 \\ b_2 \\ c_2 \end{bmatrix} = \frac{n_1}{n_2} \cdot \begin{bmatrix} a_1 \\ b_1 \\ c_1 \end{bmatrix} - \left(\frac{n_1}{n_2} \cos i_1 - \sqrt{1 - \left(\frac{n_1}{n_2}\right)^2 + \left(\frac{n_1}{n_2} \cos i_1\right)^2} \right) \cdot \begin{bmatrix} \lambda_1 \\ \mu_1 \\ \nu_1 \end{bmatrix} \tag{11}$$

with $(\lambda_1, \mu_1, \nu_1) = (0, 0, 1)$ in the chosen local coordinate system (plane normal on interface), so that the cosine of the incidence angle i_1 becomes $\cos i_1 = a_1\lambda_1 + b_1\mu_1 + c_1\nu_1 = c_1$.

The intersection of the ray with the glass/water interface is

$$\begin{bmatrix} X_{B_2} \\ Y_{B_2} \\ Z_{B_1} \end{bmatrix} = \frac{d}{c_2} \cdot \begin{bmatrix} a_2 \\ b_2 \\ c_2 \end{bmatrix} + \begin{bmatrix} X_{B_1} \\ Y_{B_1} \\ Z_{B_1} \end{bmatrix}, \tag{12}$$

and the direction cosine in water becomes

$$\begin{bmatrix} a_3 \\ b_3 \\ c_3 \end{bmatrix} = \frac{n_1}{n_2} \cdot \begin{bmatrix} a_2 \\ b_2 \\ c_2 \end{bmatrix} - \left(\frac{n_2}{n_3} \cos i_2 - \sqrt{1 - \left(\frac{n_2}{n_3}\right)^2 + \left(\frac{n_2}{n_3} \cos i_2\right)^2} \right) \cdot \begin{bmatrix} \lambda_2 \\ \mu_2 \\ \nu_2 \end{bmatrix}, \tag{13}$$

again with $(\lambda_2, \mu_2, \nu_2) = (0, 0, 1)$ and $\cos i_2 = a_2\lambda_2 + b_2\mu_2 + c_2\nu_2 = c_2$.

Having performed this ray tracing procedure (9) - (13) for two corresponding points of two images (the indices l, r denote the left and right image), a spatial intersection can be computed:

$$Z_P = \frac{Y_{B_{2(r)}} - Y_{B_{2(l)}}}{\dfrac{b_{3(l)}}{c_{3(l)}} - \dfrac{b_{3(r)}}{c_{3(r)}}}, \tag{14}$$

if the larger component of the baseline between the two cameras is in Y-coordinate direction, and

$$Z_P = \frac{X_{B_{2(r)}} - X_{B_{2(l)}}}{\frac{a_{3(l)}}{c_{3(l)}} - \frac{a_{3(r)}}{c_{3(r)}}}, \tag{15}$$

if the larger component is in X-coordinate direction. With *the* appropriate Z_P, we obtain

$$X_P = \frac{1}{2}\left(X_{B_{2(r)}} + X_{B_{2(l)}} + Z_P \cdot \left(\frac{a_{3(l)}}{c_{3(l)}} + \frac{a_{3(r)}}{c_{3(r)}}\right)\right),$$
$$Y_P = \frac{1}{2}\left(Y_{B_{2(r)}} + Y_{B_{2(l)}} + Z_P \cdot \left(\frac{b_{3(l)}}{c_{3(l)}} + \frac{b_{3(r)}}{c_{3(r)}}\right)\right). \tag{16}$$

This direct solution is considerably faster than the least squares adjustment with the collinearity condition, but it is not a strict solution due to the neglection of the redundant information, and it can only be used (optionally) for the spatial intersection in the particle positioning procedure, but not for the spatial resection in the calibration.

2.2.3 Further influences of the multimedia environment

Besides these exactly modeled effects the multimedia environment causes some problems which are not contained in the mathematical model but do have a non-negligible influence on the accuracy of results:

- Aberrations: The optical system (water - glass - air - camera lens) passed by each ray is not corrected for aberrations if lenses are corrected for the use in air. This leads to a degradation of image quality, especially as the cameras are arranged convergent.
- The dispersion in water is much larger than in air. The variance of the refractive index in water over the visible spectrum of light is 1.4% in contrast to 0.008% in air (Hoehle, 1971). This leads to colour seams at the edges of imaged objects which will appear as blur in black-and-white images.
- Inhomogenities of the refractive index (due to local pressure and temperature differences) cause deviations from the mathematical model.
- Deviations from the planeness of the glass walls falsify the incidence angles.
- The network geometry is deteriorated by the smaller intersection angle of rays due to the fact that rays are broken towards the optically denser medium. This causes larger errors in the Z-coordinates.
- Diffusion and absorption in water cause an extinction of light and reduce the image contrast.
- The diffraction is not rotational symmetric with convergent camera arrangement (Meid, 1991).

All these effects cause a degradation of image quality of multimedia images and systematic errors, which necessarily leads to larger errors in the particle coordinate determination compared with applications in air.

2.3 Handling of the mathematical model

The collinearity condition equation in combination with the additional parameters and the multimedia module equations (2) - (8) represents a functional model, which is suited to be linearized as observation equations in a least squares adjustment.

$$(x_i', y_i') = f(X_0, Y_0, Z_0, \omega, \varphi, \kappa, c, x_h, y_h, k_1, k_2, k_3, p_1, p_2, a_1, a_2, X_i, Y_i, Z_i) \ . \qquad (17)$$

In the calibration procedure (paragraph 3.) this model is being applied for a spatial resection using control points on a calibration frame with known coordinates (X_i, Y_i, Z_i) and introducing the 16 parameters of exterior and interior orientation, lens distortion and affine transformation as unknowns. After the calibration the same model can be used for a spatial intersection introducing the coordinates of a particle as unknowns (paragraph 4.3.). In a photogrammetric bundle solution (Gruen, chapter 3), calibration and point determination for an arbitrary number of cameras and object points can be solved in one system, making the solution more accurate and stable. As an option, also the refractive index of the liquid can be introduced as an unknown into the bundle solution and determined simultaneously with the camera parameters in the calibration process (Maas, 1995).

3. System Calibration

As already mentioned in paragraphs 1. and 2., the observation system has to be calibrated for the determination of unbiased 3-D particle coordinates. This calibration includes the exterior orientation parameters of all cameras as well as the interior orientation parameters, lens distortion and effects of the A/D conversion and is advantageously based on photogrammetric principles.

Four different methods for the calibration of photogrammetric systems are described by Gruen in chapter 3. Under the aspects of accuracy and user-friendliness, the reference frame method will often be an appropriate choice for the calibration of a 3-D PTV system. For this purpose, a calibration frame with a large number of targets, spanning the entire 3-D observation volume, has to be placed into object space and imaged once by all cameras of the system. With the known 3-D coordinates of the targets on the reference body and the image coordinates of these points measured with subpixel accuracy e.g. by a centroid operator (paragraph 4.1.), the 16 parameters describing each camera can be determined from a single image, using the mathematical model as described by the collinerity condition in combination with in eqs. (1)-(8). A minimum requirement for the determination of the parameters is the measurement of at least 8 points, which have to be threedimensionally distributed; if possible, a much larger number of points (50-100), well distributed over the planned observation volume and covering the whole image format, should be used.

The processing of the calibration images can be performed automatically after a user supported pre-orientation of the cameras by identifying at least three points on the reference body. Using coded targets including a binary coded point number, the calibration procedure can be completely automated. After the calibration, the 16 parameters describing exterior and interior orientation, lens distortion and affine transformation for each camera are known and can be used for the spatial intersections in the particle positioning. The large number of targets on the calibration frame allows some of the points to be used as control points and the rest as check points. If these check points are treated like unknown points (i.e. their coordinates are determined by spatial intersection using the calibration parameters determined from the resection with the other points), the rms deviations between determined coordinates and the given point coordinates of the check points are a good estimate for the accuracy potential of the method, yielding precision figures not influenced by deficits of the geometric and statistic model.

4. Algorithmic Aspects

Following the chain of automatic data processing from the digital images to 3-D particle trajectories as outlined in Figure 1, a number of algorithmic aspects have to be discussed under the aspect of the optimization of resolution, precision and reliability. Figure 4 shows the first two frames of a trinocular sequence of a digital images with a large number of particles. In these images, particles have to be automatically detected, measured and matched.

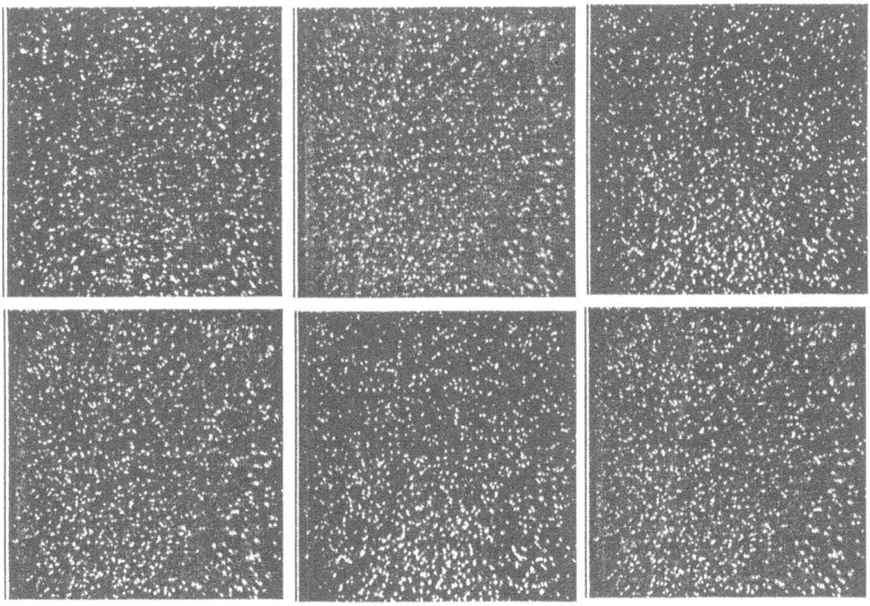

Figure 4: First two time instances of a trinocular image sequence

4.1. Image processing

Essentially the task of automatic particle identification and particle image coordinate determination in the digital images is trivial, as the particles appear as bright stains on dark background. To remove non-uniformities of the background intensity level due to reflections and the intensity profile of the illumination, the images are highpass-filtered in a first pre-processing step; highpass-filtering is performed in the spatial domain by the subtraction of an unsharp mask from the original image.

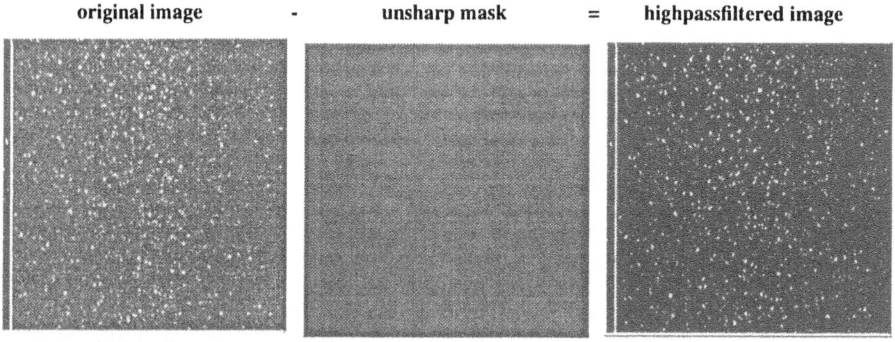

Figure 5: Highpassfiltering in the spatial domain

After that, particle images can be segmented by a simple thresholding algorithm, and image coordinates can be determined with subpixel accuracy by using the greyvalue-weighted centre of gravity of the segmented regions. Under ideal conditions, the centroid operator offers an accuracy potential of $^1/_{25}$ ... $^1/_{50}$ pixel; under realistic PTV conditions with moving tracers, multimedia environment and illumination effects, this potential may be significantly degraded.

With the large number of particles often employed in 3-D PTV applications (sometimes more than 1000 particles in an image of 512 x 512 pixel), a problem with particles optically blocking or partly overlapping each other in one or more views (Figure 6) will occur.

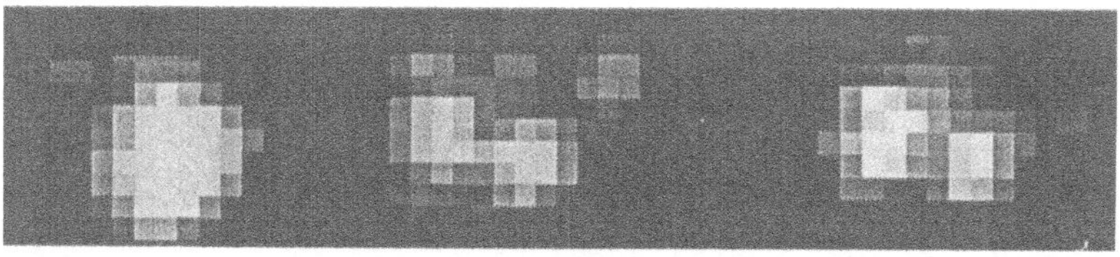

Figure 6: Single particle and examples of overlapping particle images

This problem of overlapping particles has rarely been treated in the PTV literature dealing with automatic systems. An analysis on the expectable number of overlapping particle images (N_a) as a function of the total number of particles in an image (n), the sensor area (F) and the particle image size (f) has been derived in (Maas, 1992a):

$$N_a = (n-1) + \frac{F}{4f} \cdot \left(e^{-\frac{(n-1) \cdot 4f}{F}} - 1 \right) \qquad (18)$$

In a 512 x 512 pixel image with 2000 particles each covering 10 pixel on average, almost 300 cases of overlapping particles have to be expected. This number grows linearly with particle image size and approximately with the square of the number of particles per image. This shows clearly that the problem has to be accounted for if the spatial resolution of the system is to be optimized without loss of reliability. In (Maas, 1992a), several methods of digital image analysis have been tested to detect and split overlapping images of two or more particles:

- Filtering the images with differential operators proved to be inadequate as they do well in detecting particle image borders but may fail at pixels where particles touch each other, if the contrast is low.
- Template matching (Gruen/Baltsavias, 1988), morphological operators (Ogniewicz and Kübler, 1989) and Hough transform (Hough, 1962) failed due to the small target size of only 3 x 3 pixels on average.
- Fourier analysis of the particle image border line (Kuhl/Giardina, 1982) worked well with sharp particle images but failed with slightly defocussed images of overlapping particles where the characteristic waist in the border line is blurred.
- Invariant greyvalue moments (Teh/Chin, 1988) do not show this sensitivity to defocusing, but higher order moments are rather noise sensitive. For the recognition of overlaps second order moments proved to be sufficient if only two particles were overlapping, but failed if more than two particles formed a blob.

These problems lead to the development of a modified anisotropic thresholding operator, which searches for discontinuities in the greyvalues inside segmented blobs and splits blobs if a discontinuity exceeding a certain empirically determined limit is detected (Maas, 1992a). The operator can be characterized by some rules describing a particle image:

- all pixels of a particle image have a greyvalue which is larger than a preset threshold
- a particle image shows exactly one local maximum
- greyvalues drop continuously inside a particle image
- a pixel representing a local greyvalue minimum belongs to the particle containing its neighbour pixel with the largest greyvalue

This operator can be implemented very efficiently and performed well. From comparisons with a human supervisor it can be said that in all cases where the human could meet a clear decision on where to split particle images, the operator came to the same decision. Most importantly, unlike some of the other methods mentioned above, it does not only indicate cases of overlapping particle images but splits them correctly at local greyvalue minima.

4.2. Establishment of stereoscopic correspondences

To compute 3-D particle coordinates, a spatial intersection has to be performed with corresponding image coordinates of a stereopair. The establishment of those stereoscopic correspondences, however, poses some problems. The particle images do not show any characteristic features (like colour, size, shape, etc.), which could allow for a reliable distinction of particles. The only criterion which can be applied is the geometric constraint of the epipolar line: Knowing the orientation parameters of the cameras from the calibration procedure (paragraph 3.), proceeding from a point P' in one image an epipolar line in an other image can be calculated, on which the corresponding point has to be found. In the strict mathematical formulation this line is a straight line, in the more general case with convergent camera-axes, non-negligible lens distortion and multimedia geometry, the epipolar line will be a slightly bent line (Figure 7), which can be approximated by a polygon. Its length l can be restricted if approximate knowledge about the depth range in object space coordinates is available, which is determined from the knowledge of the illuminated test section. Adding a certain tolerance width ε to this epipolar line segment (due to data quality) the search area for the corresponding particle image becomes a narrow two-dimensional bandshaped window in image space.

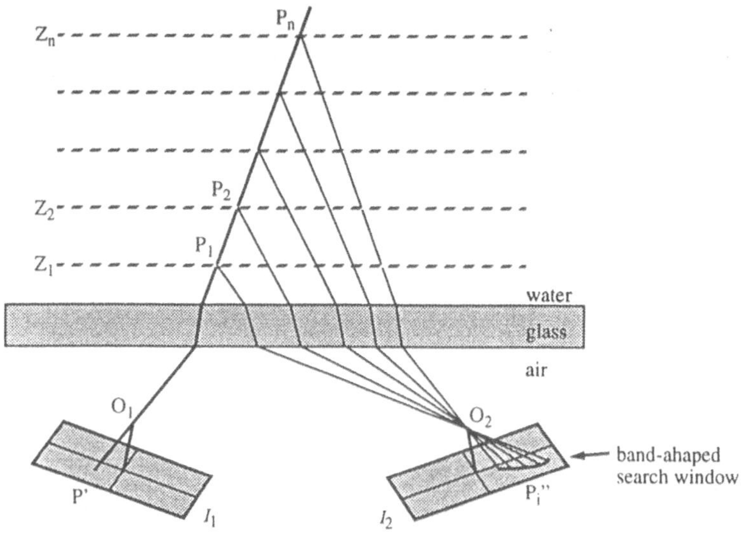

Figure 7: Derivation of epipolar lines in multimedia

Due to the large number of imaged particles a second problem of ambiguities occurs, as often two or more particle images will be found in the epipolar search area. The number of ambiguities grows (Maas, 1992b)

- approximately with the square of the number of particles
- linearly with the length of the epipolar line segment
- linearly with the width of the epipolar search window.

If the particle features do not allow a reliable distinction of particles, these ambiguities cannot be resolved by a system based on two cameras only. With realistic estimates for the number of particles per image and the dimensions of the epipolar search window in a reasonable camera setup the expected number of ambiguities becomes too large for a two-camera-system to yield a robust solution of the correspondence problem (see TABLE 1).

A consequent solution of the ambiguity problem is the use of a third camera in a setup as shown in Figure 8 with the camera projective centers forming a triangle (in the ideal case an equilateral triangle) and the camera axes approximately forming a tetrahedron (Maas, 1990).

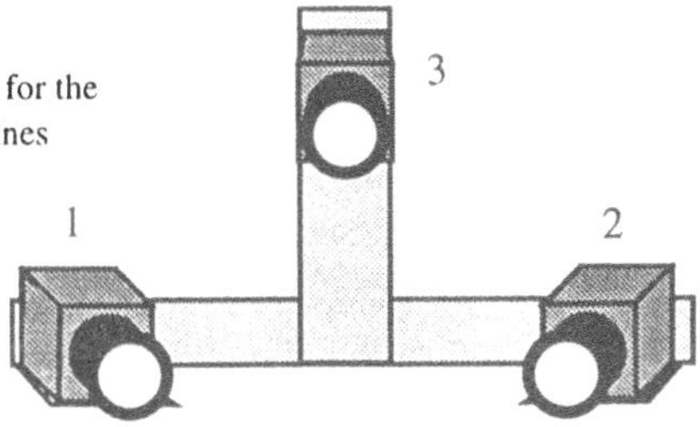

Figure 8: Arrangement of three CCD cameras for the method of intersection of epipolar lines

This setup allows the calculation of epipolar line segments intersecting in the image, which can be exploited as shown in the example in Figure 9:

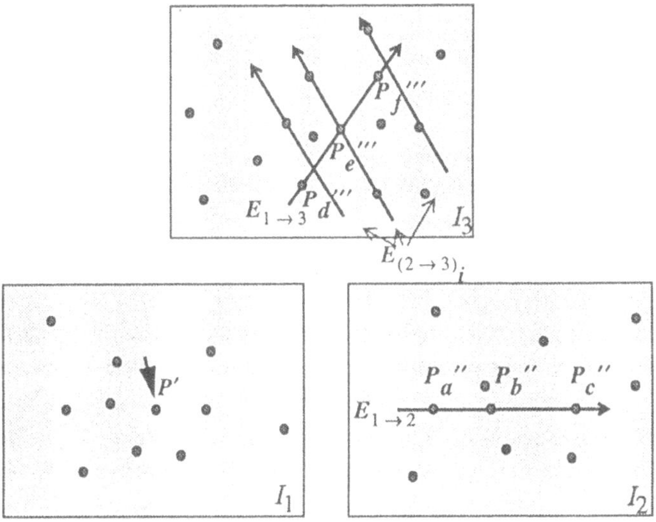

Figure 9: Principle of intersection of epipolar lines

Proceeding from a point P' in the image I_1 all epipolar lines $E_{1\rightarrow 2}$ in I_2 and $E_{1\rightarrow 3}$ in I_3 are derived, on which respectively candidates P_a'', P_b'', P_c'', and P_d''', P_e''', P_f''' will be found in Figure 9. An unambiguous determination of the particle image corresponding to P' can neither be found in I_2 nor in I_3. However if all epipolar lines $E_{(2\rightarrow 3)_i}$ of all candidates P_i in I_2 are being intersected with the epipolar line $E_{1\rightarrow 3}$, there will be a large probability that only one of the intersection points will be close to one of the candidates in I_3 (in Figure 9: P_e'''). This consideration has been implemented via a combinatorics algorithm which tries to find such consistent triplets in the three particle image coordinate datasets and rejects points which are members of more than one consistent triplet. Such an unambiguous consistent triplet is a necessary and sufficient condition for the establishment of correct correspondence. It can be shown, that the method of intersection of epipolar lines can reduce the number of remaining ambiguities by at least one order of magnitude (Maas, 1992b); these remaining ambiguities cause contradictions and inconsistent triplets and can thus be rejected.

The method described above is the most evident, but not the only way of exploiting a third camera. An approach which based on combinatorics rather than epipolar lines is shown in (Kasagi/Nishino, 1990). Employing a parametrization on the epipolar line, one can also work with three cameras which are arranged in a way that their projective centers are lying on a straight line rather than forming a triangle (Maas, 1992b). The expectable numbers of remaining ambiguities for the methods discussed above are compiled in TABLE 1 for realistic assumptions for the number of particles (n), the depth range in object space (ΔZ) and the width of the epipolar search area (ε) for a base $b_{13} = 200$ mm and a camera constant $c = 9$ mm:

TABLE 1: Remaining numbers of ambiguities in the establishment of correspondences

	2 cameras	**3 cameras collinearly**	**3 cameras in equilateral triangle**
n = 1000, ε = 10 μm, ΔZ = 40 mm	**401**	**40**	**35**
n = 2000, ε = 10 μm, ΔZ = 40 mm	**1605**	**160**	**140**
n = 1000, ε = 5 μm, ΔZ = 40 mm	**201**	**10**	**9**
n = 1000, ε = 10 μm, ΔZ = 80 mm	**802**	**40**	**35**

The number of remaining ambiguities is reduced by at least one order of magnitude for both of the three-camera arrangements compared with a two-camera arrangement. The number of ambiguities produced by a three-camera system may often be acceptable, if tracers are only considered statistical velocity probes in an Eulerian flow analysis. If the number is still considered too large it can be reduced straightforwardly by the use of a fourth camera with the aim of establishing consistent quadruplets in the four sets of particle image coordinates. The projective centers of a four-camera system can be arranged in a square (Figure 10), collinearly or in a combination of both.

The principle has been extended to the use an arbitrary number of camera stations for general matching purposes (Maas, 1996), but more than four cameras will rarely be necessary in 3-D PTV applications. A four-camera arrangement can be considered ideal: it reduces the number of remaining ambiguities to almost zero, and it even allows a reliable determination of most of those particles which are completely hidden in one of the four images. Nevertheless, a three-camera system may often be an appropriate choice, as it is optimally supported by RGB image sequence acquisition hardware components.

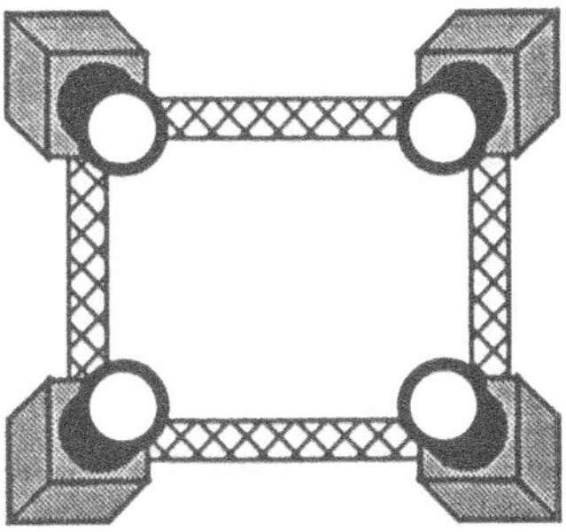

Figure 10: Example of a four camera arrangement

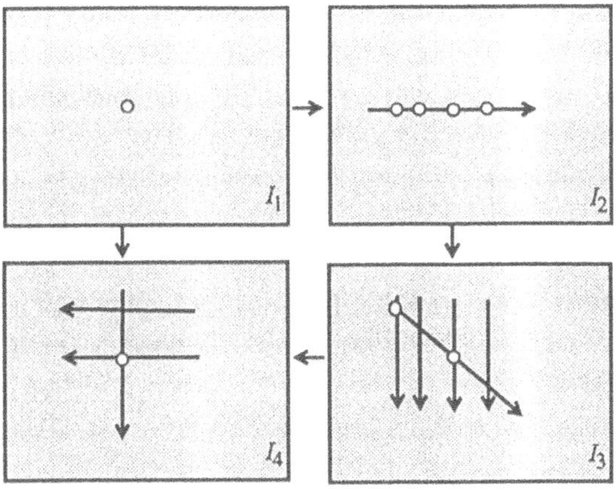

4.3. 3-D coordinate computation

After the establishment of correspondences the 3-D coordinates for the non-ambiguous particle image quadruplets (triplets, pairs) can be determined by spatial intersection using either the strict solution via least squares adjustment or bundle solution using the collinearity condition and eqs. (1) - (8), or the faster, but less accurate solution eqs. (9) - (16). Another benefit of the three- and four-camera systems as compared with a two-camera-system is improvement of the accuracy of the coordinates of a point by a factor of approximately 1.4 or 1.7, respectively.

The results of these processing steps are datasets with 3-D particle coordinates and standard deviations, which are passed to a tracking module to establish correspondences between consecutive datasets (Dracos, chapter 3). It should be noted that the precision figures are effected by locally correlated distortions and that the precision of a single particle velocity vector of limited length, computed from coordinate differences in consecutive datasets, will usually be better than the precision of 3-D particle coordinates (Maas, 1992a). Velocity errors are absolute values due to coordinate errors, the error does not depend on the length of the velocity vector; more importantly, errors in velocity vectors do not propagate along trajectories.

5. Discussion of the potential of the methods

Simulations and a number of practical experiments and performance tests have shown that with the use of three or four CCD cameras rather than two the 3-D coordinates of more than 1000 particles in a flow can be determined when standard videonorm 25 Hz CCD cameras are being used. Using large format CCD cameras with resolutions of up to 2048 x 2048 pixel, the yield may be increased to more than 10'000 particles, but at much higher system costs and at reduced temporal resolution. On the other extreme, high-speed solid state sensor camera systems (Maas, 1992c) allow for temporal resolutions of 1000 Hz and more, but at reduced spatial resolution.

Based on subpixel accuracy image measurement operators and strict geometric and stochastic modeling, the accuracy potential of the method is - although degraded by effects of the multimedia environment and the not perfectly spherical shape of tracers - in the order of 1 : 5'000 to 1 : 10'000 of the lateral observation volume dimension, with the depth coordinate usually being by a factor of two worse than the lateral coordinates (Maas, 1992a). Using redundant information in the coordinate determination process, the precision of coordinates can be estimated, and gross errors in the measurements can be detected.

The computation times of the above mentioned techniques, implemented on a SUN SparcStation 5, are listed for a typical image triplet in TABLE 2.

TABLE 2: Computation times

highpass-filtering (3 imgs)	2 seconds
particle detection and image coordinate determination	2 seconds
establishment of multi image correspondences	3 seconds
3-D coordinate determination	1 second

Considering an imaging rate of 25 Hz, the times are still by more than a factor of 100 away from realtime processing. However, the whole data processing can be performed completely automatically after an initialization of a few steering parameters by the user.

6. References:

El-Hakim, S.F., 1986: Real-Time Image Metrology with CCD Cameras. *Photogrammetric Engineering and Remote Sensing*, Vol. 52, No. 11, pp. 1757 - 1766.

Gruen, A., Baltsavias, E., 1988: Geometrically Constrained Multiphoto Matching. *Photogrammetric Engineering and Remote Sensing*, Vol. 54, No. 5, pp. 633 - 641

Hoehle, J., 1971: Zur Theorie und Praxis der Unterwasser-Photogrammetrie. Schriften der DGK, Reihe C, Heft 163.

Hough, P., 1962: Method and Means for Recognizing Complex Patterns. U.S. Patent 3069654

Kasagi, N., Nishino, K., 1990: Probing Turbulence with Three-Dimensional Particle Tracking Velocimetry. Proceedings International Symposium on Engineering Turbulence - Methods and Measurements, Dubrovnik, September 24-28

Kotowski, R., 1988: Phototriangulation in Multi-Media-Photogrammetry. International Archives of Photogrammetry and Remote Sensing, Vol. XXVII

Kuhl, F., Giardina, Ch., 1982: Elliptic Fourier Features of a Closed Contour. *Computer Graphics and Image Processing* 18, pp. 236 - 258

Maas, H.-G., 1990: Digital Photogrammetry for Determination of Tracer Particle Coordinates in Turbulent Flow Research. SPIE Proceedings Series Vol. 1395, Part I

Maas, H.-G., 1992a: Digitale Photogrammetrie in der dreidimensionalen Strömungsmesstechnik. Ph.D. Thesis No. 9665, ETH Zurich

Maas, H.-G., 1992b: Complexity analysis for the determination of image correspondences in dense spatial target fields. International Archives of Photogrammetry and Remote Sensing, Vol. XXIX, Part B5, pp. 102-107

Maas, H.-G., 1992c: High-Speed Solid State Camera Systems for Digital Photogrammetry. International Archives of Photogrammetry and Remote Sensing, Vol. XXIX

Maas, H.-G., 1995: New developments in Multimedia Photogrammetry. *Optical 3-D Measurement Techniques III* (Eds.: Gruen, A.; Kahmen, H.), Wichmann Verlag, Karlsruhe

Maas, H.-G., 1996: Automatic DEM generation by multi-image feature based matching. IAPRS Vol. 31, Part 3

Meid, A., 1991: Wissensgestützte digitale Bildkoordinatenmessung in aberrationsbehafteten Messbildern. Ph.D. Thesis University of Bonn

Netzsch, Th., Jähne, B., 1993: Ein schnelles Verfahren zur Lösung des Stereokorrespondenzproblems bei der 3D-Particle Tracking Velocimetry. *Mustererkennung 1993* (Eds. J. Pöppl, H. Handels), Springer-Verlag

Ogniewicz, R., Kübler, O., 1989: Lage- und Skalierungsinvariante Skelette zur Robusten Beschreibung und Erkennung binärer Formen. *Mustererkennung 1989*, Springer Verlag

Okamoto, A., Höhle, J., 1972: Allgemeines analytisches Orientierungsverfahren in der Zwei- und Mehrmedien-Photogrammetrie und seine Erprobung. *Bildmessung und Luftbildwesen* 2/1972, 3/ 1972.

Teh, Ch., Chin, R., 1988: On Image Analysis by the Method of Moments. *IEEE PAMI*, Vol. 10, No. 6, pp. 496 - 513

PARTICLE TRACKING IN THREE-DIMENSIONAL SPACE

TH. DRACOS
Prof. em. ETHZ
Untergasse 14
CH - 8126 Zumikon

1. Introduction

The photogrammetric procedure described in the previous chapter yields coordinates of particle positions in space at successive time steps. These coordinates have inevitably true errors of which the root mean square values $\sigma_{i,rms}$ in x, y and z-directions have been estimated in the photogrammetric part (Maas, this chapter). Hence particle positions are determined with some known *noise* . Noise depends mainly on:

- The optical properties of the media involved, e.g. index of refraction of air, water, oil, glass, etc.
- The imaging arrangement used and the quality of the particle images
- The spatial camera resolution and imaging scale
- The quality of the calibration

It is independent of the temporal resolution of the camera, i.e. the size of the time step Δt and of the flow properties. It must be taken into account in the tracking procedure.

The tracking procedure must now establish correct links between successive positions of the individual particles. To accomplish this a prediction of the expected position $e_i(t+\Delta t)$ of each particle P_i at time t+Δt must be made, knowing its position $\boldsymbol{x}_i(t)$ at time t. The real position $\boldsymbol{x}_i(t+\Delta t)$of the particle will be located with a selected statistical confidence inside a search volume centred on the expected position. The size of this volume depends on the nature of the flow, e.g. non-turbulent or turbulent flow, turbulent flow with high or low turbulence intensity, etc., on the confidence interval adopted and on the magnitude of the noise of the particle coordinates. This volume does not have to be spherical. Let r_0 be its maximum radius and Δ_0 denote the average particle spacing. If ρ_N,the average particle number density, is known

$$\Delta_0 = \rho_N^{-1/3} \tag{1}$$

The ratio

Th. Dracos (ed.), Three-Dimensional Velocity and Vorticity Measuring and Image Analysis Techniques, 209–227.

$$p = \frac{\Delta_0}{r_i} \tag{2}$$

of the average particle spacing to the maximum radius of the search volume is a measure for tracking difficulty.

For p>>1 the probability to find only one particle in the search volume is high and tracking is easy.. It reduces to the search of the nearest neighbour. One has, however, to consider that the particle number density is not uniform in the flow field. A simulation of a random flow field using a Gaussian Kinematic Inertial Model (KSIM) (Malik et al., 1993) shows that the mean value of the spacing Δ_m of particles is log $(\Delta_m / \Delta_0) = -0.6$ with standard deviation in the logarithm of ≈ 0.5 . The 99.7% confidence interval for the variation of Δ is $log(\Delta / \Delta_0) = -0.6 \pm 1.5$ or $\Delta_{max} \approx 2.5\Delta_0, \Delta_{min} \approx \Delta_0 / 8$. In real turbulent flows this span may be even larger. It follows that even for average values of p>>1 this ratio may become locally p<1 and more than one candidate is found in a search volume. In this case more elaborate principles must be applied to determine the correct link. p=1 is in practice the bound of particle path trackability. For p<1 tracking becomes difficult.

Links cannot be established for an increasing number of particles as p decreases below a certain limit. Defining the yield of links by

$$Y = \frac{\rho_{N'}}{\rho_N} \tag{3}$$

where $\rho_{N'}$ is the number density of particles for which links are established, the yield Y decreases as p decreases. There is a limit for p below which $\rho_{N'}$ starts decreasing. An increase of ρ_N beyond that limit and correspondingly the decrease of p below a limit becomes then meaningless.

Finally, for p<<1 tracking becomes impossible.

This is another limit for the particle number density ρ_N which depends on the nature of the flow investigated. A consequence of this limitation of ρ_N is that the spatial resolution decreases when the deviation of the particle position from its expected value increases.

Another tracking difficulty is caused by *drop-outs* of particle coordinates in successive time steps. There are different reasons leading to this loss of information.

- Due to non-uniformities in particle shape and illumination a particle may become too faint and drop out during image pre-processing
- Overlapping of particle images due to viewing geometry
- Lack of success in the establishment of particle image correspondences in the photogrammetric procedure.

These drop-outs are dependent on the image acquisition and the image processing. Appropriate measures can be taken to reduce this kind of drop-out to a minimum.

When the observation volume is open, i.e. is not physically bound by solid walls or a free surface, particles are inevitably advected into and out of the observation volume at every instant. Drop-outs of this kind have a physical reason and cannot be eliminated by any means. They can, however, be reduced when an observation platform is used moving with the average velocity of the particles. The loss of information on particle position data in each time step due to drop-outs will be called *data -loss* .

In the following the particle tracking procedure developed will be presented, based on the works of Papantoniou (1989) (see also Papantoniou & Dracos, 1990b) and Malik et al. (1993).

2. Basic concept

A reliable tracking procedure must have the capability of selecting the correct link between two successive positions of a particle in presence of a multiple choice of links. According to the criteria used to resolve the ambiguities arising, a three- or four-frame procedure is adopted. The current frame is denoted by [f+0]; [f-1] is the previous frame. It is assumed that all possible links between particles in frames [f+0] and [f-1] are established. This does not apply to the first step in the procedure where frame [f-1] does not exist. This case will be treated later. Frame [f+1] is the frame to which links should be established. If necessary, a fourth frame [f+2] is used.

The basic principle of particle tracking is to demarcate search volume in frame [f+1] and possibly [f+2] within which candidates for the continuation of particle trajectories will be found. If ambiguities arise, application of sorting criteria allows to choose the most probable continuation of a trajectory. If necessary, a consistency criteria based on correlation of neighbouring link vectors can be applied to eliminate spurious links. A flow chart of this procedure is presented in Fig.1.

The criteria used to resolve ambiguity in track continuation are based on the following premises:

- Particle paths are smooth. They can be described by regular functions differentiable to any order desired
- Acceleration magnitudes of particles are bound
- Acceleration change magnitudes are bound.

The last two assumptions are heuristic. One can, however, expect that molecular viscosity will damp acceleration and its change at small scales. Some support for the last two premises comes from Direct Numerical Simulation (DNS) data by Yeung & Pope (1989), who computed the acceleration magnitude autocorrelation function and conclude therefrom that the magnitude of the Lagrangian acceleration of particles changes slowly on a time scale of approximately $0.6T_L$, where T_L is the Lagrangian integral time scale. The acceleration is therefore unlikely to change much for time steps $t_k << \Delta t < 0.6T_L$. The influence of molecular viscosity is shown in the frequency spectrum of the acceleration where a sharp drop of the acceleration magnitude to nearly zero occurs at an angular frequency of $1\omega_k$. Here ω_k is the Kolmogorov angular frequency $\omega_k = \pi / t_k$, and t_k is the Kolmogorov time scale.

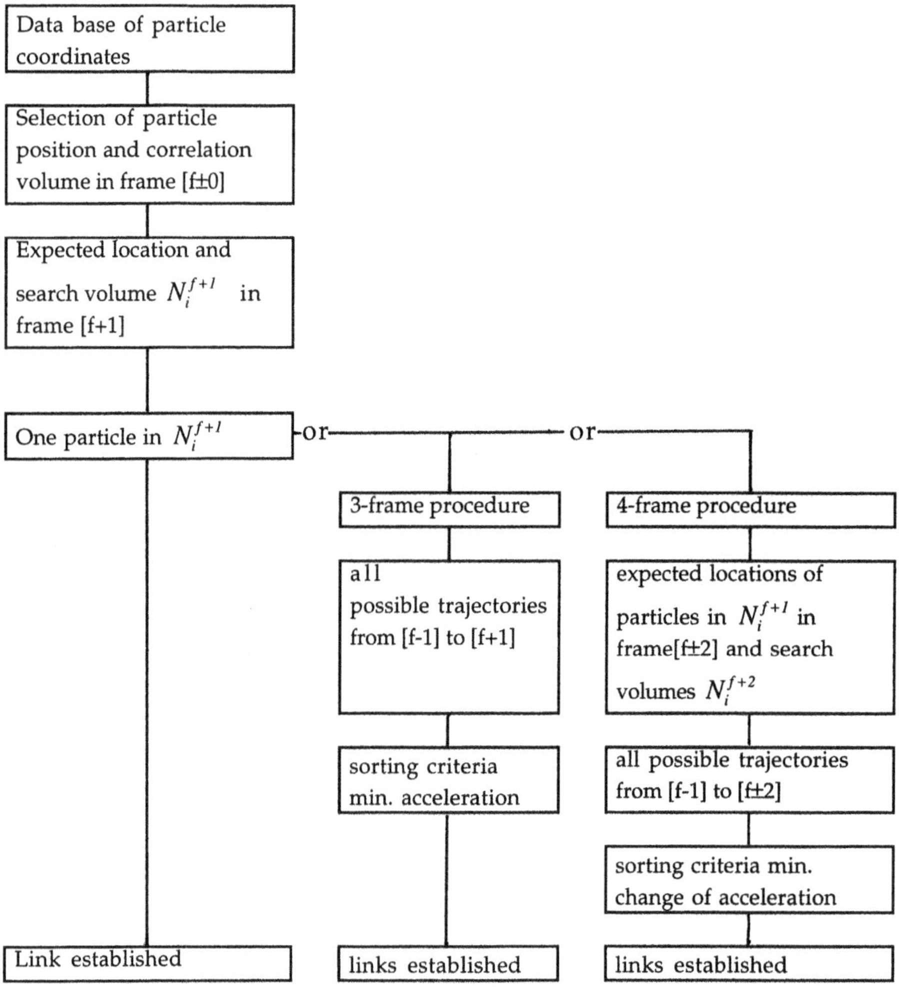

Figure 1 Flow chart of the particle tracking procedure

Yeung & Pope also state that the orientation change of acceleration changes more rapidly on time scales of the order $2t_k$. Using as criterion the magnitude of acceleration or of acceleration changes allows to follow trajectories with large orientation changes as shown in Fig.3 in Chapter Particle Tracking Velocimetry, Basic Concepts.
Criteria used by other authors, e.g. Hassan & Canaan (1991), Guezennec et al. (1994) are of the same nature.

3. Model of the tracking procedure

The tracking scheme operates on the three-dimensional coordinate data base of a sequence of a number j of frames established in the photogrammetric procedure. The time step between frames is assumed to be constant. [f+j] denotes the j-th frame and j=-1,0,1,2 is the number of frames. The location of the i-th particle in frame [f+j] is denoted by $\boldsymbol{x}_i^{f+j}$, where $\boldsymbol{x}(x,y,z)$, $i=1 \ldots I^{f+j}$. The number I^{f+j} of particles changes from frame to frame and the index i in different frames does not necessarily refer to the same particle. A link is the vector connecting the position $\boldsymbol{x}_i^{f+j}$ of a particle in the current frame [f+j] with its position $\boldsymbol{x}_i^{f+j+1}$ in frame [f+j+1]. A succession of links connecting positions of the same particle in consecutive frames form a track.

3.1. CORRELATION NEIGHBOURHOOD IN THE CURRENT FRAME

Let one select a particle with position vector $\boldsymbol{x}_i^{f+0}$ in frame [f+0]. It is expected that the movement of this particle is correlated with the movement of neighbouring particles if the distance between the particles is smaller than a correlation length r_0. A spherical correlation neighbourhood N_i^{f+0} centred on $\boldsymbol{x}_i^{f+0}$ with radius r_0 is therefore delimited. If the correlation lengths are different in x,y and z-direction, the correlation neighbourhood is an ellipsoid. To simplify the computation ellipsoids are always approximated by the circumscribed orthogonal parallelepiped. For every particle $i \in [f+0]$ there are $n \geq 1$ particles in N_i^{f+0}, with equality if particle i is isolated in N_i^{f+0}. Some of these particles, including the one selected, say n', may have links to their positions in frame [f-1] established previously. Excluding the selected particle for some of the remaining particles, say n", links may already have been established to their presumed positions in frame [f+1]. Fig.2 gives a schematic description of the situation. In this example n=4,

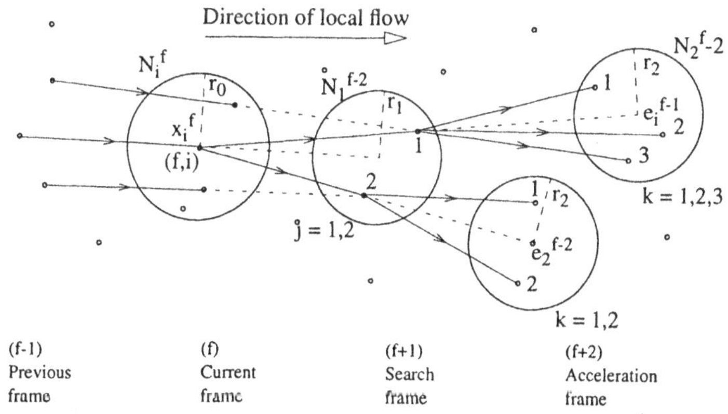

Figure 2 Scheme of particle tracking

n'=3 and n''=2. The determination of the size of the correlation neighbourhood will be given later.

3.2. PREDICTION OF THE LOCATION OF A PARTICLE IN THE NEXT FRAME

As a first step in the tracking procedure the location of a particle in frame [f+1] is roughly estimated from its location in the current frame [f+0]. The prediction of the expected location $\boldsymbol{e}_i^{f+1}$ in frame [f+1] of a particle located at $\boldsymbol{x}_i^{f+0}$ in frame [f+0] is based on the assumption that the particle moves during the time step Δt, separating the two frames, with its local velocity $\boldsymbol{u}_i^{f+0}$ in frame [f+0] without acceleration.

$$\boldsymbol{e}_i^{f+1} = \boldsymbol{x}_i^{f+0} + \boldsymbol{u}_i^{f+0}\Delta t \tag{4}$$

The problem reduces in this case to the estimation of $\boldsymbol{u}_i^{f+0}$. The following cases are distinguished:

(i) The selected particle located at $\boldsymbol{x}_i^{f+0} \in [f+0]$ has a link to its position in the previous frame [f-1]. Its local velocity is

$$\boldsymbol{u}_i^{f+0} = \frac{\boldsymbol{x}_i^{f+0} - \boldsymbol{x}_i^{f-1}}{\Delta t} \tag{5}$$

(ii) The selected particle has no link to the previous frame but a number n' of particles in the correlation volume do. The velocity of the selected particle is estimated as

$$\boldsymbol{u}_i^{f+0} = \frac{1}{n'}\sum_1^{n'} \boldsymbol{u}_j^{f+0} \quad \text{or} \quad \boldsymbol{u}_i^{f+0} = \frac{\sum_1^{n'} \boldsymbol{u}_j^{f+0}|\boldsymbol{d}_j|}{\sum_1^{n'}|\boldsymbol{d}_j|} \tag{6}$$

where $|\boldsymbol{d}_j|$ is the distance between a particle j and the selected particle i in N_i^{f+0}.and $\boldsymbol{u}_j^{f+0}$. are determined using eq.(4) accordingly.

(iii) No links exist to the previous frame. This is always the case when the tracking starts. In this case some estimate of the velocity must be made from a priori knowledge for the kind of flow investigated. In many cases an average local velocity can be determined from known relations for the flow investigated, e.g. for boundary layers, for channel flow, for jets, for shear layers, etc. If the imaging is performed from a platform moving with a constant velocity $\boldsymbol{u}_c$ and the estimated local average velocity of the flow is $\boldsymbol{u}_l$

$$\boldsymbol{u}_i^{f+0} = \boldsymbol{u}_l - \boldsymbol{u}_c \tag{7}$$

3.3 ESTABLISHMENT OF LINKS

The real position $\boldsymbol{x}_i^{f+1}$.of a particle in frame [f+1] deviates from its predicted position $\boldsymbol{e}_i^{f+1}$. The essential part of the tracking procedure is the demarcation of search volumes centred on the predicted locations of particles in the following frames. The estimation of the size of the search volume will be treated later. Once the search volume is defined, the number of particles in this volume is determined. If the size of the search volume is appropriately chosen and no particle is found in it, a data loss has occurred. If only one particle is found, the link is established. If more than one particle is found inside the search volume, links to all of them are possible. To resolve the ambiguity two options exist:

(i) *Use of the minimum acceleration criterion, three-frame procedure* (Papantoniou, 1989).
To determine the most probable link, information from the time frames [f-1], [f+0] and [f+1] is used. As shown in Fig.2 for a number of n'' particles in the correlation volume N_i^{f+0} in frame [f+0] connections may already have been established to particles in the search volume N_i^{f+1} in frame [f+1]. All possible trajectories extending over the three frames [f-1], [f+0], [f+1] containing the selected particle and the n'' particles having links to particles located in the search volume are constructed (Fig.2).

Assuming that the time steps Δt are small and that during the movement of a particle from $\boldsymbol{x}_j^{f-1} \in [f-1]$ to $\boldsymbol{x}_i^{f+1} \in [f+1]$ velocity changes are small, the acceleration along each trajectory is computed by

$$\boldsymbol{a}_j^{f+0} = \left(\boldsymbol{x}_j^{f-1} - 2\boldsymbol{x}_j^{f+0} + \boldsymbol{x}_j^{f+1}\right) / 2\Delta t^2 \tag{8}$$

$\boldsymbol{x}_j^{f-1}$, $\boldsymbol{x}_j^{f+0}$, and $\boldsymbol{x}_j^{f+1}$ are the locations of particles linked to a trajectory. Now the minimum magnitude of acceleration criterion $|\boldsymbol{a}|$ minimum is employed to select the trajectory containing the most probable link between the position $\boldsymbol{x}_j^{f+0}$ of the selected particle and one of the particle positions in the search volume N_i^{f+1}. If the trajectory with minimum acceleration magnitude does not contain this link, no link is made between the selected particle in frame [f+0] and a particle in frame [f+1]. On the other hand, if the particle to which a link is established had a previously determined link to a particle in frame [f+0], this link is cut.

For a number of particles $\boldsymbol{x}_i^{f+0} \in [f+0]$ no links exist to the previous frame [f-1]. In this case a trajectory passing through these points cannot be constructed and the acceleration criterion cannot be applied.

At this point one is left with a number of links still possible, which must be determined on the basis of information contained in the frames [f+0] and [f+1] only. A group tracking technique is optionally utilised in this case. All unmatched particles located in a correlation neighbourhood form a group and all permissible possible links to [f+1] are established for each particle of a group. All possible link combinations are formed. Each combination represents a group state. The average velocity $\boldsymbol{u}_g^{f+0}$ in the correlation volume comprising all points of a group in [f+0] is determined.

Two group tracking procedures are developed and can be used optionally. They differ at the functional minimised.

(a) Minimum of the sum of the kinetic energy of states in a group normalised by the correlation measure.

$$E_g^{f+0} = \sum_i \frac{2\left(\boldsymbol{u}_i^{f+0}\right)^2}{\left(\boldsymbol{u}_i^{f+0} - \boldsymbol{u}_g^{f+0}\right)\boldsymbol{u}_g^{f+0}} \tag{9}$$

The combination of states which minimises this sum contains the correct links.

(b) Minimum of acceleration state of the group.

$$A_g^{f+0} = \sum_i \left(\boldsymbol{u}_i^{f+0} - \boldsymbol{u}_g^{f+0}\right)^2 \tag{10}$$

The combination of states which minimises this sum contains the correct links.

In both cases a consistency test must be performed. The test will be described later.

(ii) *Use of minimum change of acceleration criterion, four-frame procedure* (Malik et al., 1993).

To apply this criterion a fourth frame [f+2] must be added.

All connections between the particle $\boldsymbol{x}_j^{f+0}$ selected in frame [f+0] and the positions of the particles $\boldsymbol{x}_j^{f+1} \in N_i^{f+1}$ in frame [f+1] are possible continuations of the trajectory through $\boldsymbol{x}_j^{f+0}$. In addition, links may already exist between n" particles in the correlation volume N_i^{f+0} and some of the particles in search volume N_i^{f+1}. Expected positions $\boldsymbol{e}_i^{f+2}$ of all particles found in N_i^{f+1} are now sought in frame [f+2]. These positions are determined by accordingly applying eq.(4) on each link anticipated between particles in N_i^{f+0} and N_i^{f+1}. Around each expected position $\boldsymbol{e}_j^{f+2}$ a search volume centred on this position is delimited and the coordinates of the particles found in each of these search volumes determined using the existing data base. All links between the particle $\boldsymbol{x}_j^{f+1} \in N_i^{f+1}$ whose expected position in frame [f+2] is $\boldsymbol{e}_j^{f+2}$ and the particles found inside the search volume centred on $\boldsymbol{e}_j^{f+2}$ are established. This way all possible trajectories from frame [f-1] to frame [f+2] containing the selected particle with position $\boldsymbol{e}_j^{f+2}$ and the n" particles having a link to particles in N_i^{f+1} are established. The principle of minimum change of acceleration along a particle trajectory is subsequently applied to select the most probable trajectory. The estimate of the acceleration (eq. (7)) is also valid for a wide range of normalised time steps $\Delta t / t_k$ when the acceleration is uniformly changing. The change in acceleration

$$\Delta \boldsymbol{a} = \boldsymbol{a}_i^{f+1} - \boldsymbol{a}_i^f \tag{11}$$

is given by

$$\Delta \boldsymbol{a}_i = \left(\boldsymbol{x}_i^{f+2} - 3\boldsymbol{x}_i^{f+1} + 3\boldsymbol{x}_i^{f+0} - \boldsymbol{x}_i^{f-1}\right) / \Delta t^2 \tag{12}$$

as it is easily shown by using eq.(8) to compute the relation given by eq.(11). The scalar quantity minimised is $|\Delta \boldsymbol{a}|$.

Links cannot be determined when the particle $\boldsymbol{x}_j^{f+0}$ has no link to frame [f-1]. However, when the four-frame procedure is used and no link to frame exists for the selected particle the minimum acceleration criterion can be used as an option for tracks in the frames [f+0] to [f+2]. Else the group tracking procedure can be applied or the link is not made, which is the conservative option often adopted.

In a second pass missing links are established if in the search volumes N_i^{f+1} in frame [f+1]. Corresponding to these particles only one particle is left without link to a particle in [f+0].

After all possible links are established a consistency test can be performed in a third pass to eliminate spurious links. It is based on a comparison of the magnitude and directions of the link vector from $\boldsymbol{x}_j^{f+0}$ to $\boldsymbol{x}_j^{f+1}$ with the magnitude and direction of all other link vectors comprised in the correlation volume N_i^{f+0} centred at $\boldsymbol{x}_j^{f+0}$. This kind of filtering is always necessary when group tracking is performed. It contains, however, some danger and is therefore optional and applied only if the magnitude and the direction deviation of the link vector exceeds some thresholds.

3.4. SIZE OF THE CORRELATION AND SEARCH VOLUMES

The size of the correlation volume in frame [f+0] and of the search volumes in frames [f+1] and [f+2] depends on the kind of flow investigated.

3.4.1. *Correlation volume in* [f+0]

If the flow is not turbulent, the velocity does not change randomly and a high correlation between velocities may exist over large distances. In turbulent flows, by contrast, where random changes of the velocity occur, finite correlation lengths are expected.

A measure for the correlation length defining a spherical volume in a turbulent flow field, in which the velocity is well correlated, is Taylor's microscale λ. Hence one can define the radius of the correlation volume in frame to be a fraction φ of λ.

$$r_0 = \phi\lambda, \qquad \phi < 1 \tag{13}$$

If the observation volume V is much larger than λ^3, $\phi = 1$ is the correct choice. For small observation volumes, i.e. for $V^{1/3}$ of $O(\lambda)$ all the velocities in the volume are well correlated and a neighbourhood with $r_0 \approx \lambda$ becomes meaningless. In fact, in this case one is seeking for a neighbourhood containing a limited number of particles n with well correlated velocities. Since the tracking scheme is only moderately sensitive to the choice of r_0 , a way to choose r_0 is to relate it to the mean particle spacing $\Delta_0 = \rho_N^{-1/3}$. If n is the number of particles wanted in N_i^{f+0} on average:

$$n = \left(r_0 / \Delta_0\right)^3 = r_0^3 \rho_N \tag{14}$$

and

$$r_0 = \left(n / \rho_N\right)^{-3}, \quad r_0 < \lambda \tag{15}$$

This is also the correct choice for r_0 when the flow is not turbulent.

As a rule the correlation volume N_i^{f+0} is assumed to be spherical.

3.4.2. *Size of the Search Volumes in* [f+1] *and.*[f+2].

The search neighbourhood is centred at the expected positions $\boldsymbol{e}_j^{f+1}$ or $\boldsymbol{e}_j^{f+2}$. Its radii r_j^{f+1}, r_j^{f+2} are related to the maximum physical distance the particle can deviate form its expected position in one time step Δt in x, y, z-direction. In a turbulent flow this is related to the decorrelation of the velocity of a particle in the same time interval and can be obtained from the Lagrangian time structure function.

$$\left\langle \delta u_i^2(t) \right\rangle = \left\langle \left[u_i(s+t) - u_i(s)\right]^2 \right\rangle = \pi C_L \varepsilon t, \quad t_k << t << T_E \tag{16}$$

or

$$\delta u_i(t) \approx \left(\pi C_L \varepsilon t\right)^{1/2} \tag{17}$$

and

$$\delta x_i = \delta u_i(t) \Delta t, \qquad i=1,2,3 \tag{18}$$

where $C_L = 2.4$ is the Lagrangian frequency spectrum constant (Novikov, 1953, Fung et al., 1992, and references therein), ε is the rate of energy dissipation per unit mass, t_k the Kolmogorov time scale and $T_E \approx L/u'$ is the turnover time scale of the large energy containing eddies, *L* is the integral length scale based upon the energy weighted scales in the flow (Batchelor, 1955). For a nearly Gaussian flow field the 99.7% confidence limit for the maximum deviation in a given direction is $3\delta x_i$.Thus

$$r_i^{f+1} = 3\delta x_i = 3\left(\pi C_L\right)^{1/2} \left(\varepsilon \Delta t^3\right)^{1/2} \tag{19}$$

An estimate for ε is:

$$\varepsilon = A\frac{u_i'^3}{L} \tag{20}$$

where u_i' is the rms value of the turbulent velocity fluctuations and A a dimensionless constant which depends on the kind of flow investigated.

The maximum distance $\delta x_{i,max}$ a particle can deviate from its present position is with 99.7% confidence $\delta x_{i,max} = 3u'_i\,\Delta t$. With this estimate eq.(19) becomes

$$r_i^{f+j} = \left(\frac{A\pi C_L}{3}\right)^{1/2}\left(\frac{\delta x_{i,max}}{L}\right)^{1/2}\delta x_{i,max}, \quad j=1 \text{ or } 2 \tag{21}$$

These radii must be increased by $3\sigma_{i,rms}$ where $\sigma_{i,rms}$ are the noise errors of particle coordinates in x,y and z-directions. Remarking that the first factor on the right hand side is a constant C depending on the kind of flow investigated, which must be adjusted but is usually of order 1.5, one can write.

$$r_i^{f+j} = C\left(\frac{3u'_i\,\Delta t}{L_i}\right)^{1/2} 3u'_i\,\Delta t + 3\sigma_{i,rms}, \quad j=1 \text{ or } 2 \tag{22}$$

The rms values u' of the velocity fluctuations can in many cases be estimated, e.g. in channel flow and boundary layers with zero pressure gradient $u'_x \approx 2u_\tau$ and $u'_y = u'_z = u_\tau$ where u_τ is the friction velocity. The integral length scale L is estimated from the geometrical features of the flow investigated. If C and L_i cannot be estimated, then

$$r_i^{f+j} = 3u'_i\,\Delta t + 3\sigma_{i,rms}, \quad j=1 \text{ or } 2 \tag{23}$$

Eqs.(22) and (23) are not always the best choice for $r_i^{f+j}, j=1 \text{ or } 2$. Recall that Δ_0 is the mean particle spacing and that some particles will be closer than Δ_0, $\Delta_n < \Delta_0$. Defining as permissible neighbourhood spacing Δ_n of particles, the mean spacing Δ_m plus one standard deviation (in the logarithm) one gets

$$r_i^{f+j} \approx e^{-1.1}\Delta_0 \approx \frac{1}{3}\Delta_0, \quad j= 1 \text{ or } 2 \tag{24}$$

r_i^{f+j},is taken as the smaller of eqs.(22) or (23) and (24). The search volumes are normally not spherical.

If the flow is not turbulent, the size of the search volumes is determined using eq.(24).

4. Validation of the tracking procedure

The tracking procedure is validated against computed trajectories in simulated flows. Papantoniou & Dracos (1990a) generated numerically a flow field with some degree of irregularity by superposing the flow field of 15 vortex line filaments of different strength and orientation randomly distributed inside a given volume. The resulting flow is not random and locally shows a high degree of organisation. A variable number of particles was injected into the simulated flow field and their tracks were computed for a sufficiently long time interval. These tracks are considered to be error free. Particle positions at discrete time steps Δt along these trajectories were determined and the tracking procedure was applied to the data base formed by particle coordinates in successive time steps. The results were excellent with almost 100% yield even for small values of the tracking parameter p. This shows that if some degree of regularity exists in the flow field, like in the near field of wakes, jets, plane sheer layers or steps, tracking is easy and the density of information is limited by the photogrammetric procedure rather than by the tracking parameter p. For this reason an unsteady random velocity flow field was generated by a Kinematic Simulation Inertial Model (KSIM) (Fung et al., 1992, Malik, 1991). In KSIM the flow field is specified as a sum of independent random Fourier modes within a limited range of wave numbers. Of course the simulated flow field is not in a strict sense turbulent, but it its sufficiently random to allow for a stringent validation of the tracking procedure developed.

In the present simulation the range of wave numbers used was 0.5<k<1200 and the number of modes was $N_k = 1000$ with one frequency mode per wave mode. The energy spectrum is taken as the Kolmogorov spectrum $E(k) = \alpha_k \varepsilon^{2/3} k^{-5/3}$ (Kolmogorov, 1941a,b). The flow field is normalised so that $\langle u_1^2 \rangle = \langle u_2^2 \rangle = \langle u_3^2 \rangle = 1$ and the integral

TABLE 1. KSIM Flow Parameters

kmin	kmax	u'	η	ε	t_k	$L_{x.y.z}$	N_k	Δ_0
05	1200	1.0	0.005	0.27	0.04	3.0	1000	0.3

TABLE 2. Tracking Parameters

p	Δt	$\Delta t / t_k$	r_0	r_0 / L	$r_{1,2}$	$r_{1,2} / \eta$
7.5	0.04	1.00	0.3	0.1	0.04	8
5.0	0.06	1.50	0.3	0.1	0.06	12
3.75	0.08	2.00	0.3	0.1	0.09	18
3.00	0.10	2.50	0.3	0.1	0.10	20
2.50	0.12	8.00	0.3	0.1	0.10	20
2.30	0.13	3.25	0.3	0.1	0.10	20

length scale is $L \approx 3$ The trajectories of 1000 particles released at random points within a volume $L^3 = 27$ were obtained over several hundred integration steps. From these 16 time frames for various time steps were extracted. The flow parameters of the KSIM are presented in Table 1, where η is the Kolmogorov length scale, and the tracking parameters are summarised in Table 2. For more details see Malik et al. (1993)

4.1. RESULTS

4.1.1. *Tracking without noise and data -loss*

Two criteria are formed to check the performance of the tracking procedure: (i) the *yield* which is the ratio of the number of a determined quantity, e.g. the number of established links or trajectories to the maximum possible number, expressed as a percentage (ii) The *quality* which is the ratio of the number of true quantity to the number of determined quantity also expressed as a percentage.

Note that in the simulated case without artificially introduced noise or data-loss the maximum number of possible quantities is equal to the number of the particles released in the flow volume. During the validation no consistency control of the established links

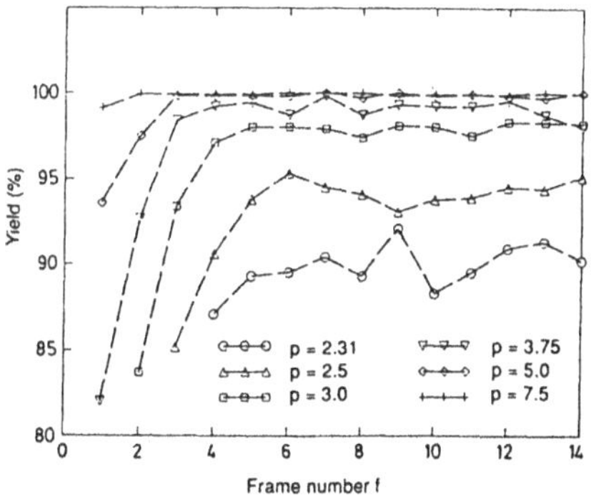

(a)

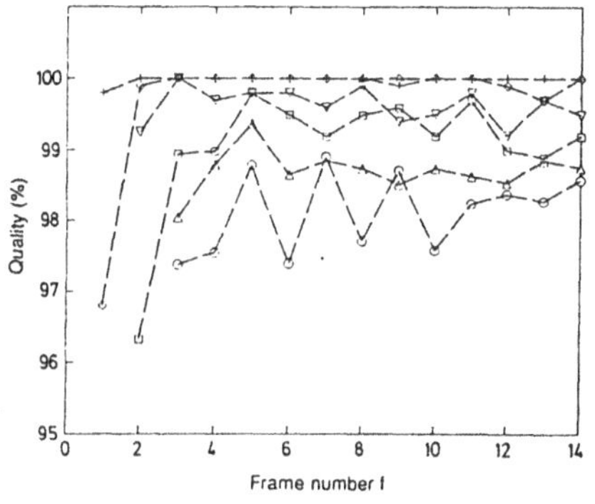

(b)

Figure 3 (a) Frame-by-frame percentage yield of links, (b) frame-by-frame percentage quality of links, for different values of the tracking parameter p as indicated in (a)

was performed. The yield and quality of links are shown frame-by-frame for different values of the tracking parameter p in Fig.3(a), (b).Both yield and quality decrease with decreasing p but the decrease in quality is less pronounced. The figures also show that yield and quality stabilise after a small number of steps at a high level.The mean percentage yield and quality of links and of trajectories after the procedure out-put stabilised is shown in Fig.4(a),(b)

Fig.4(a) shows in a more pronounced way the high quality achieved. The result presented in Fig.4(b) is one of the more important results of the validation. It shows that the yield of complete trajectories, i.e. trajectories containing the maximum possible number of links, is drastically reduced with decreasing value of the tracking parameter. This is because the increasing number of missing links, although very small, as shown in Fig.4(a), has a cumulative effect over all the frames and thus on the yield of long trajectories. However, the quality of the trajectories Fig.4(b) is consistently high. Any complete trajectory can be considered to be error free with more than 95% confidence.

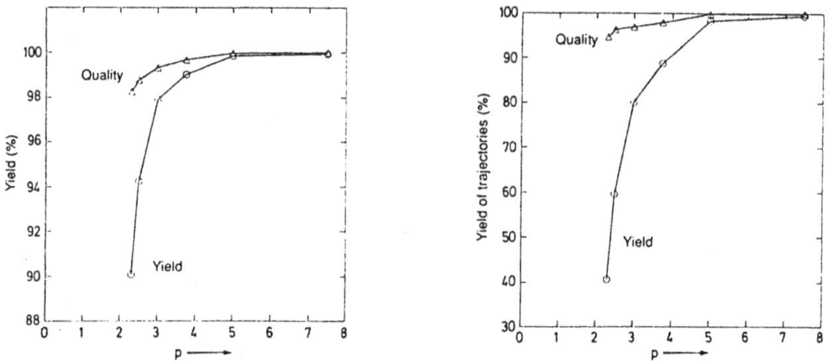

(a) (b)

Figure 4 Mean percentage yield and quality of (a) links per frame, (b) complete trajectories as a function of the tracking parameter p

4.1.2. *Influence of noise and data-loss*

Data-loss does not reduce the number of particles in consecutive frames. There is simply a number of particles in each frame which neither exist in the previous frame nor in the next frame. This condition was simulated in the KSIM data by randomly scrambling the position of n% of the particles in each frame. The percentage of data-loss was varied from n=0% to n=3%. Fig.5(a),(b) shows the influence of noise on mean percentage yield and quality for different values of the tracking parameter. The influence of data-loss is weak. Yield is reduced by a maximum of 5% and the quality by only 1%. The percentage yield of complete trajectories is on the contrary drastically reduced (Fig.6).

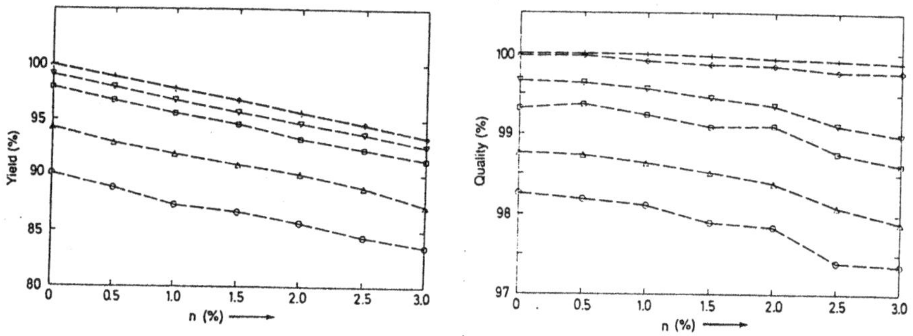

(a) (b)

Figure 5 (a)Mean percentage yield, and (b) mean percentage quality of links against data-loss (%) for different values of p. Symbols as in Fig.3(a)

Data-loss has the same effect as the reduction of yield of links with decreasing values of the tracking parameter. This is the reason that noise has the stronger effect at the largest The yield of complete trajectories decreases from 100% for n=0% to nearly 60% for n=3%. For Lagrangian statistics and for the study of particle dispersion in turbulent flows long trajectories are needed. The KSIM simulation yields results for a limited number l of links in a trajectory. However, these results can be extrapolated to any

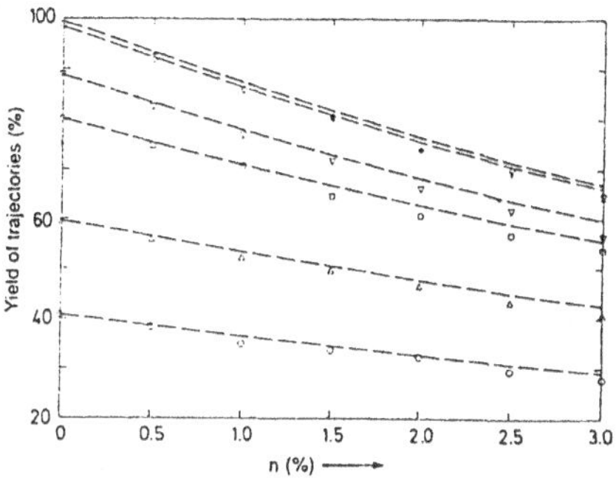

Figure 6 Percentage yield of complete trajectories against data-loss (%) for different values of p. Symbols: KSIM results as in Fig.3(a) , dashed lines, eq. (25)

number N of links. If $C_0^l(p)$ is the fractional yield of trajectories of length l links, the fractional yield of trajectories of length N is $C_0^N(p)$ Since the tracking error and data-loss error are independent processes, the probability of a link between two consecutive frames is given by the product $(1-n')^2 C_0(p)$, where n'=0.01n, is the fractional level of noise. The fractional yield of complete trajectories with N links is

$$T(N; p, n') = (1-n')^{N+1} C_0^N(p) \tag{25}$$

This result is plotted in Fig.6 and is in good agreement with KSIM results.

Tracking is also influenced by noise. To study this influence a random error with $\varepsilon_{rms} = \pm 0.05u'\Delta t$ and $\varepsilon_{rms} = \pm 0.1u'\Delta t$ was added to the x and y coordinates and double this value to the z coordinate of a particle in each frame. The search radii were accordingly adjusted. The results for p=3 and p=5 are compared with the case without noise and are presented in Table 3.
Obviously, noise does not significantly influence the yield of links per frame. The percentage of mis-links is however drastically increased. It nearly doubles in case a and more than triples in case Δa for a value of the tracking parameter of p=3 and more than triples in case a and increases by a factor of 13 in case Δa for p=5. However, for p=5 it is still very small and does not exceed 1% in the worst case.

TABLE 3. Influence of noise, a tracking by minimum acceleration, Δa tracking by minimum change in acceleration

p	data-loss %	noise % of u'Δt	yield of links per frame, %		mis-links per frame, %	
			a	Δa	a	Δa
3	0	0	97.0	97.3	1.6	0.6
3	0	5	95.6	95.5	1.9	1.0
3	1	5	93.5	93.2	2.1	1.3
3	0	10	86.8	86.7	3.0	2.1
5	0	0	99.7	99.9	0.2	0.03
5	0	5	99.6	99.9	0.3	0.1
5	1	5	97.5	97.5	0.5	0.3
5	0	10	98.2	98.1	0.7	0.4

This table also shows the performance of the two tracking criteria, minimum acceleration a, and minimum change in acceleration Δa. The influence on percentage yield of links in a frame is marginal. The percentage of mis-links is, however, considerably reduced when the second criterion is used. This goes to the expense of computational time

5. Determination of Velocity Vectors

The tracking procedure establishes links between the current position of a particle at time t=0 and its position in time t=t+Δt, where Δt is the imaging time step. A succession of such links form a trajectory. Assume that a trajectory extends from time t=t-2Δt to time t=t+2Δt and that the corresponding positions x(t),y(t),z(t) of the particle are known. Estimates are now desired of the velocity components at the current position of the particle, i.e. at time t=0. Without loss of generality only one component is considered in the following.

Let the x-component of the position vectors of a particle at different times be

$$x = x_1 \ \ at\ t\text{=-}2\Delta t, \quad x = x_2 \ \ at\ t\text{=-}\Delta t, \quad x = x_3 \ \ at\ t\text{=}0$$

$$x = x_4 \ \ at\ t\text{=}\Delta t, \quad x = x_5 \ \ at\ t\text{=}2\Delta t \tag{26}$$

Following schemes are used to estimate the velocity components at $x = x_3$, $t\text{=}0$:
A two-point first order scheme (s):

$$v_s = \frac{x_4 - x_3}{\Delta t} + O(\Delta t) \tag{27}$$

an unbiased three-point second order scheme (q), which is also used for the estimation of the acceleration:

$$v_q = \frac{x_4 - x_2}{2\Delta t} + O(\Delta t^2) \tag{28}$$

$$a_q = \frac{x_2 - 2x_3 + x_4}{\Delta t^2} + O(\Delta t^2) \tag{29}$$

an unbiased five-point fourth order scheme (f)

$$v_f = \frac{x_1 - 8x_2 + 8x_4 - x_5}{12\Delta t} + O(\Delta t^4) \tag{30}$$

These schemes are tested with KSIM generated data. This is done in the following way.

As previously explained, the release of particles in the flow field generated by KSIM allows to generate an artificial data base of particle positions on which particle tracking is applied. A sufficient number of tracks containing five consecutive particle positions are determined and the velocity **v**, here its x-component v is determined at point $\mathbf{x}_3$ of each trajectory by applying one of the eqs.(27), (28) or (30). The true velocity component in x-direction u at the same location is determined from KSIM. The rms error normalised by the rms of the x-component of the velocity fluctuations u'_{rms} is then computed for each of the estimation schemes (s, q, f) applied.

$$E_{rms} = \frac{\left\langle (v-u)^2 \right\rangle^{1/2}}{u'_{rms}} \tag{31}$$

and depicted in Fig.7(a). This figure shows that the first order two-point scheme is significantly less accurate than the two other schemes. The three-point scheme gives slightly better results than the five-point scheme, but the errors are still large. Note, however, that these errors refer to the rms values of the fluctuating part of the velocity. If u'_{rms} is p% of the local mean velocity, the error $E_{m,rms}$ in reference to the local mean velocity

$$E_{m,rms} = E_{rms} \frac{p}{100} \tag{32}$$

is much smaller. For example, for p=10% an error E_q of 25% on u' corresponds to an error of 2.5% in the mean local velocity.

One has to consider that the tracking procedure has a finite time resolution determined by the imaging time step Δt. The velocity estimates made cannot correspond to the value of the local instantaneous velocity at $\mathbf{x}_3$. They rather represent a time averaged value over a time interval proportional to Δt. If this interval is 2Δt, the true average value of the

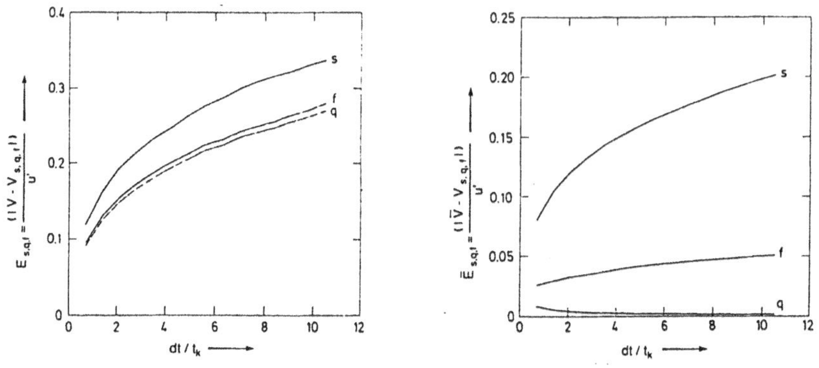

Figure 7 rms error of the estimated velocity normalised by the rms value of the velocity fluctuations u', s two-point scheme, q three-point scheme, f five-point scheme. (a) error in reference to the instantaneous velocity at $\mathbf{x}_3$, (b) error in reference to the velocity at $\mathbf{x}_3$ averaged over 2Δt, eq.(33)

velocity at $\mathbf{x}_3$ is

$$\bar{u} = \frac{1}{2\Delta t}\int_{-\Delta t}^{+\Delta t} u(t)dt \tag{33}$$

This quantity is computed from KSIM data and the error

$$\bar{E}_{rms} = \frac{\left\langle (v-\bar{u})^2 \right\rangle^{1/2}}{u'_{rms}} \tag{34}$$

is shown in Fig.7(b). The three-point scheme gives a very good estimate of the mean velocity defined by (33) and should be used if possible. The five-point scheme gives also good results whereas the results of the two-point scheme have improved but are still not satisfactory.

Particle tracking procedures give velocities at randomly distributed locations inside an observation volume. For further analysis it is often desirable to have information about the velocity in a regular grid. An interpolation scheme was developed for this purpose and checked with KSIM data. It allows to interpolate at any desired point in space the velocity from data scattered in three-dimensional space (Malik & Dracos, 1995).

References

Batchelor, G.K. 1960. *The theory of homogeneous turbulence*, Cambr. Univ. Press

Fung, J.C.H., Hunt, J.C.R., Malik, N.A., Perkins, R.J. 1992. Kinematic simulation of homogeneous turbulence by unsteady random Fourier modes, *J. Fluid Mech,* **236**, 281 - 317

Guezennec, Y.G., Brodkey, R.S., Trigui, N., Kent, J.C. 1994. Algorithms for fully automated three-dimensional particle tracking velocimetry, *Exp. Fluids,* **17**, 209 - 219

Hassan, Y.A., Canaan, R.E. 1991. Full field bubbly flow velocity measurements using a multiframe tracking technique, *Exp. Fluids,* **12**, 49 - 60

Kolmogorov, A.N. 1941a. The local structure of turbulence in incompressible viscous fluid for very large Reynolds numbers, *Rep. Roy. Soc., London, 1991*, **A434**, 9 - 13

Kolmogorov, A.N. 1941b. Dissipation of energy in the locally isotropic turbulence, *Rep. Roy. Soc., London 1991,* **A434**,, 15 - 17

Malik, N.A. 1991. Studies of turbulent dispersion using kinematic simulation, *Ph.D. Thesis, Univ. of Cambridge*

Malik, N.A., Dracos, Th. 1995. Interpolation schemes for three-dimensional velocity fields from scattered data using Taylor expansions, *J. Comp. Physics,* **119**, 231 - 243

Malik, N.A., Dracos, Th., Papantoniou, D.A. 1993. Particle tracking velocimetry in three-dimensional flows, Part B: Particle tracking, *Exp. Fluids,* **15**, 279 - 294

Novikov, E.E. 1953. Random force method in turbulence theory, *Sov. Physics,JETP,* **17**, 1449 - 1453

Papantoniou, D.A. 1989. 3-D Particle tracking code, *Intern. report,* Institute of Hydromechanics and Water Resources Management (IHW), Swiss Federal Institute of Technology (ETHZ)

Papantoniou, D.A., Dracos, Th. 1990a. Analysing 3-D turbulent motions in open channel flow by use of stereoscopy and particle tracking, *Advance in Turbulence, 2*, Springer Berlin, 278 - 285

Papantoniou, D.A., Dracos, Th. 1990b. Lagrangian statistics in open channel flow by 3-D particle tracking velocimetry, *Eng. Turb. Model Expt.,* ed. Rodi & Garnic, Elsevier, 315 - 324

Yeung, P.K., Pope, S.B. 1989. Lagrangian statistics from direct numerical simulations of isotropic turbulence, *J. Fluid Mech,* **207**, 531 - 586

ESTABLISHMENT OF A VIDEOGRAMMETRIC PTV SYSTEM

M. VIRANT, TH. DRACOS
Institute of Hydromechanics and Water Resources Management,
Swiss Federal Institute of Technology, ETH Hönggerberg,
CH-8093 Zürich, Switzerland

1. Introduction

A PTV system must be implemented according to the requirements of the hydromechanical problem studied and the image- and data processing envisaged. This will be exemplified in the following.

A two-camera PTV system and a rather low particle number density may be sufficient when one seeks the velocity field in a stationary non-turbulent flow. This is even more the case when the flow is nearly two-dimensional.

Turbulent flows, even if in the average steady, are not stationary and are intrinsically three-dimensional. If the turbulence is homogeneous and isotropic and a nearly 90 degree arrangement of the cameras is possible, a two-camera system may still be sufficient. This is particularly true when the particle number density can be kept low.

At high particle number densities and non-homogeneous, non-isotropic turbulent flows as well as when the optical access to the experimental flow field does not allow a 90 degree arrangement of the cameras, a three-camera system is advantageous. Such a system allows to increase the number of velocity vectors determined at each time step but also to improve the accuracy with which the particle coordinates are determined in space.

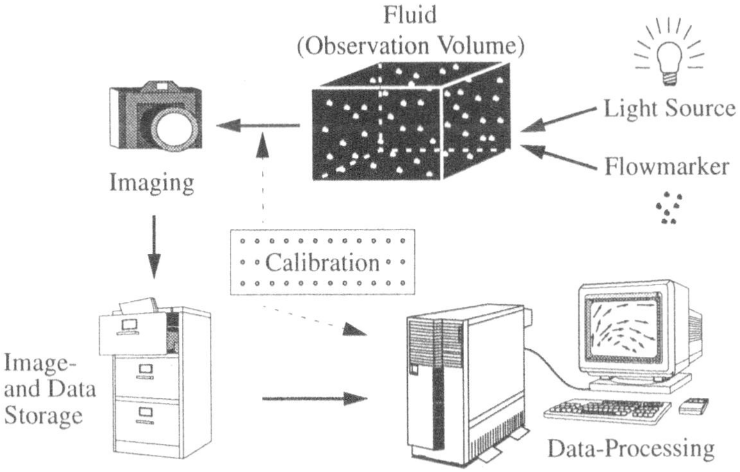

Figure 1. Basic elements of a PTV system.

Th. Dracos (ed.), Three-Dimensional Velocity and Vorticity Measuring and Image Analysis Techniques, 229–254.

When spatial derivatives have to be determined, both the particle number density and the accuracy of velocity measurements must be as high as possible. A four-camera system may in this case become necessary.

A four-camera system is almost indispensable when long particle trajectories are to be determined, as data loss must be reduced to a minimum.

The particle tracking procedure presented in the previous chapters is applicable to two-, three- or four-camera systems.

The basic elements of a PTV system are shown in Fig. 1. Detailed information and newest developments for the single elements needed are found in the product information and the operation manuals of the manufacturers of such equipment. There are, however, some basic considerations which always have to be taken into account. They will be discussed in some detail in the following, focusing on the most important aspects one must consider when planning a PTV system. It is very important to realize that a carefully established system and carefully performed measurements are the basis for a successful experimental investigation. It is usually impossible to recover from flaws made in data acquisition.

2. The observation volume

The observation volume contains the part of the flow field to be investigated. The illuminated space in 3D-PTV is a thick light slab rather than a thin light sheet. It is determined by the thickness of the illuminated spaced, by the field of view and by the depth of focus of the cameras (Fig. 2).

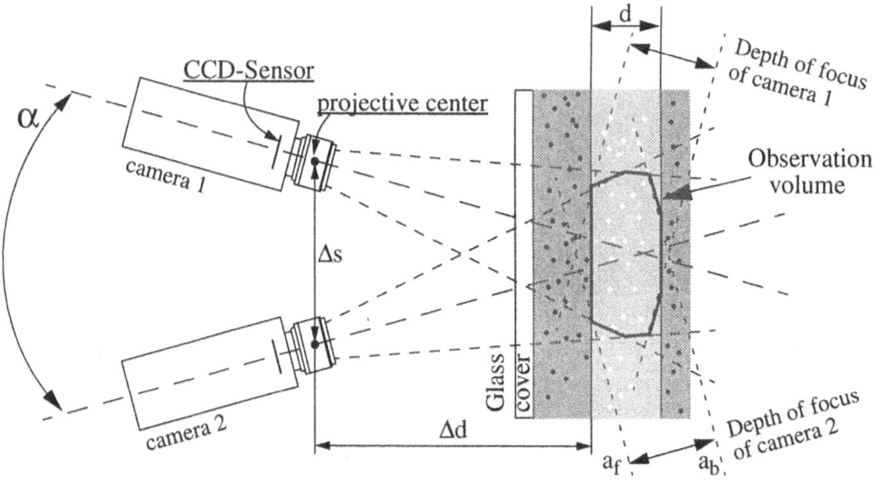

Figure 2. Determination of the observation volume.

After defining the size of the observation volume needed, the appropriate illumination must be selected, the optimal geometrical arrangement of the cameras chosen, and the lenses calculated. This calculation should be based on the premise that the depth of focus of the whole imaging system should match the thickness of the illuminated space.

Denoting by: a the distance of the object from the projective center, f the focal length of the lens system, y the object size (the projection of the diagonal of the observation volume), y' the image size (diagonal of the sensor), β' the magnification (y'/y), $\varnothing_{EP}$ the diameter of the iris, k the f-step, w the width of view, and by $u_{\varnothing}'$ the diameter of stigma at the front and rear plane when the camera is focused on the mid plane of the observation volume. following relations from geometrical optics can be used.

$$f = \frac{a}{1 + \frac{1}{\beta'}} = \frac{a}{1 + \frac{y}{y'}} \tag{1}$$

$$w = \operatorname{atan}\frac{y'}{f} \tag{2}$$

$$y = \left(\frac{a}{f} - 1\right) y' \tag{3}$$

$$y' = \frac{1}{\frac{a}{f} - 1}\, y \tag{4}$$

$$a = \left(1 + \frac{1}{\beta'}\right) f = \left(1 + \frac{y}{y'}\right) f \tag{5}$$

$$k = \frac{f}{\varnothing_{EP}} \tag{6}$$

Object distance to the front of the volume imaged

$$a_f = \frac{a\, f^2}{f^2 - u_{\varnothing}'\, k\, (a + f)} \tag{7}$$

Object distance to the rear of the volume imaged

$$a_r = \frac{a\, f^2}{f^2 + u_{\varnothing}'\, k\, (a + f)} \tag{8}$$

Eqs. (7) and (8) indicate that a small f-step k increases the depth of focus (a_r - a_f) of the imaging system. This can only be achieved by a strong illumination and good tuning of the illumination/imaging system.

3. Sensor properties and particle imaging

Since, due to their reasonable prices, off-the-shelf black and white video cameras are mainly used in PTV, the following considerations are based on properties of CCD sensors employed in such cameras. The basic principles of imaging with CCD cameras is, however, also valid when special cameras, e.g. high-resolution or progressive scan cameras, are used

A CCD sensor consists of a two-dimensional array or frame of light sensitive picture elements (pixels). The converted electric charge of each pixel is proportional to the photon flux reaching the pixel times the time the pixel is exposed to this flux. This exposure time is equal to the integration time of the sensor. At the end of each exposure, the charge of each pixel is read out. The integration time plus the read-out time determine the framing time step Δt of the sensor. This time is fixed by the video norm used. Fig. 3 shows the data transfer principles mainly used.

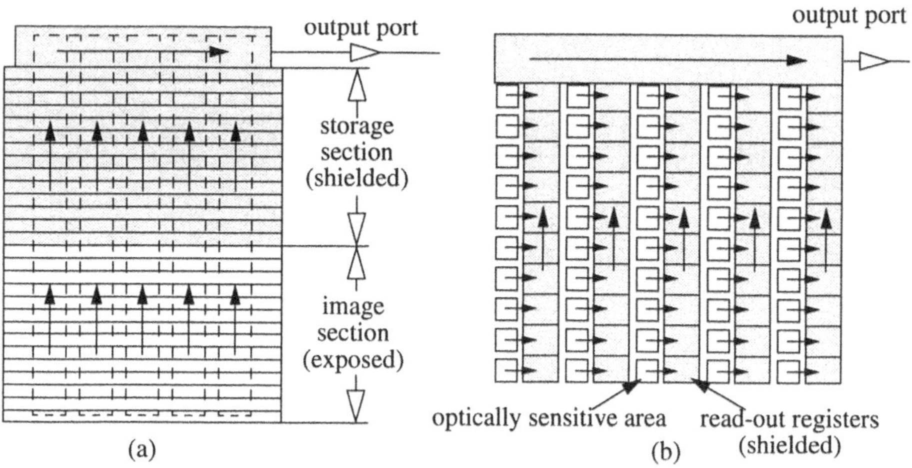

Figure 3. (a) Frame Transfer CCD sensor (FT), (b) Interline Transfer CCD sensor (ILT).

The Frame Transfer (FT) sensor consists of an imaging section and a storage section which is permanently shielded from light. After the exposure time, the charges of the pixels in the imaging section are rapidly clocked into the storage section. At the top of the storage array there is a read-out register which transports the signal sequentially out of the chip. Although the image signal charge is transported fast from the imaging area to the storage area, image smear occurs. This smearing is caused by the sustained imaging of the vertical shift register as the charge is moved through it. This impairment of original information affects mostly the lower part of the imaging section, as the charge has the longest way to the storage section. A poor charge transfer efficiency (losing charge from one location to the next) and a low video bandwidth (output amplifiers are not fast enough to convert the charge into a exactly equivalent voltage signal) increases this effect. An increased transmission time in recently developed sensors reduces these problems significantly. The major advantage of frame transfer architecture is the concentration of all pixels on a small area with a minimum of blind holes.

The pixels of an Interline Transfer (ILT) sensor are arranged in columns connected to adjacent shielded columns. These shielded shift registers are connected at the top with a horizontal shift register. The charges from the pixels are moved via the vertical to the horizontal shift register, where they are conveyed to the amplifier stage. Although smearing is avoided, an overblooming of charges between adjacent pixels is still possible, but is suppressed in modern CCD realizations. The disadvantage of this architecture are the blind areas between the pixels, which cause a loss of light sensitivity per unit area. Hyper Had techniques with a lens array over the imaging array reduce this problem.

In order to increase the vertical resolution beyond the Nyquist limit imposed by the line separation of the sensor, commercial video standards use interlacing for the read-out of the imaging frame and for the display on a video screen. Applying an appropriate read-out scheme, Interlacing can be applied to FT and ILT sensors. Each frame is accordingly subdivided in two fields. Field 1 consists of the odd number lines and field 2 of the even number lines. The read-out cycles of the two fields are shifted by half the framing time step Δt. Fig. 4 shows the integration time shift for odd and even lines. Table 1 gives the video specifications used in USA/ Japan (RS 170/ EIA) and in Europe (CCIR).

TABLE 1. Video standards.

	USA/ Japan	Europe
B/ W	RS 170/ EIA	CCIR
Color	NTSC	PAL/ Secam
Frame Rate/ Field Rate	30/ 60 Hz	25/ 50 Hz
Number of lines	525	625
Number of active lines	480	576
Field time	16 2/3 ms	20 ms
Time per line	63.49 μs	64 μs
Active line period	~ 52.5 μs	52 μs
Nominal video bandwidth	4.5 MHz	5.5 MHz
Resolution (Nyquist)	472	572
Aspect ratio	4 : 3	4 : 3

Imaging of continuously illuminated moving particles in full frame integration mode with an integration time of nearly 40 ms (CCIR) leads at higher particle velocities to streak images and a displacement of the particle images between field 1 and field 2 of one frame as shown on Fig. 5. This distortion of particle images causes errors in the determination of the particle position in the image frame. There are two possibilities to avoid these errors. One is to shorten the integration time by electronic shuttering (Fig. 4(b)). This is generally only possible with field rate, which means that the time resolution is in this case doubled (50 Hz CCIR) but the spatial resolution is decreased by a factor of two, which again reduces the accuracy with which the position of the particle can be determined in the image. Specialized cameras like progressive scan cameras have the capability to shutter with frame rate.

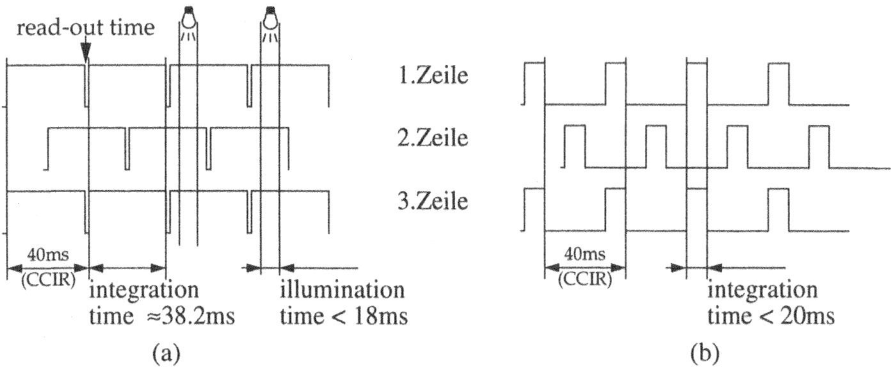

Figure 4. Interlaced Video signal in overlapping (a) Full Frame mode (b) electronically shuttered field mode.

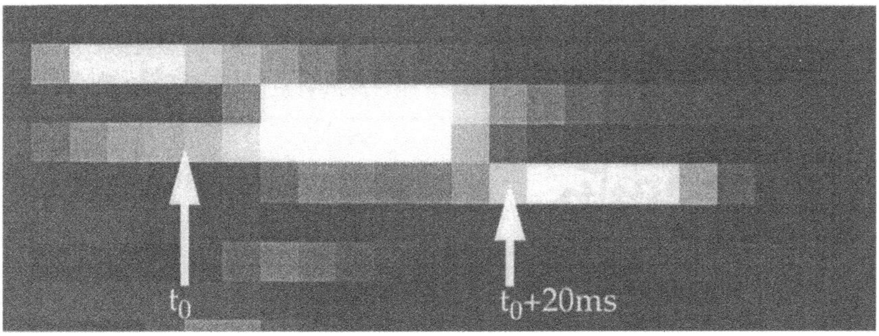

Figure 5. Image of a particle, recorded in frame integration mode by an interlaced CCD camera.

The other possibility is to use pulsed illumination. To obtain properly imaged particles with the full resolution at frame rate, the pulses must have the frame frequency of the sensor and their phase must be shifted so that the illumination time of the particles coincides with the overlapping part of the integration times of the two fields (Fig. 4(a)). Thereby the pulse length must not exceed half the integration time. Adapting the pulse length to the velocity of the particle also avoids streaky images. Correctly imaged particles are shown in Fig. 6(a). The whole field particle image in Fig. 6(b) has been high-pass filtered making use of an adaptive filter (Maas, chapter 3).

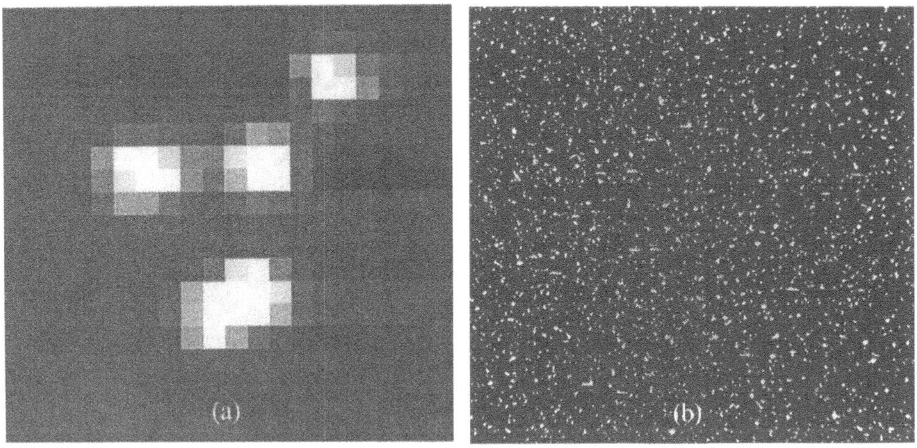

Figure 6. (a) Part of an image with properly illuminated and imaged particles, (b) high-pass filtered image with approximately 2500 particle images.

4. Illumination

In 3D-PTV a rather thick slab illumination is needed. Therefore laser light mostly used in visualization is not always the best choice. The monochromatic nature of the light avoids optical dispersion but the Gaussian intensity distribution over the beam cross-section persists in both directions of the cross-section of the slab after the optical adaptation of the light beam (Fig. 9(a)). This leads to poorly illuminated particles near the boundaries of the observation volume. For that reason other light sources have to be considered too, which often have also the advantage of being much less expensive.

The illumination should fulfill the following requirements:

- It must be intensive to allow a sufficient depth of focus and when velocity measurements are planned to allow the use of small particles which follow the flow better.
- It must be as homogeneous as possible over the entire observation volume to avoid information loss in poorly illuminated parts of the observation volume.
- It must allow to produce sharp boundaries of the illuminated slab.

Monochromatic light leads to sharply-defined particle images. Polychromatic light, on the other hand, activates a wider range of the spectral integration bandwidth of a CCD sensor. The brightness of the object on the image depends not only on the light emitted by the light source but also on the spectral response of the imaging device. Therefore these spectra should overlap over a large range.

Fig. 7 shows the relative spectral intensity of some light sources, like Argon-ion, Helium-Neon lasers and halogen short-arc lamps. The spectral response of a CCD sensor is also shown in this figure. CCD sensors have their sensitivity maximum in the red color band. The intense blue light part of the short-arc lamp emission spectrum contributes therefore little to the brightness of the illuminated object. The effectiveness of the laser light used is in the same way dependent on the relation of the emitted light to the spectral response of the CCD sensor.

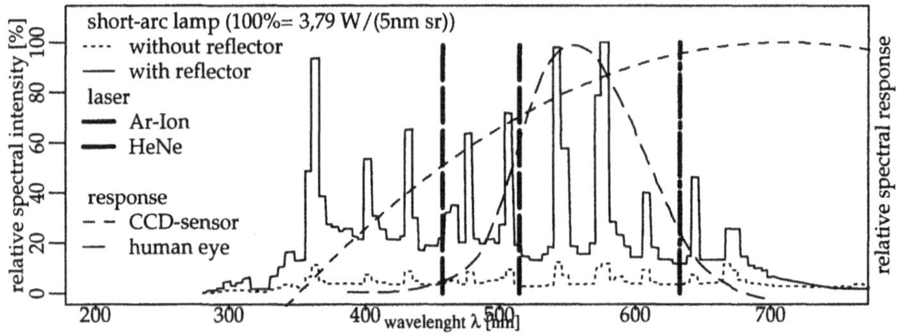

Figure 7. Relative spectral intensity of different light sources and spectral response of a CCD sensor.

As an example for the evaluation of the effectiveness of the illumination/ imaging system the response of a CCD sensor to blackbody radiation is displayed in Fig. 8. The emission spectra of a light bulb and of daylight do not cover the spectral response of the CCD sensor. The integral of their weighted spectral response is for this reason nearly identical but the maximum of this response occurs at different wavelenghts. The sensors of commercial CCD cameras are usually shielded with an infrared filter. This filter reduces the signal noise of the CCD sensor. The idealized characteristic of such an infrared filter is shown in Fig. 8(a) and (b) as a dashed line. After filtering the blackbody radiation with this infrared filter the integrals of the weighted responses of the CCD array are quite different. In this calculation the transmission of the lenses is assumed to be 100%, which is quite close to the real 90-98%. Especially at lower light source temperatures it is obviously preferable to remove the infrared filter.

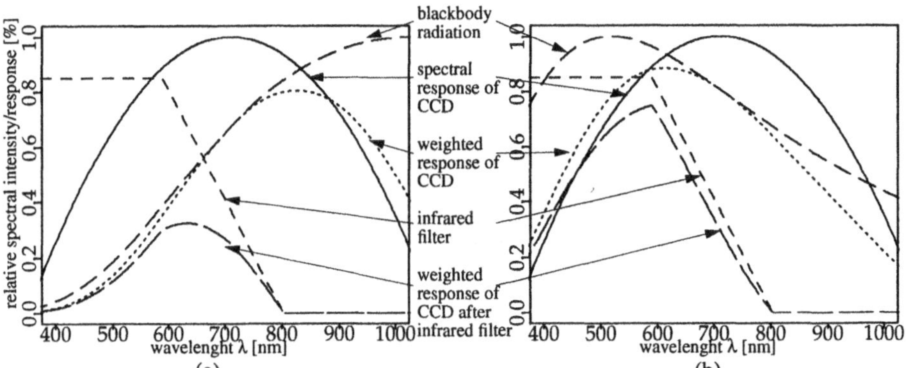

Figure 8. Relative spectral intensity of a blackbody radiation after Planck's law for a light source with the temperature of (a) 2800 K (60 W bulb) and (b) 5600 K (daylight) and the weighted relative spectral response of a CCD sensor with and without an infrared filter (idealized).

The illumination used also depends on the optical properties of the particles tracked. Mostly particles are used which either scatter the light or which fluoresce. The intensity of the scattered light varies with the viewing direction in respect to the incidence direction of the illumination (Hinsch & Hinrichs, chapter 2). As more than one camera are used in three-dimensional PT/PTV, it is impossible to place all of them in the most favorable viewing direction. This can lead to significant differences in image quality of the particles in the different cameras used. These differences can be reduced if it is possible to illuminate the particles from two or more different sides.

When fluorescent particles are used, this problem does not arise, as fluorescence is equally intensive in all directions. However, it is preferable to use a light source emitting at a wave length corresponding to the absorption wave length of the fluorescent particle. Laser light with the appropriate wave length may in this case be a favorable light source.

Another condition posed on the illumination is that it must be pulsed. Laser light can be pulsed by using a Bragg cell. Stroboscopic light is another option when a broad band light source is used. Mechanical chopping of the light is, however, the simplest and more flexible solution.

Some commonly used light sources are presented in Table 2.

Fig. 9 shows two ways to produce a light slab, using a laser and a bragg cell or a halogen-short-arc lamp with reflector and a mechanical chopper.

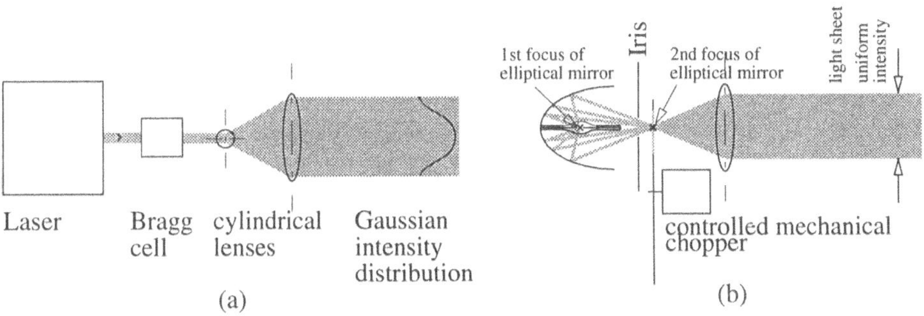

Figure 9. Light slab generated by (a) a Laser or (b) a short arc lamp.

TABLE 2. Comparisons between different light sources.

	advantages	disadvantages
Argon-ion laser	• high intensity • monochromatic light • exact positioning of the light sheet	• inhomogeneous intensity distribution over beam profile (Gaussian profile) • expensive • high energy consumption • difficult handling (dangerous)
diode laser	• extremely high intensity • monochromatic light • exact positioning of the light sheet • low power consumption	• inhomogeneous intensity distribution over beam profile (Gaussian profile) • difficult handling (dangerous) • expensive
Halogen-wire lamp	• high flexibility in combination with glass fiber arrays • nearly homogeneous intensity profile over lightsheet • very inexpensive • very easy handling	• relatively low intensity
Halogen-short-arc lamps	• high intensity • high flexibility in combination with glass fiber optics • nearly homogeneous intensity profile over lightsheet • inexpensive	• difficult handling (dangerous) • large part of the light in the blue band -> lower effectiveness
Light-emitting diodes	• very easy handling • high flexibility in combination with glass fiber optics • nearly homogeneous intensity profile over lightsheet • very inexpensive • extreme low power consumption • small band in spectrum of light (red available)	• very low intensity

5. Flow markers

When Particle Tracking is used as a velocity measurement technique, the particles must follow the movement of the fluid particle they mark closely. This means that the velocity difference $|\underline{v}_p - \underline{u}|$ between the velocity of the particle $\underline{v}_p$ and the velocity of the surrounding fluid $\underline{u}$ must be small say $0.01 \cdot |\underline{u}|$ For small particle Reynolds numbers (Adrian 1991, Malik et al. 1993) gives:

$$|\underline{v}_p - \underline{u}| = \frac{\rho_p \cdot d_p^2 \cdot |\dot{\underline{v}}_p|}{36 \cdot \rho_f \cdot \nu} \tag{9}$$

which for $|\underline{v}_p - \underline{u}| < 0.01\ |u|$:

$$|\dot{\underline{v}}_p| \le 0.36 \frac{\rho_f \cdot \nu}{\rho_p \cdot d_p^2} \cdot |\underline{u}| \tag{10}$$

or

$$d_p \le 0.6 \cdot \left(\nu \cdot \frac{\rho_f}{\rho_p} \cdot \frac{|\underline{u}|}{|\dot{\underline{v}}_p|} \right)^{1/2}. \tag{11}$$

When measuring in liquids, it is convenient to choose particles which have density ρ_p close to the density of the fluid ρ_f. For experiments performed in water with a kinematic viscosity of $\nu = 10^{-6}$ m^2/s a range of velocities from 0.1 m/s to 10 m/s cover most experimental conditions. If somehow arbitrarily one accepts that over a distance of $d_a = 0.001$ m the change in velocity does not exceed its maximum value, following acceleration time scales $t_a = d_a/|\underline{u}|$ are obtained: $t_a = 0.01$ s, $t_a = 0.001$ s and $t_a = 0.0001$ s for velocities of $|\underline{u}| = 0.1$ m/s, $|\underline{u}| = 1$ m/s and $|\underline{u}| = 10$ m/s. The corresponding accelerations $|\underline{a}| = |\underline{u}|/t_a$ are then $|\underline{a}| = 10$ m/s$^2 \approx 1$g, $|\underline{a}| = 10^3$ m/s$^2 \approx 100$g and $|\underline{a}| = 10^5$ m/s$^2 \approx 10^4$g. Neutrally buoyant particles will thus follow the flow if their diameter does not exceed $d_p = 60\mu$m, $d_p = 20\ \mu$m and $d_p = 6\ \mu$m respectively. The accelerations on which theses estimates are based are very high and the estimates for the particle sizes can be considered to be conservative.

With the equipment available at the time, velocities higher than 0.5 m/s cannot be measured reliably. Sizes of particles used in PTV extend from 30 µm to 200 µm. For flows with velocities in the range between 0.1 m/s and 0.5 m/s particles with sizes in the range $30\ \mu m < d_p < 60\ \mu m$ seem acceptable. Particles in the range between $100\ \mu m < d_p < 200\ \mu m$ may not follow the flow and if possible should not be used for velocity and even less for acceleration measurements.

Optical requirements pose on the other hand, a low bound for particle size. To determine the particle position in the image with subpixel accuracy the image of a particle must contain a sufficient number of pixels, say 8 to 16 pixels. To achieve this the particles must be sufficiently large, i.e. of order $3 \cdot \beta \cdot d_{pix}$, where β is the magnification and

d_{pix} the size of a pixel. The size of the particle image depends also on its brightness, which should be as high as possible to allow a small diaphram aperture of the camera in order to get large depth of focus and at the same time a good exploitation of the dynamic range of the grey levels.

Particles should have nearly a spherical shape to avoid fainting away during the time they are observed. It is also obvious that a small variation in size is desirable. In Table 3 properties of some conventionally used particles in Particle Tracking are listed.

TABLE 3. Properties of particles.

Material	ρ_p [kg/m3]	color/ brightness	shape	diameter [µm]	remarks
Pliolite/ Vestosint	1.02	white	nearly spherical	1-500	must be sieved
Hollow glass spheres	1.00	silver coating available	spherical	10	appropriate for thin light sheets
Fluorescent dye coated particles	0.95-1.1	absorption peak 510nm, emission 535nm	spherical	20-40 and 50-70	
Lycopodium	≈ 1	yellow	spherical	≈ 3	

When the flow field and trajectories of particles not following the flow are investigated and image size is used to separate the two particle classes, the sizes of the particles used should differ significantly.

The maximum number of particles is limited by the spatial resolution of the sensors. If for the determination of the position of a particle in a frame with 512 x 512 pixels a particle image must have in the average 12 pixels and images of particles are well distinguishable if they are in the average at least three image diameter apart, the maximum number of particles imaged is $N_I \approx 2400$. One must expect that in this case a large number of images will overlap. Nevertheless, the number of particles imaged N_I is independent of the size of the observation volume and consequently also of the particle number density. On the other hand, given the number of particles N_I imaged on a sensor, the corresponding particle number density is variable and can be determined from

$$\rho_p = \frac{N_I \cdot \beta^2}{A_s \cdot d}. \tag{12}$$

It depends on the magnification β and the thickness of the observation volume d but not on the area A_s of the sensor, which is constant. This fact is schematically depicted in Fig. 10.

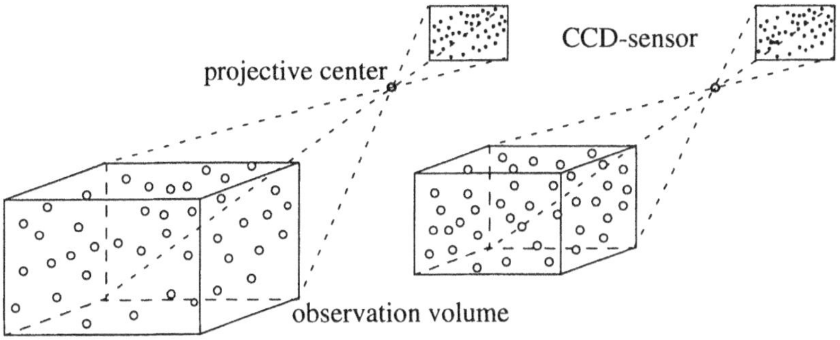

Figure 10. Influence of the magnification on the particle number density in the observation volume with a fixed sensor resolution.

6. Overlapping of particle images

Overlap of particle images is a major problem in particle tracking. When the size of the image frame A_s and of the image of a particle A_p are known, the number of particle overlaps N_a as a function of the number of particles imaged in a frame N_I can be computed from (Maas, chapter 3):

$$N_a = N_I - 1 + \frac{A_s}{A_c} \cdot \left[\exp\left(-(N_I - 1) \cdot \frac{A_c}{A_s} \right) - 1 \right]. \tag{13}$$

A_c is a critical area with the same center as the image of a particle. The critical area is essentially the area inside which the center of the image of a particle is located when overlap occurs. It is equal to four times the area of a particle image A_p. All areas can be expressed in pixels. The grey level distribution in overlapping images is the result of superposition of the grey levels of the images of the individual particles. When the centers of these images approach each other, multiple (usually two) maxima of grey levels become more and more difficult to distinguish unequivocally. As a consequence an image separation based on connectivity analysis (Maas, chapter 3) fails in an increasing number of cases. It is reasonable to assume that it becomes practically impossible when the center of the image is located inside a limiting area $A_l = 2\ A_p$. Application of eq. (13) gives the number N_a' of unresolvable image separations, which in the following will be called unresolved overlaps.

The number of overlaps and the number of unresolved overlaps are computed for a sensor area of $A_s = 512 \times 512 = 261144$ pixels, an average area of particle images $A_p = 12$ pixel and the corresponding critical and limiting areas $A_c = 4\ A_p = 48$ pixel and $A_l = 2\ A_p = 24$ pixel. The results are shown in Table 4. The number of particles in a frame is varied from 50 to 3000. The corresponding mean distances Δ of particles normalized by the mean particle diameter d_p is also given in this table. The fraction of unresolvable overlaps varies little in proportion to the number of particle images and is nearly

50%. The number of overlaps N_a, or N_a' is proportional to the square of the number of particles, i.e. $N_a^{1/2}/N_I \approx C$, and inversely proportional to the fourth power of the normalized mean distance of particles, i.e. $N_a^{1/2}/(\Delta/d_p)^2 \approx C$. The fractional number of overlaps $p_{ov} = N_a/N_I$ and of unresolved overlaps $p_{ov}' = N_a'/N_I$ is also given in Table 4.

TABLE 4. Number of overlap and unresolved overlap as a function of N_I and n.

N_I	Δ/d_p	N_a	N_a/N_I	N_a'	N_a'/N_I %	N_a'/N_a %	p_2' %	p_3' %	p_4' %	p_2'/p_3'	p_2'/p_4'
50	20.90	0.22	0.44	0.11	0.22	50	0.44	0.003	-	147	-
100	14.78	0.90	0.90	0.45	0.45	50	0.90	0.012	-	75	-
500	6.61	22	4.4	11.2	2.2	51	4.4	0.28	0.0015	16	2933
1000	4.67	86	8.6	44	4.4	51.5	8.6	1.07	0.020	8	430
1500	3.82	188	12.5	98	6.5	52	1.3	2.23	0.088	6	148
2000	3.30	325	16	172	8.6	53	16	3.74	0.23	4	70
2500	2.96	494	20	265	10.6	54	20	5.4	0.44	3.7	45
3000	2.70	691	23	376	12.5	54	23	7.2	0.70	3.2	33

Imaging of particles by more than one camera is an independent process. The probability that the images of the same particles overlap in images simultaneously taken by n cameras is therefore p_{ov}^n. The total fractional number $p_{t,n} = N_t/n\, N_I$ of independent overlaps in n cameras is consequently:

$$p_{t,n} = np_{ov} - \sum_{j=1}^{n} (n-j) \cdot p_{ov}^{j+1} . \tag{14}$$

When only two cameras are used, this number is $p_{t,2} = 2\, p_{ov} - p_{ov}^2$. As both images of a particle are needed for the determination of the spatial coordinates of a particle, this number gives also the fractional number p_2' of particles for which the coordinates cannot be determined. In the following only unresolved overlaps will be considered. The independent overlaps in three cameras are $p_{t,3}' = 3\, p_{ov}' - 2\, p_{ov}'^2 - p_{ov}'^3$. In this case, however, for each camera considered two others are left, which can be used to determine the spatial coordinates of particles having unresolvable overlaps in the first camera. If p_{ov}' is the number of overlaps in camera one, $p_{ov}'^2$ of them will have counterparts in the second and third camera. Of these p_{ov}^3 will simultaneously overlap also in camera two and three. The number of independent overlaps in camera two and three and thus the number of particles with overlapping images in camera one, for which coordinates cannot be determined with the help of the other two cameras, is $2\, p_{ov}^2 - p_{ov}^3$. Taking into account the number of simultaneous overlaps one gets in an analogous way for the second camera $2\, (p_{ov} - p_{ov}^2) - (p_{ov} - p_{ov}^2)^3$ and for the third camera $2\, (p_{ov} - p_{ov}^2 - p_{ov}^3) - (p_{ov} - p_{ov}^2 - p_{ov}^3)^3$. The fractional number of particles for which spatial coordinates cannot be determined when three cameras are used is the sum of the numbers estimated for each camera, which after some rearrangement gives

$$p_3' = 2p_{ov}'^2\left[1+\left(1-p_{ov}'\right)^2+\left(1-p_{ov}'-p_{ov}'^2\right)^2\right]-p_{ov}'^3\left[1+\left(1-p_{ov}'\right)^3+\left(1-p_{ov}'-p_{ov}'^2\right)^3\right]. \quad (15)$$

Now neglecting $p_{ov}'^2$ against 1 as $p_{ov}'^2 << 1$, one gets

$$p_3' = 6p_{ov}'^2 - 11p_{ov}'^3. \quad (16)$$

The fractional number of independent overlaps in a four-camera system is $p_{t,4} = 4\,p_{ov}'^3 - 3\,p_{ov}'^2 - 2\,p_{ov}'^3 - p_{ov}'^4$. Noting that now for each camera there is a group of 3 cameras available for the determination of coordinates of particles with overlapping images in this camera and applying the same reasoning as for three cameras one gets after simplification:

$$p_4' = 72p_{ov}'^4 - 348p_{ov}'^5. \quad (17)$$

The percent of particles for which coordinates cannot be determined when two, three or four cameras are used estimated with the help of eqs. (16) an (17) are also given in Table 4. It is seen that when only two cameras are used, the data-loss due to overlaps of particle images is significant for particle numbers $N_I \geq 500$. The addition of a third camera shifts this limit to $N_I \geq 1500$ particle images per frame whereas the addition of a fourth camera virtually eliminates the problem of data-loss due to overlapping images of particles. Data-loss is then reduced to loss due to the particle leaving or entering the observation volume (Dracos, chapter 3) or the lack of success in the establishment of image correspondences (Maas, chapter 3). This is important when long trajectories of particles are needed. As data-loss influences also the quality of the links established in the tracking procedure (Dracos, chapter 3), the addition of a fourth camera is recommended when velocity derivatives are computed.

7. Calibration

The determination of the parameters of the photogrammetric model, i.e. the inner and outer camera orientation, lens distortion, multimedia geometry etc., requires an in-situ calibration under conditions which are identical to the one prevailing during the experiment (Maas, chapter 3). The performance of the calibration is based on imaging of target points evenly distributed and precisely positioned inside the observation volume. The images of the targets should have as uniform a brightness and size as possible. This poses stringent requirements on the uniformity of size and illumination of the targets. The number of targets imaged should be sufficiently high to allow a least squares calibration and leave at the same time enough target information unused to determine the position errors of target points after calibration.

There are different ways to produce a set of reference points. They depend on the optical access to the observation volume and the possibilities feasible for the illumination of the targets used for calibration. One possibility is the use of manufactured three-dimensional calibration bodies with marked and exactly measured target points. The difficulty of an appropriate and homogeneous illumination of these points can be solved by using light fibers as sketched in Fig. 11(a).

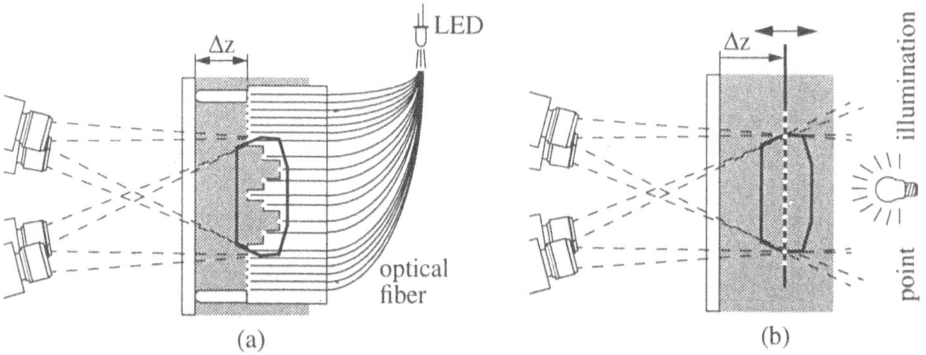

Figure 11. Imaging of reference points for calibration, (a) three-dimensional body with marked or illuminated points, (b) metal coated glass plate with photochemically etched points.

Another technical solution is the use of a metal coated plane glass plate with photochemically etched target points. By moving this glass plate parallel to itself by a high precision drive the required accuracy in the point determination can be reached. The size of the target points must be such that their images contain about 16 pixel at least. It thus depends on the magnification factor. The brightness of the targets should give images which exploit as well as possible the dynamic range of grey levels. This can be achieved by dimming the illumination of the target points. Both solutions are not appropriate when a nearly 90 degree arrangement of the cameras is used. Here a solution must be found appropriate to the experimental set-up.

8. Control, storage and processing

The topics discussed so far concern the imaging part of the experimental set-up. The next points which have to be taken into account are the control of the imaging system, the digitalization of the images and the processing of the data.

As more than one camera are used, a synchronization of the cameras is necessary. It was previously shown that to achieve a proper imaging of the particles the illumination must be pulsed. The illumination pulses must also be synchronized with the camera field phase. It is also advantageous in some cases to label the video frames with a binary code for control reasons. A central control unit is therefore necessary which provides a triggering signal with which the cameras, the pulsing equipment of the light and the binary code are synchronized.

To enable digital image processing the output of the CCD cameras must be digitized and appropriately stored. An 8 bit resolution of the greyscale is common for the digitalization of b/w images. The grey level information of one pixel can be stored in one byte and the information of a frame digitized by a 512 x 512 pixel frame grabber needs 262144 byte of storage. The data rate of a camera with 25 Hz framing frequency (European norm) is ≈ 6.6 MB per sec. and for a camera with 30 Hz framing frequency (US/Japan norm) is ≈ 7.9 MB per sec. These rates must be multiplied by the number of cameras used.

One way to cope with this high rate of data is to store the output of the cameras on analog video tapes and subsequently digitize the video signals. This solution excludes an on-line data processing. The intermediate analog storage introduces some degradation of the video signal. It is therefore preferable to directly digitize the output of the CCD cameras and store the information in digital form.

Digitalization is possible at sufficiently high frame rates. The bottleneck is the transfer and storage of the data. Fast buses used at present in Personal Computers and Workstations can handle the data rates of even a four-camera system. Storage must, however, occur in this case on the RAM of the computer. When long image sequences are needed the storage requirements become excessively high. They can be reduced if the data are compressed before they are stored. Data compression will reduce the data quality. However, a compression by a factor of five affects the data quality less than an intermediate analog storage.

Another solution is to store the data on hard discs. At present data transfer on discs is too slow, so data compression is necessary in this case also. This solution is more flexible and allows the storage of a large amount of data.

The different possibilities of data acquisition and storage are shown in Fig. 12.

The quality of the digitized image depends on the way it is digitized. Off-the-shelf video cameras with European norm have sensors with 575 lines and 739 pixels per line. A digitalization with 512 pixel in the rows is asynchronous, as shown in Fig. 13(b). This leads to an additional degradation of the signal. It can be avoided in directly digitized images if the digitizer is synchronized by the pixel clock signal of the camera. Such digitizers are now available. As the number of rows do not match the number of rows of the CCD sensor, a number of rows at the top and bottom of the image are not digitized. Similar conditions also prevail where cameras with US norm are used. Digital read-out of the CCD signal is becoming more and more common. This will in the future open the way to preprocess the image and determine the two-dimensional particle coordinates in real time and store only these coordinates for further processing. This will reduce the storage requirements by at least one order of magnitude.

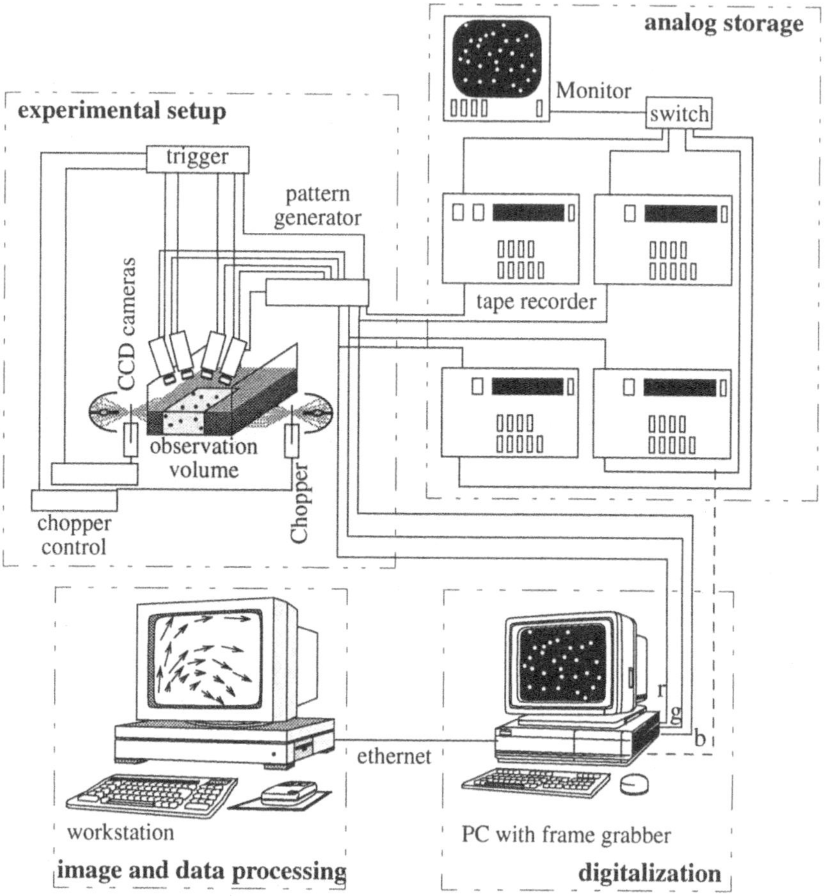

Figure 12. Scheme of hardware setup.

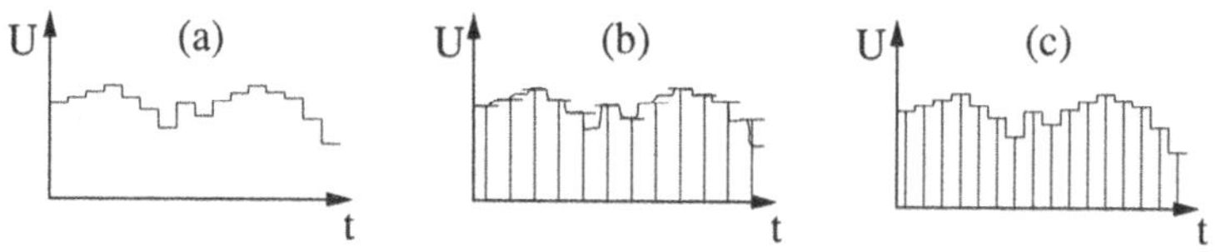

Figure 13. (a) Original video signal, (b) asynchronous digitalization of the video signal smoothed by the analog storagewith -> degradation, (c) direct digitalization triggered with pixel clock with a minimum of degradation.

9. Accuracy and performance tests

Estimates of the accuracy and of the capability of the method to follow the movement of particles have been given by Maas (chapter 3) and Dracos (chapter 3). Here experimental checks will be presented which confirm to a large extend the predictions and show the capabilities and limits of the method.

9.1 ACCURACY TESTS

The overall accuracy of the method depends on the accuracy with which the spatial coordinates are determined at each time step and the precision of the time-step Δt. The latter is very high and has little influence on the overall accuracy. Tests can also be performed which allow to estimate the precision with which the velocity component u and w in x_1- and x_3-directions can be determined Lagrangian observations of trajectories enable the direct check of the accuracy of all three components in the flow investigated.

9.1.1 Accuracy of particle coordinates

As previously mentioned, if a sufficient number of target points are available for calibration, a number of them can be treated as unknown. After calibration the images of this points are used to compute their spatial coordinates using the photogrammetric model. The RMS of the differences between known coordinates and computed coordinates gives an estimate of the error of the coordinates computed (Maas, chapter 3). One must, however, keep in mind that this check takes place in situ but with stationary cameras and targets.

In the experiments conducted calibration was performed with metal coated glass-plates with etched target points and the cameras used were of European video norm ($\Delta t = 0.04$ s). The size of the plates and of the target points was adapted to the size of the observation volume. With an average size of target images of 22 pixels the coordinates of a target point in an image are determined in the average of the four cameras with an accuracy of $\approx 0.6\ \mu m \pm 0.07\ \mu m$. The size of a pixel in the sensors used is $15.6\ \mu m$ x $11.0\ \mu m$ and its linear dimension is $\sqrt{15.6 \cdot 11.0} \mu m = 13.1 \mu m$. Target point coordinates are accordingly determined with approximately an accuracy of 1/20 of the linear dimension of a pixel. This accuracy is an evidence for the quality of the photogrammetric model and of the calibration procedure and facility. Two different sizes of observation volumes have been used. One has the dimensions length $x_1 = 427$ mm, depth $x_2 = 90$ mm width $x_3 = 229$ mm and the other $x_1 = 100$ mm, $x_2 = 90$ mm, $x_3 = 73$ mm. In both cases the depth of the volumes and thus also the depth of focus is extremely large and limits the convergence angles of the cameras. This affects mainly the accuracy of the x_2 coordinates of particles. The coordinates errors determined with a four-camera system are in the large volume: $\sigma_{x1} = 0.097$ mm, $\sigma_{x2} = 0.224$ mm and $\sigma_{x3} = 0.096$ mm and in the small volume: $\sigma_{x1} = 0.038$ mm, $\sigma_{x2} = 0.110$ mm and $\sigma_{x3} = 0.038$ mm. As expected, in both cases the position error in x_2-direction is more than two times larger than in x_1-

and x_3-direction. In these two directions the errors are nearly equal.

9.1.2 Accuracy of the velocity components

An estimate of the velocity errors in the u_1 component of the velocity can be found by using the stationary calibration plate and tracking images taken with video-rate from cameras moving with known velocity. In the following experiments only the large calibration plate was used. The velocity of the carriage on which the cameras were mounted was measured independently. Its value was 134.26 mm/s or 5.37 mm/Δt. The RMS value of its variation was $\sigma_{uc} = 1.168$ mm/s or $\sigma_{uc} = 0.047$ mm/Δt and expressed in percent of the average velocity $100 \cdot \sigma_{uc}/u_c = 0.745\%$. Tracking produced the following results: $u_1 = 134.30$ mm/s, the average velocity was thus measured with an accuracy of ≈ 0.3‰ or 1.6 mm/Δt, $\sigma_{u1} = 0.047$ mm/Δt, $\sigma_{u2} = 0.092$ mm/Δt and $\sigma_{u3} = 0.017$ mm/Δt. These results are produced under nearly ideal imaging conditions. There is therefore little difference in the precision of particle coordinates in the different cameras. In addition, all points have the same velocity. The interesting result of this test is that under these ideal conditions the errors of u_1 found by tracking can be entirely contributed to the velocity variations of the carriage. An estimation of the velocity errors based on error propagation of the coordinate errors gives for the errors of the distance Δx_i travelled by the particle in one time-step Δt $\sigma_{\Delta xi} = \sigma_{ui(p)} = \sqrt{2} \cdot \sigma_{xi}$. The error on Δt is very small and can be neglected. The error of the velocity component in x_i-direction should in this case be $\sigma_{u1} = \sqrt{2\sigma_x^2 + \sigma_{uc}^2} = 0.145\frac{mm}{\Delta t}$ or about 3 times larger than the measured. In analogy, assuming that $\sigma_{uc2} = \sigma_{uc3} = 0$ the errors in x_2- and x_3-directions should be $\sigma_{u2} = 0.318$ mm/Δt and $\sigma_{u3} = 0.136$ mm/Δt. The errors experimentally determined are again much smaller and the biggest part of them can most probably be contributed to variations of the carriage position in x_2- and x_3- directions during its movement. Obviously the error of the coordinates of a particle in consecutive time-steps have in addition to the true error a correlated (systematic) error, which when differences are computed cancels out.

The experiments related to these calibrations were performed in an open channel. For more details see Virant (1996). Long trajectories of particles were determined and the Lagrangian autocorrelation functions for the local velocity fluctuations $u_{p,i}'(\underline{x},t) = u_{pi}(\underline{x},t) - u_{pi}(\underline{x})$ computed. These functions allow to determine the errors of the velocity component which occurred during these measurements.

Assume that $u_{pi,m}'(\underline{x},t)$ is measured with some true error $\delta u_{pi}'(\underline{x},t)$, i.e $u_{pi,m}'(\underline{x},t) = u_{pi}'(\underline{x},t) + \delta u_{pi}'(\underline{x},t)$, where $u_{pi}'(\underline{x},t)$ is the true value. The first value of the autocorrelation function is proportional to $\overline{[u_{pi,m}'(\underline{x},0)]^2}$ or $\overline{[u_{pi}'(\underline{x},0)]^2} + \overline{2u_{pi}'(\underline{x},0)\,\delta u_{pi}'(\underline{x},0)} + \overline{[\delta u_{pi}'(\underline{x},0)]^2}$. As $u_{pi}'(\underline{x},0)$ and $\delta u_p'(\underline{x},0)$ an statistically independent, the second term vanishes and

$$\overline{[u_{pi,m}'(\underline{x},0)]^2} = \overline{[u_{pi}'(\underline{x},0)]^2} + \overline{[\delta u_{pi}'(\underline{x},0)]^2} \tag{18}$$

In all other ordinates of the autocorrelation function the products

$$\overline{[u_{pi,m}'(\underline{x},t)\,u_{pi,m}'(\underline{x}+\Delta\underline{x},t+\tau)]} = \overline{u_{pi}'(\underline{x},t)\,u_{pi}'(\underline{x}+\Delta\underline{x},t+\tau)} + \overline{u_{pi}'(\underline{x},t)\,\delta u_{pi}'(\underline{x}+\Delta\underline{x},t+\tau)} +$$

$$\overline{u_{pi}'(\underline{x}+\Delta\underline{x},t+\tau)\,\delta u_{pi}'(\underline{x},t)} + \overline{\delta u_{pi}(\underline{x},t)\,\delta u_{pi}(\underline{x}+\Delta\underline{x},t+\tau)} \tag{19}$$

appear. Due to the independence of the errors all terms on the right hand side except the first one vanish. These terms are thus not affected by measurement errors of the velocity components. With the help of these unaffected values the true value $[u_{pi}'(\underline{x},0)]^2$ can be estimated (Virant, 1996) and the RMS

$$\sqrt{\overline{[\delta u_{pi}'(\underline{x},0)]^2}} = \sqrt{\overline{[u_{pi,m}'(\underline{x},0)]^2} - \overline{[u_{pi}'(\underline{x},0)]^2}} \tag{20}$$

of the errors of the Lagrangian velocity fluctuations determined.

Note that the individual particles sample with time the whole velocity field. The two terms on the right hand side of eq. (20) are thus averaged over time and space.

The channel in which the experiments were performed is 600 mm wide and the flow depth was 90 mm, i.e. equal to the thickness of the observation volume. the average velocity of the flow was 234 mm/s and the RMS value of the Lagrangian velocity fluctuations in x1-direction $\sqrt{\overline{u_1'^2}}$ =26.25 mm/s or 11.2% of the mean velocity. The RMS errors of the RMS values of the velocity fluctuations determined from Lagrangian autocorrelations are $\sigma_{RMS1} = 0.0496$, $\sigma_{RMS2} = 0.5293$ and $\sigma_{RMS3} = 0.0383$. Relative to the mean flow velocity these errors are $\sigma_{u1} = 0.56\%$, $\sigma_{u2} = 5.93\%$ and $\sigma_{u3} = 0.43\%$ or σ_{u1} = 1.31 mm/s = 0.052 mm/Δt, σ_{u2} = 13.88 mm/s = 0.555 mm/Δt and σ_{u3} = 1.01 mm/s = 0.040 mm/Δt.

The RMS of the carriage velocity fluctuations does not change considerably. The error in x2-direction again can be explained to a large extend by the error introduced by variation of the carriage velocity.

Part of the error in x_2-direction can be explained also by the error introduced by the carriage. The remaining part is still considerably larger than expected. Large uncorrelated errors must occur in the determination of the x_2 particle coordinates in consecutive time-steps. This is due to the extreme thickness of the observation volume combined with its size in the other two directions which are also large. This leads to unfavorable convergence angles of the cameras. In x_3-direction the error can to a big part be explained by the errors due to the carriage and is considerably smaller than expected.

These results show that the particle tracking technique developed yields results of very high accuracy in the x_1- and x_3-directions. To improve the results in x_2-direction more favorable arrangements of the cameras must be chosen if the geometrical conditions in the experimental facility permit.

9.2 Performance tests

The accuracy of the measurement being satisfactory, the next step is to find to what extent PTV is capable of measuring mean quantities of the velocity and their fluctuations in turbulent flows. First, one has to check if the criteria used in the tracking procedure influence the measured data. Then it will be shown that PTV is capable of measuring the quantities mentioned with satisfactory accuracy.

9.2.1 Acceleration tests

Ambiguities in tracking are resolved by using criteria based on minimum acceleration or minimum change of acceleration along the path of a particle. This may bias the results towards paths with low acceleration. Fig. 14(a) shows pdf's of the acceleration measured in open channel flow with the PTV method presented. For comparison a numerically obtained pdf of the acceleration for grid-turbulence (Yeung & Pope, 1989) is also presented in this graph. The shape of these functions are similar. A bias toward small accelerations in the PTV measurements is not observed.

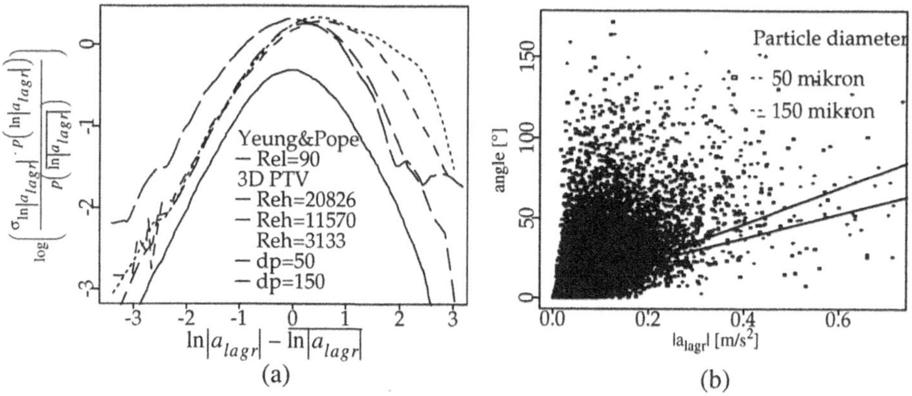

Figure 14. (a) Pdf of the normalized magnitude of acceleration, (b) scatter diagram of the directional change of acceleration vector.

The directional change of the acceleration vectors determined by PTV in the same flow is shown as scatter diagram in Fig. 14(b). A bias caused by the sorting criteria is not introduced, either. In Fig. 14(b) the directional change of the acceleration vector for large particles is also depicted. The increase of inertia of the larger particles decreases this change and shows that larger particles do not follow the flow faithfully.

9.2.2 Measurements in turbulent flow

The experiments conducted in the laboratory flume were used to test the performance of PTV against other methods. For more details see Virant (1996). As also Lagrangian analysis of the data was planed, the observation volume was chosen as large as possible. Its dimensions are previously given. It covers about 38% of the width w of the channel and the whole depth d of flow. At an aspect ratio $w/d \approx 6.7$ it is expected that in the central part, where the observations are made, the mean flow is nearly two-dimensional. This is important because averaging is performed over time and space, and flow homogeneity must in this case persist in x_1- and x_3-directions inside the observation volume. Flows with three different Reynolds numbers based on depth and mean velocity $Re_1 = 3133$, $Re_2 = 11570$ and $Re_3 = 20826$ were investigated. Results are presented for Re = 20826. The results for the other two Reynolds numbers are very similar. The number of vectors available is in this case 550'000. Assuming homogeneity and ergodicity in horizontal layers, the average velocity was calculated in 0.5 mm layers. 180 velocity values were thus obtained in a vertical. On an average ≈3055 velocity vectors are available per layer. The real distribution of vectors is shown in Fig. 15.

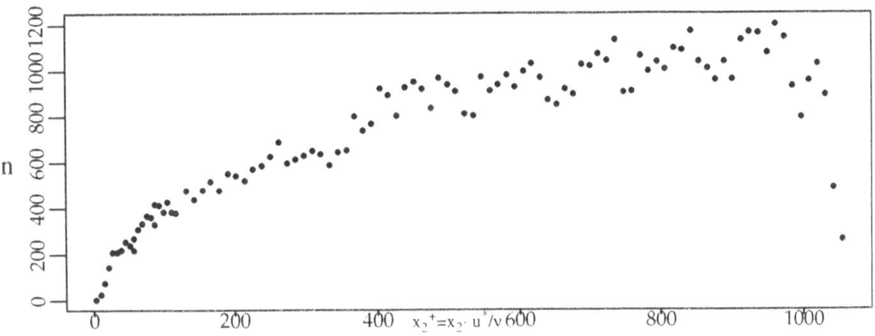

Figure 15. Number of velocity vectors over a flow depth, normalized with friction velocity u*=12.2 mm/sec and kinemastic viscosity ν, $Re_h = 20826$.

Near the bottom the number of velocity vectors in a layer drops rapidly. Averages near the bottom are therefore less reliable. The velocity distribution normalized with inner parameters is presented in linear plot in Fig. 16.

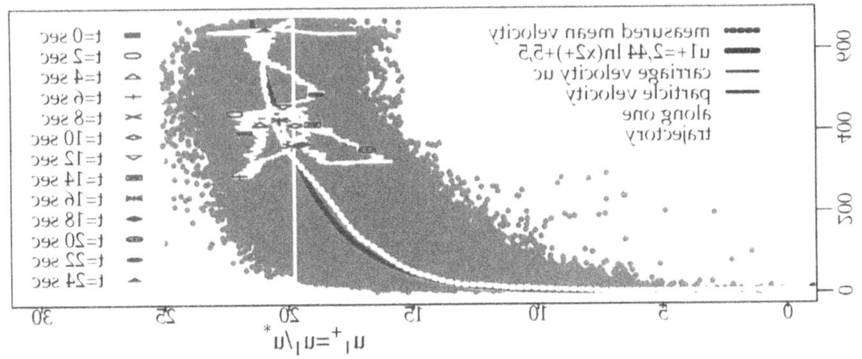

Figure 16. Statistics over 550'000 velocity components over a flow depth of h=9 cm, normalized with friction velocity u*=7.5 mm/sec and kinemastic viscosity ν, $Re_h = 11570$.

The variation span of the velocity is shown in this figure too. A semilogarithmic plot is depicted in Fig. 17. For comparison LDA measurements (Steffler et al., 1983) and Prandtl's law

$$u^{+} = 5.5 + 2.44 \cdot \ln x_2^{+} \tag{21}$$

are also presented in this figure. The agreement with the LDA measurement and with theory is satisfactory. For y+ > 400 the PTV data are considerably better approximated when the law of the wake (Coles, 1956) is introduced. This indicates intermittency in the upper part of the flow. At the top the velocity is forced back to the carriage velocity by a glass plate touching the water surface necessary for undisturbed optical access to the observation volume. The poor agreement near the bottom is attributed to lack of a sufficient number of data in this range. The velocity profile is elsewere determined with an accuracy comparable to LDA measurements.

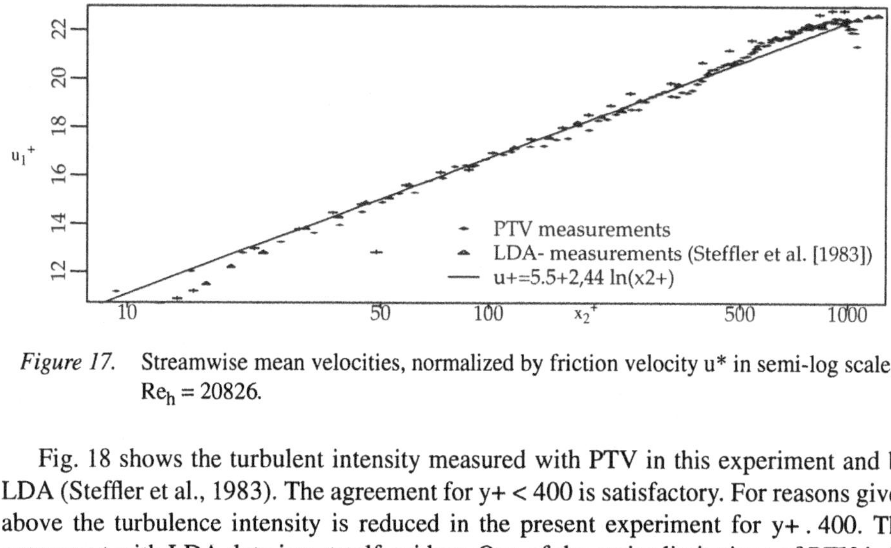

Figure 17. Streamwise mean velocities, normalized by friction velocity u* in semi-log scale, $Re_h = 20826$.

Fig. 18 shows the turbulent intensity measured with PTV in this experiment and by LDA (Steffler et al., 1983). The agreement for y+ < 400 is satisfactory. For reasons given above the turbulence intensity is reduced in the present experiment for y+ . 400. The agreement with LDA data is not self-evident. One of the major limitations of PTV is its poor spectral resolution. The imaging frequency of 25 Hz imposes an upper bound of frequency resolution of 12.5 Hz. High-frequency fluctuations are consequently filtered.

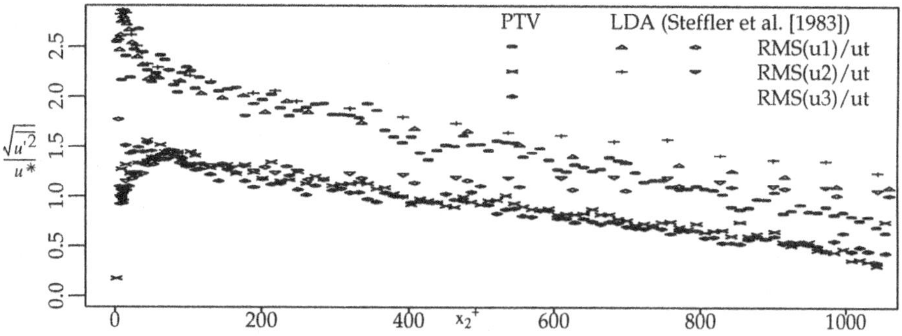

Figure 18. Turbulent intensities, normalized by friction velocity u*, $Re_h = 20826$.

10. Present limitations and future developments of PTV

The major limitations stem from the off-the-shelf equipment available at present. One of these limitations is the low imaging frequency of common CCD cameras. Specialized cameras with 512 x 512 pixel sensors capable of taking 500 images per second are available. They have the big advantage of delivering the image information in digital form and storing it directly on a build-in RAM. At maximum 3072 images can be stored. As a PC is integrated in the camera hardware, a preprocessing of the data is feasible. The price of such cameras is of course accordingly high. Nevertheless, they allow to increase the spectral resolution of the measurement and to perform measurements at higher velocities. If the accuracy of the displacement of a particle during a time-step depends on the precision with which the spatial positions of particles is determined, which in the present case is not changed, an increase of the frame rate and thus a decrease of the time-step will reduce the accuracy of the velocity. The accuracy analysis shows, however, that this problem is minor as errors in particle positions cancel out at a large extend when displacements are computed.

Another disadvantage of PTV is the relatively low number of vectors one can determine per time-step.Using a four-camera system the upper limit is at present close to 1500 vectors per time-step. When derivative depending quantities like vorticity or rate of strain tensor are wished the spacing between vectors may be too large for accuracy computations of derivatives. Interpolation is one way of overcoming this difficulty (Malik & Dracos, 1995), however, at the expense of accuracy. Another way is to reduce the size of the observation volume and thus increase the number density. As the average distance among particles and thus also vectors depends on the -1/3 power of particle number density (Dracos, chapter 3), a reduction of the volume by a factor of eight reduces the distance by only a factor of two. The spatial spectral resolution is at the same time also reduced by a factor of two.

It is expected that CCD cameras with 1024 x 1024 pixels and 100 Hz video frequency will be available in the future. If the size of the images of particles can still be kept at an average of 12 pixels, the number of vectors determined per time-step will be increased to a maximum of nearly 6000 and at the same time the spectral resolution to 50 Hz. This resolution is sufficient for channel flows with Reynolds numbers up to 30'000. Smaller particles and more intense illumination will be necessary in this case.

The problem of data transfer and storage will become more acute with increasing spatial and temporal resolution of the cameras. Nevertheless, these developments will make PTV a very competitive measuring technique for the investigation of flows in liquids. Applications in air on the contrary, will always be restricted to low-velocity flows.

References

Adrian, R. J., 1991, Particle-imaging Techniques for experimental Fluid Mechanics, Ann. Rev. Fluid Mech., Vol.23, pp. 261-304.

Coles, D., 1956, The law of the wake in the turbulent boundary layer, JFM, Vol. 1, pp. 191-226.

Malik, H.-G., Dracos, Th., Papantoniou, D.A., 1993, Particle tracking velocimetry in three-dimensional flows, Part II: Particle Tracking, Exp. in Fluids 15, pp. 279-294.

Malik, H.-G., Dracos, Th., 1995, Interpolation Schemes for Three-Dimensional Fields from Scattered Data Using Taylor Expansions, J. of Computational Physics, 119, pp.231-243.

Steffler, P. M., Rajaratnam, N., Peterson, A. W., 1983, LDA Measurements of mean velocity and turbulence distribution in a smooth rectangular open channel, Water Resources Eng. Rep. WRE 83-4, University of Alberta, Edmonton, Canada.

Virant, M., 1996, Anwendung des dreidimensionalen "Particle-Tracking-Velocimetry" auf die Untersuchung von Dispersionsvorgängen in Kanalströmungen, Ph.D. thesis, ETH Zürich.

CHAPTER IV

LASER INDUCED FLUORESCENCE VELOCIMETRY

LASER INDUCED FLUORESCENCE VELOCIMETRY
Basic Concepts

TH. DRACOS
Prof. em. ETH Zürich
Untergasse 14
CH - 8126 Zumikon

1. Introduction

In Particle Image Velocimetry (PIV) and its special form of Particle Tracking Velocimetry (PTV) presented in Chapters II and III of this volume the fluid particle are marked by small inert particles, which hopefully follow their movement. A thin sheet (PIV) or thick slab (PTV) illumination is used and the incident light is scattered by the particles, nearly according to Mie's scattering theory for spherical particles. The number density of particles is in both cases kept small enough to ensure that individual particles can be imaged on one or more plane optical sensors. The images of particles contain as a rule more than one pixel (Adrian 1991).

Instead of using particles one can use fluorescent molecules (dye) to mark fluid particles. Thin sheet illumination preferable by laser light of appropriate wavelength excites the molecules of the marker which emit light with a certain red shift (Stokes shift) and same intensity in all directions. A pixel of a sensor with optical axis perpendicular to the sheet of light receives light emitted by all molecules inside the small volume $\Delta V = \Delta x \Delta y \Delta z$ called a voxel mapped on this pixel. $\Delta x \cdot \Delta y$ is a small area in the light sheet and Δz the thickness of the light sheet. If the response of the sensing pixel is proportional to the number of fluorescent molecules in this volume, its measurement allows to determine the "local" concentration of dye. In the last two decades this principle has been extensively used for the measurement of concentration of conservative scalars and its distribution in fluid flows (Dewey 1976, Dimotakis et al. 1983, Mc Daniel 1983, Koochesfahani & Dimotakis 1985, Kychakoff et al. 1987, Papanicolaou & List 1988, Papantoniou & List 1989, Arcoumanis et al. 1990, Dahm & Dimotakis 1990, Gaskey et al. 1990, Komori et al. 1991, van Cruyningen et al. 1991, Yoda et al. 1992, Simoens & Ayrault 1994, a.o.). Most of these visualisations are either line observations during a certain time or planar observations at a fixed time.

Concentration distributions give insight into mixing processes, particularly in turbulent flows. However, turbulent mixing is a typically three-dimensional process. The need for three-dimensional imaging of concentration fields was soon recognised. It was achieved

Th. Dracos (ed.), Three-Dimensional Velocity and Vorticity Measuring and Image Analysis Techniques, 257–263.

by moving the light sheet quickly parallel to itself in small steps (Faris & Byer 1986, Yip et al. 1987, Prasad & Sreenivasan 1989, Dahm et al. 1991, Merkel et al.), thus producing tomographically a quasi instantaneous three-dimensional image of the concentration field.

Gradients of the concentration distribution allows to localise areas where high chemical activity is expected to occur in reactive flows. Such areas are as a rule folded and twisted surfaces in three-dimensional space nearly identical with material surfaces if the molecular diffusion of the dye molecules is much smaller than the molecular diffusion of momentum, i.e. if the Schmidt number

$$Sc = \nu / D >> 1 \tag{1}$$

ν is the molecular diffusion coefficient of momentum or the kinematic viscosity of the fluid and D the molecular diffusion coefficient of the dye molecules. This is more generally true for isoconcentration surfaces. The volume dV is the fluid particle observed at a given time t. Its size depends mainly on the edge length l_p of a pixel in the sensor used, the mapping scale factor M and the thickness of the light sheet Δz. Or for square pixels

$$\Delta V = l_p^2 M^2 \Delta z \tag{2}$$

If cubic voxels are asked for, Δz must be

$$\Delta z = M \cdot l_p \tag{3}$$

For large Schmidt number and short observation time the number density of fluorescent molecules in the fluid volume dV imaged on a pixel will change with time mainly due to advection of fluid in and out of this volume. Consecutive tomographic imaging of the concentration field gives an insight into the evolution of the concentration distribution in the observation volume with time and allows to determine time derivatives of concentration. When such derivatives are determined, one must consider the mode of observation, which can be either Eulerian or Lagrangian.

Laser Induced Fluorescence (LIF) and its extension to Laser Induced Fluorescence Tomography (LIFT) proved to be, if properly applied, a precise measuring technique for concentration distributions of passive scalars and their derivatives.

It is desirable to also extract some information on the flow field from such observations. Two different approaches have been developed recently in this respect. One is based on the solution of the inverse problem using the balance equation for a conservative scalar tracer (Dam et al. 1992), the other is a tracking method of small fluid patches based on least square matching techniques (A. Grün, Chapter III of this volume)

2. Basic principle of the inversion method

The balance equation for a conservative scalar in an incompressible fluid for which $\nabla \boldsymbol{u} = 0$ is

$$\frac{\partial C(\boldsymbol{x},t)}{\partial t} + \boldsymbol{u}(\boldsymbol{x},t)\nabla C(\boldsymbol{x},t) = D\nabla^2(\boldsymbol{x},t) \tag{4}$$

The influence of the velocity field is expressed by the term giving the advective change of concentration. In this term the velocity appears explicitly. The scalar product of the velocity vector and the gradient vector of the concentration of scalar can also be written

$$\boldsymbol{u} \cdot \nabla C = u_{//}|\nabla C| \tag{5}$$

where

$$u_{//}(\boldsymbol{x},t) = \boldsymbol{u}(x,t) \cdot \nabla C(x,t) / |\nabla C(\boldsymbol{x},t)| \tag{6}$$

is the component of $\boldsymbol{u}(\boldsymbol{x},t)$ in direction of the gradient of the scalar $\nabla C(x,t)$.
If the first and second order derivatives appearing in eq.(1) can be determined from tomographic LIF imaging this equation can be solved for $u_{//}(\boldsymbol{x},t)$, which in dimensionless form is

$$u_{//}(\boldsymbol{x},t) = \left[\frac{1}{R_e S_c}\nabla^2 C(\boldsymbol{x},t) - \frac{\partial C(\boldsymbol{x},t)}{\partial t}\right]\left(|\nabla C(\boldsymbol{x},t)|\right)^{-1} \tag{7}$$

The change of $u_{//}$in space is expressed by

$$\nabla u_{//} = \boldsymbol{u} \cdot \nabla \boldsymbol{e}_{\nabla C}^T + \nabla \boldsymbol{u}^T \cdot \boldsymbol{e}_{\nabla C} \tag{8}$$

where $\nabla \boldsymbol{u}$ is the gradient field of $\boldsymbol{u}$, i.e. a second rank tensor and T stand for the transposed matrix. Discrete three-dimensional information available from LIF is used to solve iteratively for $\boldsymbol{u}$ and the gradient field of $\boldsymbol{u}$ making use of eqs. (7) and (8). The derivatives appearing in these equations must be determined correctly and precisely.

This procedure has been used by Dahm et al. (1992) and was further developed by Su & Dahm (1995) for the determination of the velocity field and its derivatives.

The problems inherent to such an approach in particular concerning the uniqueness of the solution are discussed by Pearlstein & Carpenter (1995), who show that n-1 different scalar fields are needed to determine a n-dimensional flow fields uniquely in incompressible fluids. Note that the solutions of this problem depend on derivatives of the concentration of the scalar. Derivatives are very sensitive to noise and other inevitable errors in LIF measurements. This is particularly true for second order derivatives. Eq. (5) shows, however, that the term containing the Laplacian of concentration has little influence when Re and Sc are large.

3. Videogrammetric Concept

The videogrammetric approach to the problem of flow field measurements using LIF tomography is based on tracking the movement and transformation of small cuboids in time. The Least Square Matching used for this purpose is explained in principle by A. Grün in Chapter III of this volume and its application on three-dimensional LIF-Velocimetry will be presented to some extent by H.-G. Maas in this chapter. Here the basic requirements for its application will be discussed briefly. Before doing so reference is made to a similar approach presented by Tokumaru & Dimotakis (1995).

The approach used is basically Lagrangian. A small cuboid containing n^3 voxels, where n is of order 10, is cut at time *t.* out of the three-dimensional image optained tomographically by using LIF The aim of the method is to find the location of this cuboid and determine its transformed shape in the next three-dimensional LIF image at time $t+\Delta t$. The following assumptions are made:

(i) The concentration of dye in a voxel whose movement is followed and thus the corresponding grey-levels do not change significantly during the time step Δt. This implies that the time step Δt is small compared to the diffusion time it takes dye molecules to pass the boundaries of a voxel

$$\Delta t << \left(l_p \cdot M\right)^2 \cdot (4D)^{-1} = \left(l_p \cdot M\right)^2 \cdot Sc \cdot (4\nu)^{-1} \qquad (9)$$

It is easily seen that this condition is fulfilled if *Sc* is sufficiently large.

(ii) Inside a cuboid the concentration of dye and consequently the grey-levels in its obtained image change significantly and form well recognisable patterns. In other words, significant concentration respectively grey-level gradients exist inside such a cuboid.

Only under this condition the correlation between the grey-levels in the template cuboid at time t and its displaced and transformed form at time $t+\Delta t$ can be determined unambiguously. If condition (i) is also fulfilled, the correlation must be very high.

Translation and rigid body rotation do not change the shape of the cuboid followed in time. Depending on the size of the cuboid and on the time increment Δt the deformation of the cuboid can be approximately linear or of higher order (see also Tokumaru & Dimotakis 1995). Although theoretically higher order deformation can be modelled, computation efficiency considerations demand that linear deformation can be assumed. This is only possible if the relative deformations are small. This poses restrictions on the size of the cuboid and on the size of the time increment, which anyhow must be as small as possible for reasons explained before.

4. Applicability of LIF-Velocimetry Methods

Both methods briefly introduced are only applicable if derivatives of scalar concentration are present. In free shear flows like jets ambient often irrotational fluid is entrained into the flow field investigated. The process of entrainment starts by engulfment of ambient fluid by large eddies (Hussain 1986, Papantoniou & List 1989, Yoda et al. 1992), which then is mixed by the action of turbulence with the jet fluid. If the jet fluid is marked with fluorescent dye, the probability of encountering patches of ambient undyed fluid in the interior of the jet is high. In such patches gradients of grey-levels are absent and both methods cannot be applied.

Gradients of scalar concentration can also be absent or too weak when well mixed conditions prevail. This may be the case in flows in which entrainment does not occur, for instance in channel flow. In such cases only part of the fluid must be marked by fluorescent dye. Turbulence will produce a mixing zone in which concentration gradients will be present. Measurements of the flow field are restricted to this zone.

The cuboid tracking method fails when the velocity is tangential to isoconcentration surfaces, i.e. when $\boldsymbol{u} \cdot \nabla C = 0$.

The cuboid tracking procedure is Lagrangian in nature. Observations should therefore be made from a platform moving with the local average velocity of the flow. This is also commendable when the inversion method is used to avoid errors in the determination of derivatives in z-direction.

The matching method has been applied by Papantoniou et al. (1990) and by Tokumaru & Dimotakis (1995) on two-dimensional LIF data. Such applications must be restricted to strictly two-dimensional flow fields. In the case of three-dimensional flow two-dimensional applications yield results of some reliability only for the translation, i.e. the velocity components in the illuminated plane. Vorticity and rate of strain results are not reliable. First of all, continuity is not necessarily fulfilled in two dimensions. In other words, the area of the square template at time t and of its advected and transformed form at time $t + \Delta t$ are not necessarily equal. Secondly, the rotation and deformation are not projections of the real ones on the observation plane, which is fixed in space, whereas the cuboid of which the template and its transformed form are cuts, moves, rotates and deforms in the three-dimensional space. This method yields reliable results for three-dimensional flow fields only when it is applied in its three-dimensional form.

References

Adrian, R.J., 1991. Particle imaging techniques for experimetal fuid mechanics, *Annu. Rev. Mech.*,**23**, 261

Arcoumanis, C., Mc Guirk, J.J., Palma, J.M.L.M. 1990. On the use of fluorescent dyes for concentration measurements in water flows, *Exp. in Fluids,* **10**, 177

Cetegen, B.M., Mohamad, N. 1993. Experiments on liquid mixing and reaction in a vortex, *J. Fluid Mech,* **249**, 391

Dahm, W.J.A., Dimotakis, P.E. 1990. Mixing at large Schmidt number in the self-similar far-field of turbulent jets, *J. Fluid Mech.,***217**, 299

Dahm, W.J.A., Lester, K.S., Southerland, K.B. 1992. A scalar imaging velocimetry technique for fully resolved four-dimensional vector velocity field measurements in turbulent flows, *Phys. Fluids, A***4** (10), 2191

Dahm, W.J.A., Southerland, K.B., Buch, K.A. Direct high resolution, four-dimesnional measurements of the fline scale structure of Sc>>1 molecular mixing in turbulent flows, *Phys. Fluids,* A **3** (5), 1151

Dewey, C.F. 1976. Qualitative and quantitative flow field visualisation utilizing laser-induced fluorescence, *Proc. AGARD Conf. Appl. of Non-Intrusive Instrumentation in Fluid Flow Research,* AGARD-CP, 193

Dimotakis, P.E., Miake-Lye, R.C., Papantoniou, D.A. 1983. Structure and dynamics of round turbulent jets, *Phys. Fluids,* **6**, 3185

Faris, G.W., Byer, R.L. 1986. Quantitative optical tomographic imaging of a supersonic jet, *Opt. Lett.,* **11**, 413

Gaskey, S., Vacus, P., David, R., Villermaux, J. 1990. A method for the study of turbulent mixing using fluorescence spectroscopy, *Exp. Fluids,* **9**, 137

Husain, A.K.M.F. 1986. Coherent structures and turbulence, *J. Fluid Mech.,* **193**, 303

Komori, S., Kanzaki, T., Murakami, K. 1991. Simultaneous measurements of instantaneous concentrations of two reacting species in a turbulent flow with a rapid reaction,*Phys. Fluids,* A **3** (4), 507

Koochesfahani, M.M., Dimotakis, P.E. 1985. Laser Induced fluorescence measurements of mixed fluid concentration in a liquid plane shear layer, *AIAA J.,***23**, 1700

Kychakoff, G., Paul, P.H., van Cryningen, I.L.,Hanson, R.K. 1987. Movies and 3-d images of flow fields using planar laser-induced fluorescence, *Appl. Opt.,***26**, 2498

McDaniel, J.C. 1983. Quantitative measurements of density and velocity in compressible flows using Laser-Induced Iodine Fluorescence, *AIAA-83-0049, AIAA, Twenty-first Aerospae $Science Meeting, Reno, N.V.*

Merkel, G.

Papanicolaou, P.P., List, E.J. 1988. Investigations of round vertical turbulent buoyant jets, *J. Fluid Mech.,***195**, 341

Papantoniou, D.A., Bühler, H., Dracos, T. 1990. On the internal structure thermals and momentum puffs, *Int. Conf.onn Phys. Modeling of Transport and Dispersion, MIT,* 7 - 10 Aug.

Papantoniou, D.A., List, E.J. 1989. Large-scale structure in the far-field of buyoant jets, *J. Fluid Mech.,***209,** 151

Pearlstein, A.J., Carpenter, B.N. 1995. On the determination of solenoidal or compressible velocity fields from measurements of precise or reacitive scalars, *Ohys. Fluids,* **F** (4), 754

Prasad, R.R., Screenivasan, K.R. 1990. Quantitative three-dimesional imaging and the structure of passive scalar fields in fully turbulent flows, *J. Fluid Mech.,* **216**, 1

Simoens, S., Ayrault, M. 1994. Concentration flux measurements of a scalar quality in turbulent flows, *Exp. Fluids,* **16**, 273

Su, L.K., Dahm, W.J.A. 1995. Scalar Imaging Velocimetry and its application in measurements of the strucutre and dynamics of the complete velocity gradient tensor in turbulent flows, Rep. No. 026779 - 14, Dep. of Aerospace Eng., Univers. of Michigan, Ann Arbor, MI, 48109 - 2118, USA

Tokumaru, P.T., Dimotakis, P.E. 1995. Image correlation velocimetry, *Exp. Fluids,* **19**, 1

van Cruyningen, I., Lozano, A., Hanson, R.K. 1991. Quantitative imaging of concentration by planar laser-induced fluorescence, *Exp. Fluds,* **10**, 41

Yip, B., Lam J.K., Winter, M., Long, M.B. 1987. Time-resolved three-dimensional concentration measurements in a gas jet, *Science,* **235**, 1209

Yoda, M., Hesselink, L., Mungal, M.G. 1992. The evolution and nature of large-scale structures in the turbulent jet, *Phys. Fluids A,***4** (4),

ADAPTIVE LEAST-SQUARES-MATCHING FOR ACCURATE TRACKING OF PATTERNS IN VOXEL DATA SEQUENCES

H.-G. Maas
Institute of Geodesy and Photogrammetry,
Swiss Federal Institute of Technology, ETH Hoenggerberg,
CH-8093 Zurich, Switzerland

1. Introduction

The tomography data acquisition system as described in (Merkel et al., chapter 4) delivers sequences of voxel datasets as shown in Figure 1. In contrast to the marking with discrete tracers as described in the 3-D PTV chapter of this book, the visualization of the flow can be considered a continuous visualization here. Thus the determination of velocity fields in these datasets cannot be based on the tracking of discrete tracers; instead, volume- or feature-based approaches have to be applied. In the following, an adaptive correlation technique for the simultaneous determination of velocity fields and deformation parameters of fluid elements in sequences of voxel-datasets will be described.

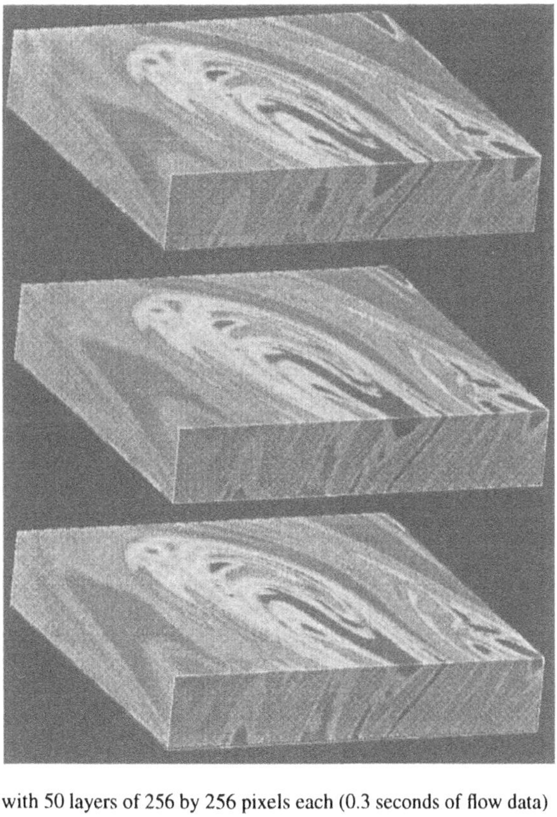

Figure 1: Three consecutive volume datasets with 50 layers of 256 by 256 pixels each (0.3 seconds of flow data)

Th. Dracos (ed.), Three-Dimensional Velocity and Vorticity Measuring and Image Analysis Techniques, 265–276.

2. Least-squares-matching for tracking of flow patterns

Least squares matching (LSM) is a correlation technique which is often being in photogrammetry, mainly for the establishment of stereoscopic or multi-image correspondences and for the determination of image coordinates of signalized points using artificial templates (Gruen/Baltsavias, 1988). The basic principle of LSM is to take a patch P_1 of a certain size in an image I_1 and transform it into a patch P_2 in image I_2 in a way that the sum of the squares of greyvalue differences between P_1 and P_2 reaches a minimum; see (Gruen, chapter 3) for an outline of 1-D and 2-D LSM. Usually an affine transformation is used for the deformation of the patch, assuming a linear deformation. The patches may be:

- Corresponding regions in two or more images taken from different viewpoints (establishment of multi-image correspondences).
- A signalized target to be measured in an image using an artificial template (least squares template matching).
- Corresponding locations in consecutive images of an image sequence (tracking).

LSM has also been proven to be well suited for the establishment of temporal correspondences through structure tracking in 2-D image sequences (Papantoniou et al., 1990). An example for the application of LSM for LIF flow pattern tracking in two consecutive images of an image sequence is shown in Figure 2. The shift parameters of the affine transformation applied to a selected patch describe the velocity vector for the fluid element covered by the patch, while scales, rotation and shears contain information on the transformation of the fluid element.

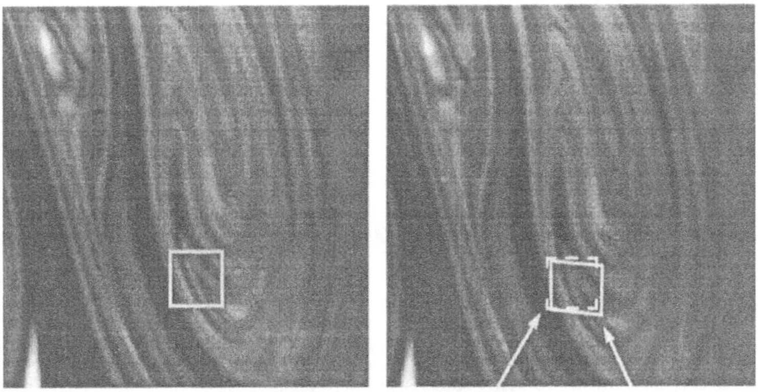

Figure 2: LSM for flow pattern tracking in 2-D

By defining a regular grid over an image and applying LSM to each individual grid mesh, a 2-D velocity field can be computed similarly to PIV techniques.

It is obvious that the application of LSM (like other tracking techniques) to 2-D image sequences may lead to biased information on the velocity field, if structures show velocity components in the third dimension.

3. 3-D least squares matching

Least squares matching can be extended to a 3-D algorithm, working with 3-D volume data and voxels rather than 2-D images and pixels (Maas, 1993). To match two discrete cuboids of voxel intensities $g_1(x, y, z)$, $g_2(x, y, z)$ taken from two consecutive volume datasets $\mathcal{V}_1$ and $\mathcal{V}_2$, cuboid g_1 has to be transformed into g_2 in a way that

$$g_1(x, y, z) = g_2(x, y, z) . \tag{1}$$

To better accommodate noise and various radiometric and geometric dissimilarities in the imaging conditions of two 3-D datasets, which are not covered by eq. (3), a true error vector has to be added

$$g_1(x, y, z) - e(x, y, z) = g_2(x, y, z) . \tag{2}$$

To express the geometric relationship between conjugate cuboids, the 6-parameter affine transformation used to relate conjugate patches in 2-D least squares matching has to be substituted by a 12-parameter 3-D affine transformation, as

$$\begin{aligned} x_2 &= a_0 + a_1 x_1 + a_2 y_1 + a_3 z_1 \\ y_2 &= b_0 + b_1 x_1 + b_2 y_1 + b_3 z_1 \\ z_2 &= c_0 + c_1 x_1 + c_2 y_1 + c_3 z_1 \end{aligned} \tag{3}$$

with (a_0, b_0, c_0) being shifts, (a_1, b_2, c_3) scales, and $(a_2, a_3, b_1, b_3, c_1, c_2)$ rotations and shears. Equation (2) can be linearized to yield observation equations as

$$g_1(x, y, z) - e(x, y, z) = \overset{0}{g_2}(x, y, z) + \frac{\partial \overset{0}{g_2}}{\partial x} dx + \frac{\partial \overset{0}{g_2}}{\partial y} dy + \frac{\partial \overset{0}{g_2}}{\partial z} dz . \tag{4}$$

With

$$dx = \frac{\partial x}{\partial p_i} dp_i , \quad dy = \frac{\partial y}{\partial p_i} dp_i , \quad dz = \frac{\partial z}{\partial p_i} dp_i \tag{5}$$

$$p_i \in \{da_0, \dots, da_3, db_0, \dots, db_3, dc_0, \dots, dc_3\}$$

and by defining the numeric greyvalue derivatives at the location of the initial values $\overset{0}{g_2}(x, y, z)$ of the transformation

$$\frac{\partial \overset{0}{g_2}}{\partial x} = g_x \qquad \frac{\partial \overset{0}{g_2}}{\partial y} = g_y \qquad \frac{\partial \overset{0}{g_2}}{\partial z} = g_z$$

the observation equations become

$$\begin{aligned} g_1(x, y, z) - e(x, y, z) = \overset{0}{g_2}(x, y, z) &+ g_x da_0 + g_x \overset{0}{x_2} da_1 + g_x \overset{0}{y_2} da_2 + g_x \overset{0}{z_2} da_3 \\ &+ g_y db_0 + g_y \overset{0}{x_2} db_1 + g_y \overset{0}{y_2} db_2 + g_y \overset{0}{z_2} db_3 \\ &+ g_z dc_0 + g_z \overset{0}{x_2} dc_1 + g_z \overset{0}{y_2} dc_2 + g_z \overset{0}{z_2} dc_3 \end{aligned} \tag{6}$$

and they can be grouped in a Gauss-Markov estimation model as

$$-e(x, y, z) = Ax - l \quad ; \quad P \tag{7}$$

$$E(e) = 0 \qquad E(ee^T) = \sigma_0^2 P^{-1} \tag{8}$$

where the coefficient matrix A, observation vector l and vector of unknowns x are

$$A_i = [g_x, g_x x_2^0, g_x y_2^0, g_x z_2^0, g_y, g_y x_2^0, g_y y_2^0, g_y z_2^0, g_z, g_z x_2^0, g_z y_2^0, g_z z_2^0]_i \tag{9}$$

$$l = g_1(x, y, z) - g_2^0(x, y, z) \tag{10}$$

$$x^T = [da_0, da_1, da_2, da_3, db_0, db_1, db_2, db_3, dc_0, dc_1, dc_2, dc_3]\,. \tag{11}$$

The values g_x, g_y, g_z are numeric greyvalue first derivatives, calculated either in g_l or in g_2 or averaged over both; x_2^0, y_2^0, z_2^0 are voxel coordinates in g_2, at the location of the initial transformation values. In contrast to typical photogrammetric applications, a radiometric correction should be avoided here, as it would interfere with the original radiometric calibration of the system.

The vector x of unknown parameters is determined by minimizing the sum of squares of the estimated residuals $v(x, y, z)$ in the Gauss-Markov model

$$\hat{x} = (A^T P A)^{-1}(A^T P l) = Q_{xx} A^T P l \; ; \; \hat{\sigma}_0^2 = \frac{v^T P v}{r} \tag{12}$$

with
- P: weight matrix (usually P = unit matrix)
- $\hat{x}$: estimated solution vector
- Q_{xx}: (12 x 12) covariance matrix
- $\hat{\sigma}_0$ standard deviation of weight unit
- r redundancy.

For the procedure of linearization of the observation equations, inversion of the normal equation system, parameter estimation etc., see e.g. (Koch, 1980) or (Slama, 1980, pp. 88 ff). Besides the estimations of the 12 unknown affine transformation parameters in the solution vector $\hat{x}$, the Gauss-Markov model gives information on the precision and reliability of the determined parameters in the covariance matrix and the standard deviation of weight unit. The standard deviation of transformation parameters can be obtained from the standard deviation of weight unit and the covariance matrix:

$$\hat{\sigma}_{x_i} = \hat{\sigma}_0 \cdot Q_{ii} \tag{13}$$

Non-significant parameters can be excluded from the estimation process by analyzing their estimated standard deviations and the correlations with other parameters, obtained from the covariance matrix. It has to be noted that the actual observations for the determination of the affine deformation parameters are greyvalues; greyvalue first derivatives are used only in the design matrix A, thus making the method robust against noise in the greyvalue data.

The fact that $g_2(x, y, z)$ does actually contain stochastic quantities is neglected here (as usually done in LSM) to allow the use of the Gauss-Markov model (Gruen/Baltsavias, 1988).

Velocity field components are deduced with subpixel accuracy from the adjusted shift parameters as

$$u = \left(\frac{a_0}{\Delta t}, \frac{b_0}{\Delta t}, \frac{c_0}{\Delta t}\right) \tag{14}$$

while information on the deformation behavior of fluid elements is included in the remaining 9 transformation parameters $(a_1, a_2, a_3, b_1, b_2, b_3, c_1, c_2, c_3)$.

4. Additional constraints

Experiences with 2-D matching have shown that there exist several greyvalue patterns for which some of the affine transformation parameters cannot be determined due to lack of sufficient contrast or high correlation between parameters (Gruen/Stallmann, 1991). An example id shown in Figure 3.

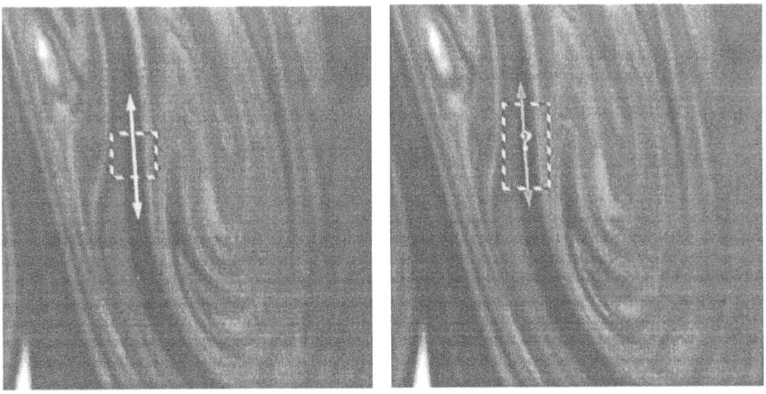

Figure 3: 2-D example with uncontrolled affine parameters in vertical direction

Similar situations can occur in 3-D voxel matching. In such cases, either certain undeterminable parameters have to be excluded from the vector of unknowns, or additional constraints have to be introduced to strengthen the system. In the following some constraints, which may be used for 3-D cuboid matching, will be explained. It has to be mentioned that they arise from the special task of matching flow tomography sequences and are not necessarily valid for general applications of least squares image matching.

4.1. Incompressibility Constraint

A fundamental property of liquids is incompressibility. That means that a cuboid, while changing it's shape due to deformations, will have a constant volume. This fact can be exploited to prevent cuboids from uncontrolled growing or shrinking in the matching process.

An incompressibility constraint can be formulated as

$$vol_1 = vol_2 \tag{15}$$

where vol_1 is $n_x \, x \, n_y \, x \, n_z$ and vol_2 is calculated as the volume of the affine transformed vol_2. This yields an additional equation for the Gauss-Markov model

$$-e_v = A_v x_v - l_v \ ; \ P_v . \tag{16}$$

With the definition

$$\frac{\partial vol_2^0}{\partial x} = vol_x \qquad \frac{\partial vol_2^0}{\partial y} = vol_y \qquad \frac{\partial vol_2^0}{\partial z} = vol_z$$

and the cuboid center pixel coordinates (x_c, y_c, z_c) the coefficient matrix A and the observation vector l become

$$A_v = [0, vol_x x_c, vol_x y_c, vol_x z_c, 0, vol_y x_c, vol_y y_c, vol_y z_c, 0, vol_z x_c, vol_z y_c, vol_z z_c] \tag{17}$$

$$l_v = vol_1 - vol_2 . \tag{18}$$

The derivatives $\partial vol/\partial p$ are calculated numerically. The weight P_v should be determined empirically; a weight of 0 invalidates the constraint, while a weight ∞ forces strict volume conservation.

4.2. Intensity Constraint

A highly specialized constraint, stemming from the particular nature of the application at hand, forces the greyvalue sums sg in the volumes covered by the cuboid g_1 and the deformed cuboid g_2 to be equal. This constraint can only be valid in 3-D, since for 2-D applications fluorescein might simply move into the third dimension and violate this criterion. In 3-D it is only valid when conservation of fluorescence and linearity between fluorescein content and voxel greyvalues can be assumed. The basic formulation of the condition

$$sg_1 = sg_2 \tag{19}$$

yields an additional equation for the Gauss-Markov model

$$-e_t = A_t x - l_t \ ; \ P_t \tag{20}$$

with the coefficient matrix A and the observation vector l

$$A_t = [sg_x, sg_x x_c, sg_x y_c, sg_x z_c, sg_y, sg_y x_c, sg_y y_c, sg_y z_c, sg_z, sg_z x_c, sg_z y_c, sg_z z_c] \tag{21}$$

$$l_t = sg_1 - sg_2 , \tag{22}$$

where sg_x, sg_y, sg_z are the numeric first derivatives of greyvalue sum and (x_c, y_c, z_c) is the transformed central voxel of the g_2 cuboid. Note that sg_2 is the total intensity over the metric volume covered by the transformed cuboid g_2, not over the matrix g_2. The weight P_t should be determined empirically.

4.3. Smoothness of Velocity Field

This constraint is based on the assumption that the sampling frequency is sufficiently high, ensuring that consecutive datasets are correlated to a certain extend and Lagrangian acceleration is relatively low. This implies that the matching is not only performed between two datasets at time instances T_i->T_{i+1}. At least a third dataset T_{i+2} has to be included and a correlation between the parameters of the affine transformations T_i->T_{i+1} and T_{i+1}->T_{i+2} has to be assumed. The constraint can be formulated as additional observation equations for the parameters. Considering for example the shift parameter in x-direction, the associated equation becomes

$$a_{0_{(0\to 2)}} = 2\cdot a_{0_{(0\to 1)}} \quad => \quad -v = 2\cdot a_{0_{(0\to 1)}} - a_{0_{(0\to 2)}} \tag{23}$$

The remaining 11 equations can be formulated accordingly. The smoothness constraints produce 12 additional equations

$$-e_s = A_s x - l_s \quad ; \; P_s \tag{24}$$

where

$$A_s = [2I_{12} \quad -I_{12}] \tag{25}$$

with I_{12} being the *12x12* unit matrix, and

$$l_s = \begin{bmatrix} a_{0_{(0\to 2)}} - 2\cdot a_{0_{(0\to 1)}} \\ a_{1_{0\to 2}} - 1 - 2\cdot(a_{1_{0\to 1}} - 1) \\ a_{2_{(0\to 2)}} - 2\cdot a_{2_{(0\to 1)}} \\ a_{3_{(0\to 2)}} - 2\cdot a_{3_{(0\to 1)}} \\ b_{0_{(0\to 2)}} - 2\cdot b_{0_{(0\to 1)}} \\ b_{1_{(0\to 2)}} - 2\cdot b_{1_{(0\to 1)}} \\ b_{2_{0\to 2}} - 1 - 2\cdot(b_{2_{0\to 1}} - 1) \\ b_{3_{(0\to 2)}} - 2\cdot b_{3_{(0\to 1)}} \\ c_{0_{(0\to 2)}} - 2\cdot c_{0_{(0\to 1)}} \\ c_{1_{(0\to 2)}} - 2\cdot c_{1_{(0\to 1)}} \\ c_{2_{(0\to 2)}} - 2\cdot c_{2_{(0\to 1)}} \\ c_{3_{0\to 2}} - 1 - 2\cdot(c_{3_{0\to 1}} - 1) \end{bmatrix}. \tag{26}$$

By manipulating the associated weight matrix P_s, the constraint can be invalidated ($P_s = 0$), or strictly enforced ($P_s \to \infty$) in which the parameter sets of consecutive transformations are forced to be identical. Eqs. (24)-(26) show the implementation for three datasets T_i->T_{i+1}->T_{i+2}. It can be straightforwardly extended to more consecutive time steps T_i->T_{i+1}-> ... ->T_{i+n}.

4.4. Multi-Patch Matching

With the solution as shown in paragraphs 3. and 4.1.-4.3., each cuboid is being treated independently from its neighboring cuboids. This may lead to certain gaps or overlaps between neighboring cuboids after the transformation. To prevent this, another geometric constraint can be formulated forcing neighboring cuboids to remain fit together after their boundaries are reshaped by the matching adjustment. An analogous approach has been formulated for DTM generation applications by (Gruen, 1985) and (Rosenholm, 1986). For the 3-D implementation it is realized by adding observation equations for the coincidence of the corner pixels of neighboring elements (7 neighbors at 8 corner voxels) in the 3-D affine transformation (Maas, 1993).

This multi-patch constraint implies that parameters of single cuboids cannot be solved independently anymore. In the extreme case, the transformation parameters of all cuboids of the complete dataset have to be estimated simultaneously in one system. Combined with the smoothness constraint which ties multiple datasets together, the resulting equation system can become extremely large. Despite its sparsity, it will impose immense computational burdens. In a realistic experiment with 256x256x64 voxels per dataset and a cuboid size of e.g. 17x17x17 voxels with smoothness constraints applied only for consecutive datasets, a total of 24576 unknowns per volume dataset would have to be solved simultaneously. It is clear that such a system would be computationally too expensive and therefore the equation system has to be split into smaller, manageable parts. A promising idea is to search for cuboids with sufficient contrast for a independent solution in a first step and then triangulate the volume dataset with a tetrahedra structure, within which the multi-patch constraint and the smoothness criterion will have to be applied.

4.5. Global Solution

A global matching solution is obtained for a single pair of cuboids by the introduction of the incompressibility and intensity constraints as

$$\hat{x} = (A^T P A + A_v^T P_v A_v + A_t^T P_t A_t)^{-1} (A^T P l + A_v^T P_v l_v + A_t^T P_t l_t)\,. \tag{27}$$

Considering more than two consecutive cuboids, a multi-temporal global matching solution is obtained by further introducing the smoothness of velocity field constraint as

$$\hat{x} = (A^T P A + A_v^T P_v A_v + A_t^T P_t A_t + A_s^T P_s A_s)^{-1} (A^T P l + A_v^T P_v l_v + A_t^T P_t l_t + A_s^T P_s l_s)\,. \tag{28}$$

5. STRATEGIES

Considering the large amount of data associated with 3-D image sequences in general, and the multi-directional velocities involved in this specific application, it is evident that the decision for an efficient matching strategy can optimize the application of the matching tool. Matching strategies should mainly concentrate on the selection of locations where matching can/should be applied, on cuboid sizes, and on the provision of approximate values of sufficient quality.

Matching locations may be chosen after the following criteria:

- Definition of a regular grid with a grid element size (r_x, r_y, r_z), and a cuboid size (c_x, c_y, c_z)
- Locations selected by an interest operator (e.g. Förstner, 1986)
- Locations chosen after criteria posed by fluid dynamics
- Locations selected interactively in a graphical user interface.

Three competing demands influence the choice of the cuboid size:

- The larger the cuboids, the more signal will be contained and the more parameters will be significantly determinable in the least squares estimation process.
- Large cuboids introduce a smoothing effect on the velocity field.
- Least squares matching is characterized by a limited convergence radius. The size of the convergence radius for the shift parameters a_0, b_0 and c_0 depends on the signal contents of the patches/cuboids and will usually be in the order of half the cuboid size. Therefore, the selection of the cuboid size will depend on the expected velocity spectrum.

In particular, the task of initial value selection for the matching parameters (especially for the three shifts a_0, b_0 and c_0) is crucial since erroneous selection will cause matching to not to converge (matching will fail, but give an indication on the failure). The use of a hierarchical strategy can greatly facilitate this task. Multigrid hierarchical strategies proceed in a number of levels, from coarse to fine, by employing successively refined grid structures (Li, 1989). Applied to the matching problem at hand, the implementation of a 3-D multigrid hierarchical strategy proceeds by selecting a 3-D grid structure, defined by its grid element size (r_x, r_y, r_z), to cover a 3-D dataset, and a corresponding cuboid size (c_x, c_y, c_z). For an initial configuration (*level 1*) matching is performed using cuboids of relatively large size on a sparse grid. Depending on the cuboid and grid element sizes, neighboring cuboids may share common borders ($g_i = c_i$ where $i \in \{x, y, z\}$), overlap ($g_i < c_i$), or show gaps ($g_i > c_i$). Obviously, the multi-patch constraint can only be applied in the first case, and overlapping cuboids will produce velocity data correlated to a certain extent. For sufficiently large cuboid sizes trivial initial approximations can be provided and a matching solution can still be achieved. The matching solution $\hat{x}^1$ from this first configuration provides the necessary information to update the conjugate positions. Using these results we proceed on the next level, at which the grid element and/or cuboid sizes become smaller. At this level, approximations for the new adjustment solution are based on the matching solution $\hat{x}^1$ of the previous level. These approximations need not be restricted solely to the obvious choice of the three shifts, but could include the scale and shear parameters as well.

However, to alleviate computational burdens, it may be advisable in a low, coarse levels to solve only for the shift parameters which are typically more crucial for successful matching. The procedure continues with approximations for *level i+1* provided directly or indirectly (via interpolation) from the matching adjustment solutions of *level i*. Thus is devised a globally iterative process for which different levels correspond to a change in the number and/or size of the cuboids, and minor iterations (local adjustment solutions) form single iterative steps of the complex multigrid solution procedure. Matching results from coarse grid levels correspond to low frequency components of the velocity field, while in finer levels the effect of high frequency velocity components is also considered.

6. SOME RESULTS

Some results of the application of 3-D LSM to the sequence of voxel datasets shown in Figure 1 are illustrated in Figure 4 and Figure 5: Figure 4 shows a cuboid of a size of 15x15x15 voxels (which has turned out to be a reasonable cuboid size for this type of data) and the corresponding cuboid in the consecutive voxel dataset resampled with the parameters of the 3-D affine transformation in 8 iterations. Only the incompressibility constraint (paragraph 4.1.) at moderate weight was introduced in this example. Below the cuboids, the 12 parameters after the last iteration and their standard deviations are listed; the precision of the shift parameters in this example is $^1/_{20}$ ... $^1/_{40}$ voxel, and all parameters are significant, although correlations of ~0.9 occurred between a_0 and c_0 and between a_2 and c_2.

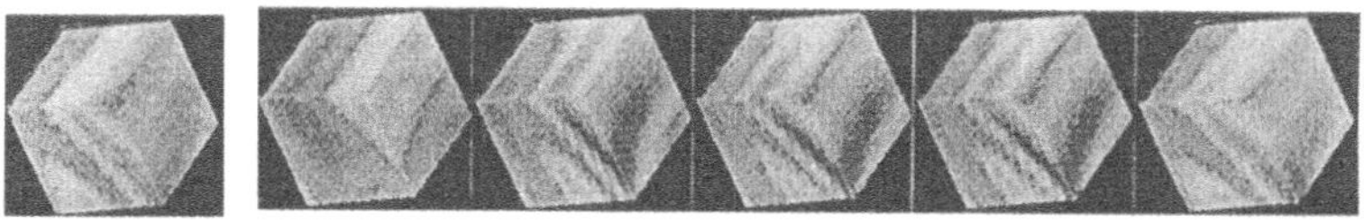

cuboid 1 cuboid 2: 0. 1. 2. 5. 8. it.

σ_0 = 3.1 grayvalues

a0: -4.778 +- 0.023 voxel	a1: 1.140 +- 0.006	a2: -0.097 +- 0.005	a3: -0.118 +- 0.005
b0: 5.652 +- 0.046 voxel	b1: 0.382 +- 0.012	b2: 0.799 +- 0.010	b3: -0.148 +- 0.010
c0: -1.698 +- 0.031 voxel	c1: 0.124 +- 0.008	c2: -0.104 +- 0.007	c3: 0.864 +- 0.006

Figure 4: Deformation of a 15 x 15 x 15 voxel cuboid through 8 iterations (Maas et al., 1994)

Figure 5 shows the 2-D projection of one slice of the velocity field, which consists of a total of ~8000 velocity vectors. Note that the velocity field was oversampled (by defining a grid with mesh size smaller than the cuboid size), but that no neighborhood constraints were introduced.

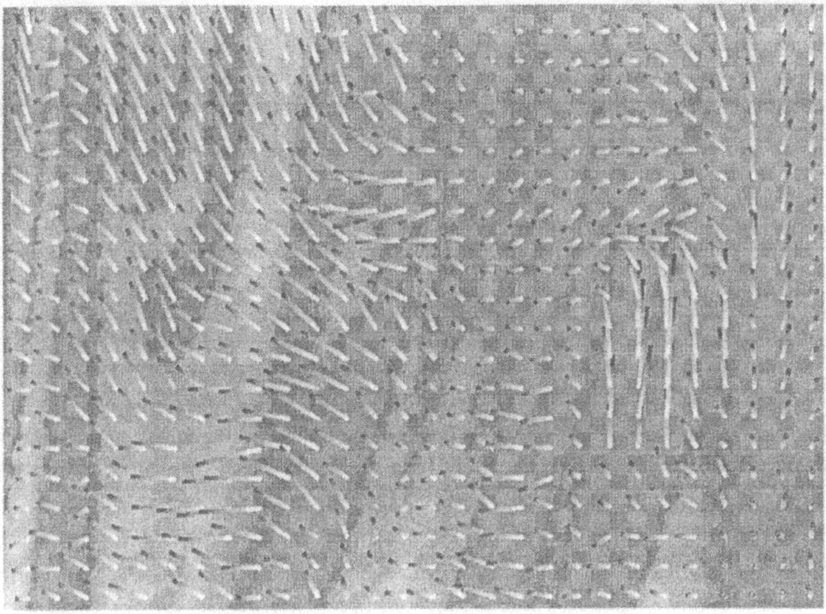

Figure 5: Slice of a velocity field determined by 3-D LSM (2-D projection)

In some regions of the datasets shown in Figure 1 no velocity vectors could be determined, in other regions only certain parameters of the 3-D affine transformation could be determined. The decision, whether a vector could be determined at all and which transformation parameters could be determined, was met after a contrast check in the cuboid and reliability tests.

7. CONCLUSION

3-D least squares matching has proven to be a very powerful technique for tracking fluid elements in tomography datasets of mixing flows marked with fluorescein. Major advantages of the technique are its inherent analysis of the quality of results, its adaptivity to the determinable transformation parameter sets, and the fact that information on the deformation behaviour of fluid elements is determined simultaneously with the velocity field information.

8. REFERENCES

Förstner W., 1986: A Feature Based Correspondence Algorithm for Image Matching, IAPRS, Vol. 26, Part 3/3, pp. 150-166.

Gruen A., 1985: Adaptive kleinste Quadrate Korrelation und geometrische Zusatzinformationen. *Vermessung, Photogrammetrie und Kulturtechnik* 9/85, pp. 309-311.

Gruen A., Baltsavias, E., 1988: Geometrically Constrained Multiphoto Matching. *Photogrammetric Engineering*, Vol. 54, No. 5, pp. 633-641.

Gruen A., Stallmann, D.(1991: High-accuracy edge-matching with an extension of the MPGC matching algorithm, Proceedings Industrial Vision Metrology, SPIE Volume 1526, pp.42-55.

Koch, K.R., 1980: *Parameterschätzung und Hypothesentests in linearen Modellen.* Dümmler Verlag, Bonn.

Li. M., 1989: *Hierarchical Multi-Point Matching with Simultaneous Detection and Location of Breaklines.* Ph.D. Thesis, Department of Photogrammetry, The Royal Institute of Technology, Stockholm.

Maas, H.-G., 1993: Determination of velocity fields in flow tomography sequences by 3-D least squares matching. *Optical 3-D Measurement Techniques II* (Eds.: Gruen, A, / Kahmen, H.), Wichmann Verlag, Karlsruhe

Maas, H.-G., Stefanidis, A., Gruen, A., 1994: From Pixels to Voxels - Tracking volume elements in sequences of 3-D digital images. IAPRS Vol. 30, Part 3/2

Papantoniou D., H. Bühler, T. Dracos, 1990: On the Internal Structure of Thermals and Momentum Puffs. International Conference on Physical Modelling of Transport and Dispersion, MIT, 7.-10. August.

Rosenholm D., 1986: Accuracy Improvement of Digital Matching for Evaluation of Digital Terrain Models. IAPRS, Vol. 26, Part 3/2, pp. 573-587.

Slama, C., 1980 (Ed.): *Manual of Photogrammetry.* American Society of Photogrammetry, Falls Church, Virginia

ESTABLISHMENT OF A TOMOGRAPHIC LASER INDUCED FLUORESCENCE SYSTEM

S. Deusch[1], Th. Dracos[2], P. Rys[1]
[1)] Dept. of Industrial Chemistry and Chemical Engineering
[2)] Prof. em.
Swiss Federal Institute of Technology, CH-8092 Zurich

1. Introduction

The accurate tracking of patterns in finite elements of a volume sequentially imaged in time by using Laser Induced Fluorescence (LIF) tomography requires high quality images of the dye distribution. Special attention must therefore be paid to the selection of the dye, the illumination system and the recording camera, keeping in mind the requirements imposed by the tracking procedure and the scales one wishes to resolve in a specific experiment.

Successful and accurate tracking is achieved by using finite elementary volumes, which in the following are called cuboids with a size of order 10x10x10 voxels. If the maximum translation of a cuboid expected in transverse direction per time-step Δt is equal to half its edge length, the minimum number of tomographic cuts forming a volume is of order twenty. Usually one wishes to determine the velocity field in more than one plane. The number of cuts will therefore in most cases be somewhere between fifty and one hundred. The resulting relatively large thickness of the volume, which requires a large depth of focus, demands careful choice of objectives for the camera used. Some basic information necessary for the establishment of a LIF system will be given in the following. For more details the reader is referred to the relevant literature on light and optics (Handbook of Optics 1995, Hecht & Zajak 1974, Hecht 1992), and fluorescence (Slavik 1994, Wolfbeis 1993). Helpful information is also found in catalogues of manufacturers of optical elements and instruments.

2. Basic Considerations

Fluorescence is the spontaneous emission of photons from electronically excited molecules or atoms. The fluorescence properties of a chemical compound depend on the electron structure of the molecule, the dynamics of the energy exchange between the

Th. Dracos (ed.), Three-Dimensional Velocity and Vorticity Measuring and Image Analysis Techniques, 277–297.

different energy state levels, and the interaction of the fluorophor with the surroundings. The preceding, and therefore underlying basic process leading to electronically excited species is the absorption of laser light. Lasers provide high photon density fluxes in narrow frequency bands. For the case considered here, the molecular transition is assumed to be induced by a single photon absorption. The interacting laser photon must provide the appropriate energy for the transition, which will only take place if the selection rules for the quantum states are satisfied. The orientation of molecules in gases and liquids is random, but upon excitation with polarized light (lasers emit light with usually vertical polarization), fluorophor molecules whose absorption transition dipole is parallel to the electric vector of the excitation field are selectively excited (photoselection). However, this fluorescence anisotropy is diminished by rotational diffusion which occurs during the lifetime of the excited state and rotates the emission dipole of the fluorophor. To reduce polarization dependence on the exciting laser, broad-band detection of the emitted light is recommended, which also enhances the signal intensity. The fluorescence light thus detected is random in momentum and phase and collected only in a fraction of the whole 4π steradians.

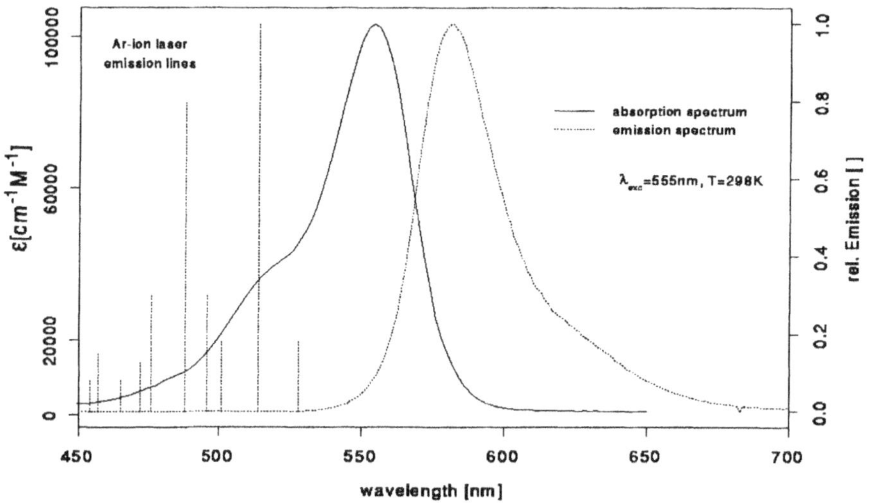

Figure 1. Measured spectra of absorption and emission of rhodamine-B in tap-water

In Fig. 1. the absorption and emission spectra of rhodamine-B are given together with the emission lines of an all-line Ar-ion laser which could serve as excitation source for the fluorescence of rhodamine-B. They are measured in tap-water such as used in the experiments. The measured ε_{max} value is 103,500 [$M^{-1}cm^{-1}$] at λ=555nm. This value is ~ 10% smaller than the value found in literature, which is due to the use of tap-water as solvent and of commercial-grade dye. In the same figure the emission lines of an Ar-ion laser are also presented. Two important features are recognizable: (i) the emission spectrum is red-shifted to lower energies, characterized by the so-called Stokes-shift (the wavelength gap between the maxima of absorption and emission), (ii) absorption and emission spectra are in a certain way like a mirror image of each other. Typical Stokes-shifts are in the range of 20 nm and more. In the present case the Stokes-shift is ~ 34 nm. An important feature of a fluorescent compound to be used in an analytical application is its fluorescent quantum efficiency Φ_f defined as the ratio between the number of emitted photons to the number of absorbed photons. The higher the Φ_f, the lower the concentration of dye in the application of interest can be made. Fluorescence quantum yields can be as high as 90% or more for very strong fluorescent compounds. Typical lifetimes of the involved processes are: excitation ~ 10^{-15}s, internal conversion and vibrational relaxation ~ 10^{-12}s, fluorescence decay τ ~ 10^{-9}s to 10^{-8}s. If the excitation is caused by a rectangular pulse of light of duration t_i, the stationary rate of emission is reached after ~ 10^{-15}s, and after the end of irradiation the fluorescence emission I_f decays from the stationary intensity I_f^{st} with τ the mean fluorescence lifetime following

$$I_f(t) = I_f^{st}\exp(-t/\tau) \tag{1}$$

τ allows for any of the competing deactiviation processes of the excited state

$$\tau = (k_f+k_i+k_x)^{-1} \tag{2}$$

where k_f denotes the fluorescence rate constant, k_i internal conversion, and k_x inter-system crossing. The raise time is negligible for all practical applications in this context. The decay time must be taken into account when the irradiation time t_i is of the same order of magnitude as τ, i.e. when $t_i \leq 5 \cdot 10^{-7}$s. For sensor integration times longer than this limit, the sensor yields stable means of fluorescence emission. The quantum efficieny Φ_f of fluorescence can be expressed in terms of the kinetics of the involved processes. Assuming a two-level electron energy system, a condition for excitation saturation and linear excitation fluorescence can be specified. The direction of an electron transition is indicated by the order of the indices, e.g. B_{ij}, where B_{12} is the Einstein coefficient of absorption and A_{21} the Einstein coefficient of spontaneous emission. Q_{21} is any type of deactivating quenching process. The rate coefficient for single-photon laser-stimulated absorption is $k_0 = B_{12}I_v$, with I_v the laser spectral intensity in [J cm^{-2}]. For the two laser-coupled energy levels the steady-state fluorescence rate R_p in [number of photons/cm^{3}s] at constant illumination intensity I_v is given by, Seitzman (1993),

$$R_P = n_1^0 B_{12} I_v \frac{A_{21}}{A_{21}+Q_{21}} \cdot \frac{1}{1+I_v/I_v^{sat}} \tag{3}$$

where n_1^0 is the initial number density n_1 of fluorophor molecules (i.e. in [molecules/

/litre]) in the solution, which can be expressed with the help of concentration as follow

$$n_1^0 = c \cdot N_A \tag{4}$$

where N_A is Avogadro's constant $N_A = 6.0221 \cdot 10^{23}$ mol^{-1}, and c the molar concentration [moles/l]. I_v^{sat} is the intensity of saturation given by

$$I_v^{sat} = \frac{A_{21} + Q_{21}}{B_{12}(1 + g_1/g_2)} \tag{5}$$

g_i is the degeneracy level of state i. Eq.(3) can be simplified for two limiting cases, corresponding to either low or high laser intensity. The first ratio on the right hand side of eq.(3) is essentially the fluorescence quantum yield Φ_F expressing the competition between emission (A_{21}) and emissionless processes (Q_{21}). For low laser intensities, $I_v \ll I_v^{sat}$, eq.(3) reduces to the linear fluorescence equation, eq.(6), with the fluorescence intensity linearly dependent on the excitation light intensity as well as on the total species molecule number density n_1^0.

$$R_P = n_1^0 B_{12} I_v \frac{A_{21}}{A_{21} + Q_{21}} \tag{6}$$

For high laser intensities, $I_v \gg I_v^{sat}$, R_p in eq.(3) becomes independent of laser intensity and quenching effects and is proportional to species concentration only. Saturated fluorescence would be a desirable goal in LIF imaging since inhomogeneities in the illuminating light sheet and unspecific quenching need not be accounted for. But saturation is difficult to achieve with the lower intensities of the expanded light sheets, especially with continuous wave (cw) lasers. Strong pulsed lasers are more likely to reach saturation. Thus the linear fluorescence equation, eq.(6), will be valid for the LIF applications envisaged. Quenching processes are a major hurdle to quantitative LIF measurements in multi-component solutions. Molecular quenching rates depend on temperature, pressure, and concentration. For this reason it is difficult to predict values for Q_{21} in reacting flows thus making it impossible to determine the quantum efficiency Φ_f other than by experimental measurement. Many chemical substances are known to act as fluorescence quenchers. Among them are halide ions, heavy metals, dioxygen, and many organic compounds. Basically, there are three types of quenching mechanisms: (i) Static quenching: In this type a parasitic side reaction takes place in which the fluorophor A reacts with Quencher Q to give a non fluorescing complex AQ, while the fluorescence lifetime of free excited A* remains unchanged. At high concentrations A may act as a quencher itself. (ii) Dynamic quenching: in this case the quencher molecule Q interacts with the excited fluorophor A* to deactivate it (e.g. heavy metal effect or collisional quenching). (iii) Energy transfer between excited donor molecules of A close to acceptor molecules of Q (Förster mechanism). Another unexpected process of upper-state decay may be laser-stimulated photo-ionization with unstable fluorescent compounds.

In the linear fluorescence limit of eq.(6) the absorption along the excitation pathway must be taken into account because it influences the fluorescence signal intensity. For a given species concentration the intensity of fluorescent light, I_f, is proportional to the absorbed intensity of excitation irradiation, where the initial intensity I_0 is either transmitted, I_t, or absorbed, I_a.

$$I_f = \Phi_f I_a = \Phi_f(I_0 - I_t) = \Phi_f I_0 (1 - I_t/I_0) \qquad (7)$$

with Φ_f the fluorescence quantum yield, the emitted light intensity may be derived using the Lambert-Beer law. This states that light propagating through an absorbing medium is attenuated exponentially according to

$$\frac{I_t}{I_0} = e^{-k\int_0^x c(x)dx} = 10^{-\varepsilon\int_0^x c(x)dx} \qquad (8)$$

where k is the logarithmic coefficient of absorption, c(x) is the local molar concentration of the chemical species, and x is the pathway of absorbance in [cm]. Designating the average concentration of c(x) along x by c and taking the first three terms of the Taylor expansion of eq.(8) , eq.(7) can be written, as

$$I_f = \Phi_f I_0[2.303\varepsilon cx - (2.303\varepsilon cx)^2/2 + (2.303\varepsilon cx)^3/6] \qquad (9)$$

For small arguments of the exponential function all higher order terms in eq.(8), except the linear one, can be neglected. Hence the fluorescence intensity may be approximated by

$$I_f = \Phi_f I_0 \, k \, c \, x \qquad (10)$$

Using instead the more common decadic extinction coefficient ε [$cm^{-1}M^{-1}$], the total fluorescence intensity at low concentrations and/or short pathway is given by

$$I_f = 2.303 \, \Phi_f I_0 \, \varepsilon \, c \, x \qquad (11)$$

To keep the LIF response nearly linear with respect to concentration, the maximum average concentration permitted should be such that no more than 5% of the transmitted light is absorbed within the pathway of length x, which means that at the end of the pathway there will be only 2% deviation from the linear relation. If this constraint cannot be fulfilled in LIF experiments, the imaged concentration must be corrected iteratively pixel by pixel along x, e.g. by a predictor-corrector procedure. In addition, corrections for an inhomogeneous illumination intensity $I_0(x,y)$ must be performed. $1/(\varepsilon \cdot c)$ has the dimension of a length (it is also called optical depth) and can be compared to x directly. If the illumination pathway x is much smaller than the optical depth, the absorption along this pathway can be considered negligible and the linear fluorescence limit is assumed to be valid. To avoid self-quenching and non-linear absorption the fluorophor concentration must be kept low. An increase of concentration is, therefore, not a remedy for weak fluorescence and faint LIF images.

The distance x is the pathway of the light ray from the beginning of absorbing solution to the outmost location of the imaged region. For well-known experimental flows, such as the turbulent free jet, one can estimate flow dimensions and scalar concentration from published data (e.g. Dahm & Dimotakis, 1990). With x and ε given, the maximum average concentration permitted can be determined. Similar arguments apply to the LIF signal attenuation from the locus of measurement up to the imaging sensor if absorbing material resides in between. The attenuation along this pathway must be taken into account. This is easy if it is constant. Alternatively, a low-pass cut-off filter may be used in front of the sensor. The longer emission wavelengths, which are outside the absorption spectrum of the molecule, can then be recorded unperturbed. The same filter also eliminates scattered light noise due to e.g. small particles.

3. Fluorescent Tracers for LIF Imaging

Technical flow visualisation applications involving fluorescence are made with a limited number of fluorescent compounds of which the most commonly used in water are fluorescein and its disodium salt, also called uranin, rhodamine-B, and rhodamine-6G. Actually, a palette of diversely coloured fluorescent tracers exists for the various fields of visualisation applications, see (Lakowicz 1991, Wolfbeis 1993). For practical reasons, the fluorescent compounds presented herein are intended for water only as solvent. A few tracers for passive LIF flow visualisation as well as ion-sensing fluorescent tracers are presented. They are described in terms of their absorption coefficient ε, their quantum efficiency Φ_f, and their wavelengths of maximum absorption and emission.

ε is a given property of the fluorescent compound and said to be independent of concentration in highly diluted solutions. The value of ε is specified at the maximum absorption peak essentially indicating ε_{max}.

The constraint of optical transparency favours chemical compounds with high molar absorption coefficient ε since the same attenuation may be reached at very low concentrations. An additional demand is a high fluorescence quantum yield which also permits imaging at low concentrations. Low tracer concentrations are generally favourable, on the one hand because the properties of the solution are kept close to those of the pure solvent, and on the other hand, because Φ_f and ε are constant. The requirements for an ideal, water-soluble fluorescent tracer compound for LIF imaging may be summarized as follows:

- good solubility in water
- high molar absorption coefficient ε
- high quantum efficiency Φ_f to provide good signal to noise ratio
- absorption and emission preferably at long wavelengths because Rayleigh scattering is dependent on $1/\lambda^4$
- Stokes-shift should be as large as possible to keep re-absorption of the emission signal small
- the absorption spectrum should match the laser emission lines for LIF
- the emission spectrum should match the characteristics of the existing instrumentation, e.g. the sensor sensitivity spectrum (see Fig.7 in Virant & Dracos chap.three)
- stability against photo-decomposition and fast quenching by other chemicals

TABLE 1: Some properties of passive tracers for LIF imaging in water

Fluorescent tracer	Φ_f	λ_{exc} [nm]	λ_{em} [nm]	ε [cm^{-1} M^{-1}]
disodium fluorescein	0.92	491	515	88,000
rhodamine-B	0.60	555	574	115,000
rhodamine- 6G	0.78	538	564	116,000

Many fluorescent compounds have been devised to enable ion concentration mapping (Lakowicz 1991, Slavik 1994, Wolfbeis 1993). A selection of pH-sensing and Ca^{2+}-specific fluorescent indicators is given in Table 2.

TABLE 2: Some properties of common ion-sensitive tracers in water

Ion sensitive tracer	Φ_f	λ_{exc} [nm]	λ_{em} [nm]	ε[cm^{-1} M^{-1}]	pKa's
		pH Probes			
fluorescein	0.92	490	520	63,900	2.1;4.4;6.8
c-SNAFL-1	0.29	510	539	31,000	7.5
pyranin	0.99	455	515	18,400	7.2-7.8
		Ca Probes			
FURA-2	0.49	335	505	35,000	-
Fluo-3	0.18	507	526	83,000	-

The spectra of absorption and emission of rhodamine-B in tap-water, a widely used dye system and well characterized in literature, are illustrated in Fig. 1. Many other absorption and emission spectra were compiled by Berlman (1971), however most spectra were not recorded in water solutions.

3.1 LASER LIGHT SOURCE

Some effort should be invested into the correct choice of the laser system for a particular experiment because it is the most expensive element of the system. A primordial issue in LIF imaging is the choice between pulsed lasers and cw-lasers. Pulsed lasers offer higher photon densities at very short times (ns range). In 2D imaging this allows for highly time-resolved, sharp image sequences of moving objects. cw-lasers deliver constant illumination intensity, which can be used to scan dynamically the focused laser beam over the region of interest while recording an image simultaneously. The other two most important questions in picking up the proper laser light source for a given LIF application are: (i) Matching of the laser wavelength with the absorption spectrum of the fluorescent tracer, (ii) sufficient light power to illuminate the whole area of interest of the experiment. With this in mind, the choice of a laser system for a particular application is one of picking the most suitable type from one of the many buyer's guides (e.g. Laser Focus World Buyer's Guide). The most important technical features which are treated therein as well as in laser literature are laser wavelength, power, linewidth, and pulsewidth. Because it is nearly impossible to keep track with the latest developments in the vast area of laser technology, the beginner is best advised to more general introductory surveys (Hecht 1992, Handbook of Optics 1995). In terms of the lasing material one can distinguish between three well developed laser systems: (i) Gas lasers (Kr-,HeNe-, Ar-ion-, CO_2-lasers), (ii) solid state lasers (ruby, Nd:YAG, Nd:glass), (iii) dye lasers.

Dye lasers have the advantage of being quasi tunable. New solid state systems of tunable lasers for the visible region are developed and commercialized now, too. One group are transition metal ion vibronic lasers, like the titanium:sapphire system, another new group are diode pumped solid state lasers. For the application envisaged cw-lasers are the appropriate choice at present.

3.2 BEAM SHAPING OPTICS

Because cubical voxels are usually desired, the thickness of the light sheet should be equal to the edge length of a voxel. The emerging laser beam must be focused to the appropriate size. Any combination of optical devices such as lenses, prisms, and mirrors may be used to achieve this. The egressing laser beam has a diameter of typically 1-2 mm. In most applications the thickness of the sheet is much smaller, e.g. 60-100 μm. To achieve best possible focusing, which is near the diffraction limited spot size, the beam must be expanded first. A beam expander is essentially a telescope used backwards. It is commercially available for any type of common laser. Optical element manufacturers offer a broad variety of lenses specifically designed for use with lasers. For Gaussian laser beams it is convenient to specify an optimized lens aperture D_{opt}, which is the diameter of the 99% throughput contour corresponding to 3/2 times the $1/e^2$ irradiance contour diameter.

The minimum focal spot diameter d_0 in mm can be estimated from

$$d_0 = 2w_0 = 4\,\lambda\,F\,/\,(\pi\,D) \tag{12}$$

with w_0 the beam radius, F the effective focal length of the focusing lens(es) [mm], D the beam diameter at the focusing unit [mm], and λ the characteristic wavelength of the laser light [mm]. Eq.(12) presupposes that the TEM_{00} mode of the laser is adjusted. To make focusing more flexible, especially for far field focusing, a set of lenses can be combined to give a zoom lens with variable focusing length. The focal length F of a two-lens system spaced at distance d, where focal lengths f_1 and f_2 of either sign are permitted, is

$$F = f_1\,f_2\,/\,(f_1 + f_2 - d) \tag{13}$$

The half beam waist at a location x from the focal point is given by eq.(14) (see Fig. 2.).

$$w(x) = w_0 \sqrt{\left[1 + \frac{\lambda \cdot x}{\pi \cdot w_0^2}\right]} \tag{14}$$

Generally, a long focusing length F combined with a large beam diameter D is favourable to obtain a beam diameter minimally varying along the dimension parallel to the beam. The image should be centred around the (flat) focal point. Usually, the flow field is imaged perpendicularly to the illumination sheet. The variation of beam diameter across the imaged region can be optimized by using eqs.(12) to (14).

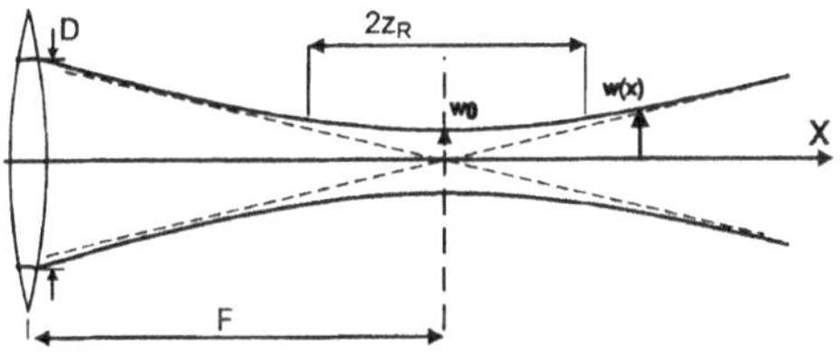

Figure 2. Gaussian laser beam focusing

The condition that w(x) does not vary by more than a factor of $\sqrt{2}$ on an interval of $2z_R$ around the focus is

$$2z_R = 2\,\pi\,w_0^2\,/\,\lambda \tag{15}$$

To obtain a 2D image of the flow the laser beam must be expanded to a laser sheet. Two possibilities to produce a sheet of laser light may be distinguished, the one being static, the other dynamic. The most common static device is a cylindrical lens which magnifies in one direction only, i.e. it expands a point image into a line image. This is the simplest way to generate a planar laser light sheet. The other method is to scan the beam in its full intensity over the illumination plane while recording. This poses the additional problem of synchronisation of illumination and imaging. Beam scanning methods can be divided into such in which the laser beam is deflected, e.g. on a rotating polygonal mirror, and such in which the beam is refracted, for instance by a rotating, transparent prism.

3.3 CYLINDRICAL LENS LIGHT SHEETS

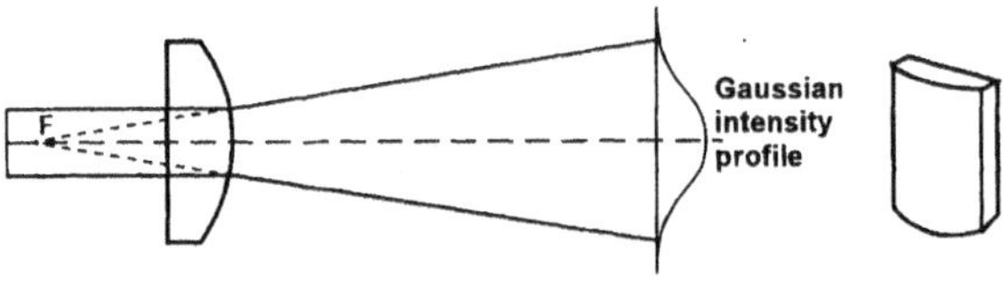

Figure 3. Schematic of a laser light sheet generated by a cylindrical lens; side view of lens

Plano-convex, see Fig. 3., or plano-concave lenses can be used to expand the beam. Their basic advantage is robustness, easy handling, and availability. Their major disadvantage is the non-uniformity of the expanded laser light sheet. The nearly Gaussian intensity profile of the laser beam is preserved, which makes a uniform light intensity difficult to achieve. A high degree of uniformity of illumination intensity within the laser light sheet is desired for LIF imaging. With cylindrical lens light sheets the problems of over-exposure in the central part of the flow image and under-exposure in the wings may arise. However, the Gaussian wings can be cut off by use of an iris aperture at the expense of partial loss of the light emitted by the laser. An evaluation of a LIF flow visualisation experiment with a cylindrical lens can be found in van Cruyningen et al. (1990), where all necessary image corrections are discussed. It should also be kept in mind that the intensity of light diminishes in radial direction proportional to 1/R, with R the distance between the focus of the lens and a point in the imaged part of the sheet. This intensity reduction can be neglected in image corrections, only if the distance between the front and rear side of the image, ΔR, is small relative to R, i.e. $\Delta R/R \ll 1$. To avoid this radial intensity reduction a more sophisticated arrangement with two cylindrical lenses can be used. The first spreads the beam while the second collimates it.

3.4 LASER SHEETS BY ROTATORY SCANNERS

According to the definitions given by Beiser et al. (1995) we can discern between objective, preobjective, and postobjective scan. Objective scan is transverse translational scanning, i.e. for example shift of the plane of focus along the optical axis by moving the lens. Preobjective scan consists of an angular scanner located ahead of the focusing unit. Postobjective scanning generates angular scanning but preserves its radial symmetry, thus engendering a circular scan locus. The focusing lens is situated ahead of the scanner. The following discussion is limited to postobjective scanners. A survey of laser scanning techniques in particle imaging velocimetry, a related experimental method, is given by Rockwell et al. (1993). Therein, rotating polygon mirrors, oscillating mirrors and acousto-optic deflectors are compared and the basic design equations are given. The concept of scanning beam illumination is advantageous over expanded beam techniques in that the former produces better quality images in high-speed flows because the scanning light beam illuminates a certain fluid particle for shorter time but with higher light intensity.

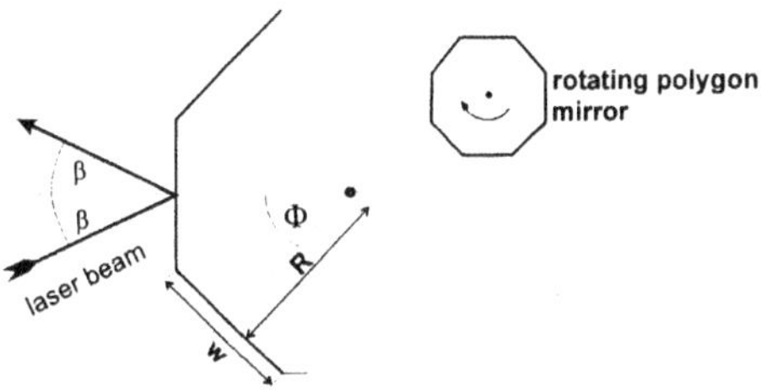

Figure 4. Beam Deflection from a rotating polygon mirror

Furthermore, the scanning beam illumination method gives a superior sheet intensity profile. Optimization of scanning techniques demands careful consideration of the concepts of illumination time and exposure time, scan rate relative to the time scale of the speediest flow domains, and the retrace time of the beam to its initial position. The various laser scanning methods yield a different exposure function E(x) which must be accounted for in the case of unsaturated, linear fluorescence for image post-correction. The intensity of light in the image plane of the camera is $I(\mathbf{x},t)$. The local exposure function $E(\mathbf{x},t)$ is the integral of the intensity over the effective time t_e of laser illumination.

$$E(\mathbf{x},t) = \int_{t_e} I(\mathbf{x},t)dt \qquad (16)$$

For a nearly instantaneous LIF imaging of a continuum moving with a mean velocity u on a pixel array sensor with a resolved dimension Δx in object space, the imposed scan time t_s is constrained by

$$t_s << \Delta x / u \qquad (17)$$

In order to freeze the motion of the flow during imaging, a rotatory scanner with a radius R and a scan angle Θ must satisfy the additional condition

$$u \cdot t_s << R \cdot \Theta \qquad (18)$$

With t_s specified by eqs.(17) to (18) the scan speed is given by

$$v_{scan} = R \cdot \Theta / t_s = R \cdot \omega \qquad (19)$$

The selection of a rotating polygon mirror, an oscillating mirror, or an acousto-optic deflector depends also on the scan frequency and magnitude of scan angle. Accordingly, each method has its trade-offs.

3.4.1 Polygonal Mirrors

Detailed design equations for a polygon laser scanner are given by Beiser (1991 & 1995). The angle between adjacent facet normals being Φ, see Fig.4, the resulting total scan angle of one polygon facet is 2Φ. For a polygon with n facets, the angle Φ is $2\pi/n$. Therefore, the total scan angle of the output beam is

$$\Theta = 2\Phi = 4\pi/n \tag{20}$$

Since the beam has to be focused down to the sub-mm-range in LIF imaging, which is only possible with a collimating lens, a polygonal mirror scanner can only be run in post-objective arrangement. The incident beam diameter may be considerable at the polygon surface, which actually reduces the usable scan angle Θ of eq.(20), if double-illumination at the edges of the facets must be avoided. Beiser (1991) shows how to calculate the facet width w taking into account the beam diameter. The diameter of the polygon mirror, $D = 2R$, follows from the trigonometric relation

$$\sin(\pi/n) = \sin(\Phi/2) = w/2R \tag{21}$$

A strategy for the implementation of a polygon scanner in a LIF experiment with a given flow geometry is basically as follows:

- determine the width of laser sheet
- the distance between scanner axis and imaged locus gives Θ
- $\Phi = \Theta/2$ ideally, but the finite beam diameter has to be accounted for, which reduces the effective scan angle and determines the facet width w
- the number of facets is $n = 2\pi/\Phi$
- beam diameter, Φ, w, and n determine the polygon diameter

A major advantage of polygonal mirror scanners in LIF imaging is the nearly constant scan velocity which gives a nearly constant exposure function, eq.(16). However, the sheet of light has a triangular form. The scan time is changing with R, which varies from R_1 at the front edge of the imaged section of the sheet to R_2 at the rear side of this section. This produces a non-uniform exposure function. The scan time t_s depends on R as

$$t_s(R) = \frac{\operatorname{atan}(y/2R)}{\omega}, \; R_1 \le R \le R_2 \tag{22}$$

where y is the width of the imaged section, and ω the angular speed. This is only negligible if the scan angle is small. If the mean of the deflection of the polygon scanner is 90°, simple geometric reasoning shows that the incident beam must be directed at a distance $\sqrt{2}R/2$ from the spin axis onto the polygon, see Fig.5. This has to be accounted for in the construction and experimental set-up. Besides, polygonal scanners can reach very high speeds of rotation with proper bearings. Up to several tenthousand Hz scanning rate can be achieved. Performance characteristics may be found in Marshall et al. (1991). A difficulty with the construction of a rotatory scanner is to avoid the wobble of the rotational axis as well as the alignment of the mirrors with the axis, which introduces an oscillation of the sheet location. Consideration of this kind of error is found in Horikawa et al. (1991).

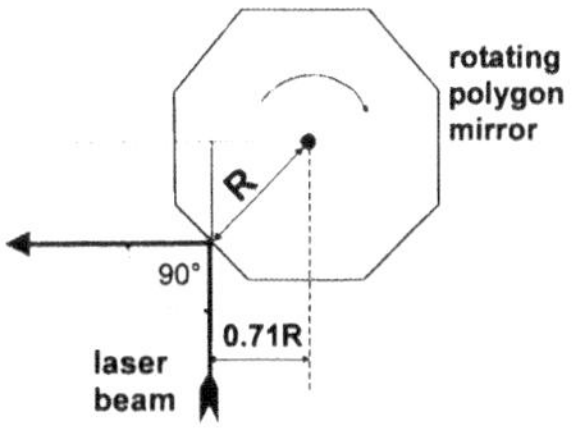

Figure 5. 90°- deflection of laser beam with polygonal scanner

A survey of bearings for rotary scanners is given in Preston et al. (1991).

A special parallel scanning beam optical arrangement with a rotating polygon mirror is presented by Gray et al. (1991). A rotating mirror is used in conjunction with a stationary parabolic mirror which converts a constant rotational into a parallel scan. In general, when w is large and Θ small, the diameter of the polygonal scanner becomes prohibitively large.

3.4.2 Oscillating Mirrors

For small deflection angles oscillating mirrors are advantageous as compared to rotating mirrors. Fig.6a shows the incident laser beam deflected at its maximum scan angle. Similarly to the polygon mirror, a mirror rotation by Φ yields a scanned region of aperture 2Φ with the incident beam fixed in position. Fig.6b depicts the scanner motion, i.e. the angle of rotation against time. The flat upward ramp corresponds to the actual scanning process taking place during t_s. The steep downward ramp represents the retrace time t_r which should be as short as possible to avoid secondary imaging. A synchronized shutter can interrupt illumination during t_r either at the laser beam or at the camera. The mirror itself must be small and thin (~0.5mm) in order to be of low inertia. The mirror width

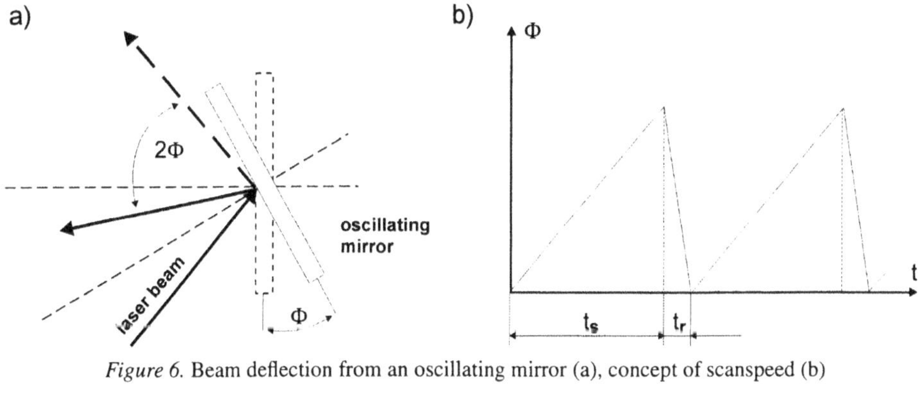

Figure 6. Beam deflection from an oscillating mirror (a), concept of scanspeed (b)

should be such that the incident beam is not cut off even at the most extreme angle β of incidence, i.e. the width $w > D_{beam}/\cos\beta$. With postobjective scanning and far-field focusing in LIF imaging the incident beam diameter may be too large for common oscillatory mirror scanners, especially if the mirror is situated right behind the collimating and focusing optics. The driving system for oscillatory scanners is preferentially a galvanometric coil which can produce constant linear scan in freely programmable forms such as sawtooth or triangular.

For LIF imaging the scan should be run in a single sweep direction in order to obtain images with equal time increments. Thus, a triangular wave form is necessary with a retrace time t_r as short as possible, as shown in Fig.6b. Mechanically these scanners are limited to frequencies of several hundred Hz which cannot compare to polygonal scanners, Rockwell et al. (1993). Resonant driving systems oscillate in harmonic motion and are therefore not recommended for LIF illumination, although they can be operated at higher frequencies and larger angles. The scan direction of two subsequent images would be anti-parallel.

The advantages of an oscillatory scanner are the variable scan angle, unlike the polygon mirror with fixed scan angle, and a programmable scan sequence. Disadvantages are the relatively low scan frequency compared to rotatory systems (~ 100 to 200 Hz), and the fact that only tiny mirrors with low inertia can be driven at high frequencies, i.e. the beam diameter must be small or the scanner close to the focusing point, which in most cases is not possible. In practice, oscillating mirrors have not be used beyond a frequency of 150 Hz in LIF tomography. A review of galvanometric and resonant low-inertia scanners is given by Montagu (1991). The state-of-the-art in oscillatory optical scanners is assessed by Marshall & Montagu (1995). Dahm et al. (1991) performed their experiments with galvanometrically-driven oscillating mirrors. An alternative to galvanometric driving of oscillating mirrors is piezo-crystal driving. This will be discussed later.

3.4.3 Prism scanners

Forming of a parallel laser light sheet by means of a rotating transparent prism is an alternative technique to the above methods. It produces a highly uniform, energetic light sheet, which can be used for LIF excitation as well as in Particle Image Velocimetry (PIV). The laser beam entering the rotating prism is refracted to the normal of the surface and when leaving the prism is again refracted back at the same angle thus shifted parallel to the incident ray, see Fig.7.

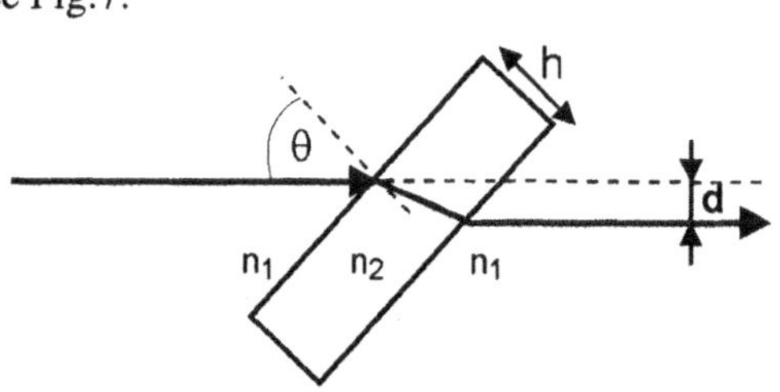

Figure 7. Parallel displacement of a light beam passing through a parallel plate.

The prism is supposed to consist of an even number of sides with parallel opposite sides. As a net effect, the laser beam is shifted parallel to itself on a line perpendicular to the prism's spin axis. The rotation of the prism thus produces a linear unidirectional scan.

The parallel shift d can be derived using Snell's law of refraction

$$d = h \cdot \sin\theta \left[1 - \frac{\cos\theta}{\sqrt{\left(\frac{n_2}{n_1}\right)^2 - (\sin\theta)^2}} \right] \tag{23}$$

A laser beam directed onto the spin axis of a rotating quadratic prism scans a width 2d as the angle θ varies from $-\pi/4$ to $+\pi/4$ between neighbouring edges, see Fig. 8. With $\theta = \pm\pi/4$, $n_2 = 1.5$, which is typical for glass, and $n_1 = 1$ for air, the ratio of the scan-width of the laser sheet to the edge length h of the prism is fixed equal to

$$2d = 0.658 \cdot h \tag{24}$$

The edges of the prism give rise to beam splitting onto adjacent sides, as is the case also with rotating polygonal mirrors. To avoid this double exposure at the fringes, the prism's edges can be covered by black light absorbing stripes of which the projected width equals one beam diameter. The usable incidence angle, i.e. its duty cycle, is reduced, thereby giving a smaller scanwidth.

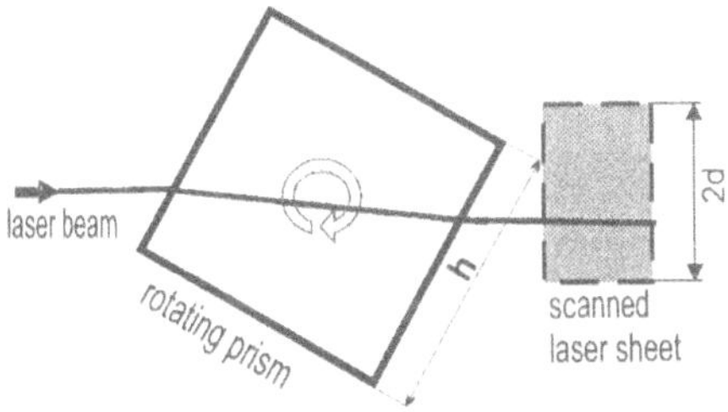

Figure 8. Schematic of a rotating prism scanner with theoretically obtainable light sheet.

The most important considerations for the construction of the rotating prism scanner are listed briefly:

- the material must be of optical quality
- tolerances of angle and surfaces must be as small as possible
- wobble-free rotation
- high strength of the material used

The framing rate in Hz of the camera is imposed by the revolution speed of the prism. For a prism with four sides, see Fig.8, it is equal to 4n/60, with n the prism revolution speed in RPM. The synchronized triggering may be achieved by means of an optical switch which is mounted on axis with the rotating prism. The local time dependent exposure function can be expressed as

$$E(\mathbf{x},t) = I(\mathbf{x},t)l_v / \mathbf{v}(\mathbf{x},t) \tag{25}$$

with $I(\mathbf{x},t)$ the illumination intensity at location $\mathbf{x}$ off the centerline, l_v the linear dimension of a voxel, and $\mathbf{v}(\mathbf{x})$ the local scanning speed. With $\mathbf{v}(\mathbf{x})$ higher near the corners of the prism, illumination time will be shorter at the edges of the sheet. But, to some degree, this effect may be counter balanced by making use of the proper laser polarization plane. According to Fresnel's law of reflection of polarized light on plane surfaces, for light with its electromagnetic field oscillating in the plane of the surface's normal (in Fig. 9. ||-polarized), reflection losses decrease with increasing incident angle until Brewster's angle is reached where reflection vanishes. This angle is larger than $\pi/4$ for optical materials used in prism construction and, therefore, is never reached during prism rotation. The $\perp$-polarized light reflection increases progressively for incident scan angle θ_1. Therefore, it is advisable to take advantage of the (usually) vertical polarization of the laser light source and direct it with the minimum number of deflecting mirrors onto the rotating prism such that the lower reflection curve in Fig.9 is valid. Accounting for the reflection losses of the || - polarized beam on the outer and inner sides of the prism, the exposure function, eq.(25), can be calculated for an ideal light sheet.

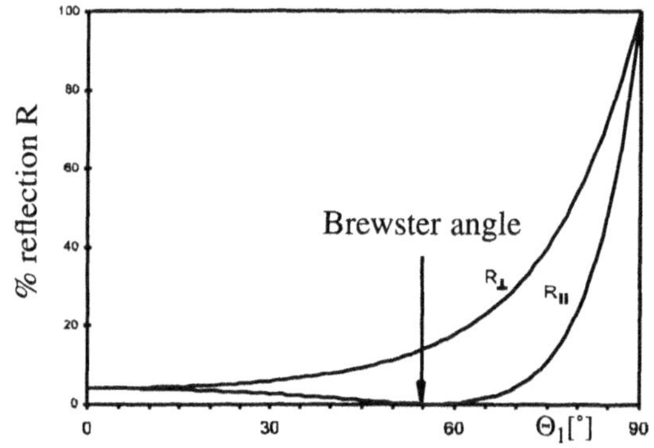

$$R_\perp = \theta\left[\frac{\sin(\theta_1 - \theta_2)}{\sin(\theta_1 + \theta_2)}\right]^2$$

$$R_{||} = \left[\frac{\tan(\theta_1 - \theta_2)}{\tan(\theta_1 + \theta_2)}\right]^2$$

Figure 9. Fresnel's law of reflection of polarized light at glass, $n_2 = 1.5$; $n_1/n_2 = \sin\theta_2 / \sin\theta_1$

An example of an exposure function for a 65mm prism with refractive index 1.5, exploiting the || - polarization of the laser, is given in Fig.10. Note that the variation of light exposure function does not fall below 80% of the maximum value, for a completely parallel light sheet without conical expansion as is the case with a single cylindrical lens. The cylindrical lens sheet intensity profiles falls off much steeper following the expanded Gaussian profile of the beam. The light intensity profile in the sheet can also be measured if necessary. In all cases discussed previously the exposure function must be determined to perform the appropriate corrections of the grey level distributions in the

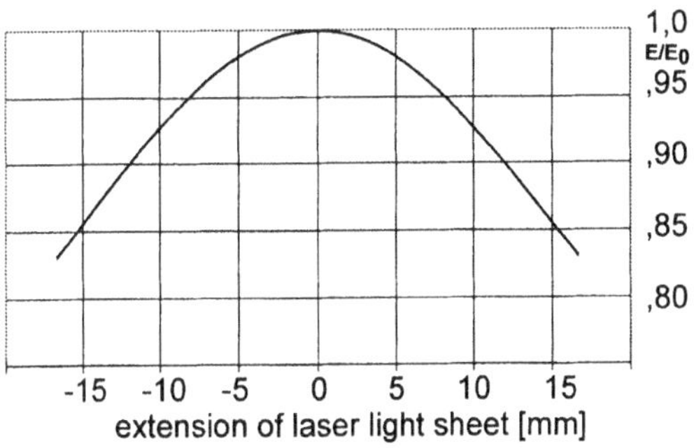

Figure 10. Exposure function of laser light sheet obtained by a rotating prism with h=65mm, n=1.5

unsaturated fluorescence regime. This is especially important when precise determination of the gradients of the grey levels is necessary. An advantage of the Adaptive Least-Squares Matching procedure used for LIF velocimetry, see Maas this chapter, is its small sensitivity to errors in grey level gradients.

3.5 Z-SHIFT OF LASER SHEET FOR 3D LIF

To obtain 3D images by LIF the laser beam or sheet must also be scanned into the third dimension. One way to achieve this is via a mirror which is mechanically coupled to the spinning axis of the rotating prism. More approved ways are by a second scanning mirror either electrogalvanic or piezo crystal controlled. This scanning can be considered parallel if $\tan\theta \approx \theta$. Both types of angular scanners are available commercially. Their driving function should be ideally stepwise as depicted in Fig.11. Φ denotes a driving voltage or scan angle. During the horizontal halt the mirror is fixed in position. This halt should last the time of one xy-scan, which should be equal to the integration time of the camera.

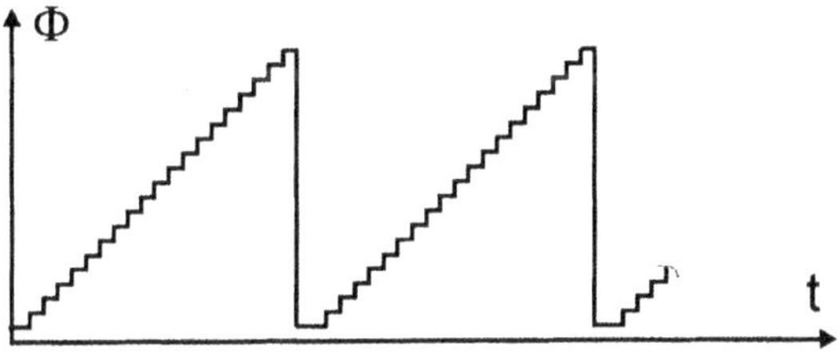

Figure 11. Step-function ramp for z-shift driver system scanning a volume of 20 planes

The downward jump to the initial z-scan position represents the retrace of the z-scan which is necessary in order to scan a 3D volume unidirectionally.Again, a small mirror may not take up the whole beam diameter, especially with postobjective scanning. The synchronization of the third scan direction with the xy-scanner producing the light sheet and with the camera must be carefully adjusted. Principally, any scanner with two adjustable mirrors can cover a 3D region of physical space, see for instance Marshall & Montagu (1995).

3.6 HIGH QUALITY OPTICS VERSUS LASER POWER

At each interface where the index of refraction changes reflection losses occur. In an optical system containing n such transitions the attenuation is

$$I_L/I_0 = 1 - (1-p)^n \tag{26}$$

where I_L is the intensity loss, I_0 is the intensity at the light source, and p the fractional loss of intensity at an interface. For normal incidence at a glass surface (e.g. a lens) p is about 4%. As an example, if ten such interfaces are present, the total loss is 34%. Using high quality optics with anti-reflective coating where reflections are reduced to below 1% at least, say p=0.5%, the same optical arrangement would attenuate the illumination light by 5% only. Therefore, investing in high quality optic components such as lenses, mirrors, filters, and the like, pays off because a laser with lower power can be employed for the same purpose and the laser is the most expensive component in the optical chain.

4. Imaging Devices

The LIF signal is collected by an imaging lens usually at right angle to the illuminated plane onto a two-dimensional visible light array detector or onto an image intensifier coupled to such an array. An image sensor's task is to convert incident light into a proportional amount of electric charge. Silicon is ideally suited for this purpose, at least for the green and red part of the visible spectrum. Since the invention of the charge-coupled device principle (CCD) in 1971 solid state imaging has made big advances in the wake of silicon semiconducter technology for computer chips. Recent developments of electronic imaging sensors are discussed by Seitz et al. (1995). The authors predict that the constantly decreasing chip dimensions of RAM technology will not be followed by image sensor manufacturers because optical limits constrain the pixel size. Fraunhofer diffraction blurs the image (besides lens aberrations) such that the expected pixel size of image sensors will not be made smaller than 2-3μm. Technical performance of a CCD image sensor is discussed in terms of the full well capacity Q_{max} , the equivalent r.m.s. readout charge noise σ, the spectral quantum efficiency, the charge transfer efficiency, the output sensitivity, the maximum readout speed, the number of readout channels, and the dark current. The pixel number of an image sensor is of course also decisive for a good resolution of the flow structure. Seitz et al. (1995) report that 5k x 5k pixel sensors are commercially available, the dynamic range of video CCD sensors is up to 94 dB. A typical CCD-sensor's spectral response is shown in Fig.7 in Virant & Dracos, chapter three. A cheaper solution, of course, is a standard CCD-video camera, with the trade-off of constant resolution of 739h x 575v and 25 Hz frame rate only (European standard), the upcoming HDTV sensors have a resolution of 2048h x 1152v. For an overview of visible array detectors, the reader is referred to Tredwell (1995), and also to the contributions of Grün and Virant et al. in this volume. Fig.12 shows the digital high-speed motion analysis system Phantom 2.0 (Vision Research) which is capable of taking 500 pictures per second with a 512x512 back-illuminated CCD sensor. Quantum efficiency at 600 nm is 68%, the pixel size is 18x18μm. The camera read-out is highly parallel (16 ports) and the pixel content is mapped into built-in RAM. The Phantom 2.0 system has a variable speed up to 500 fps, which can be triggered externally and memory gated. Care should be taken that the camera is adaptable to a widely used lens system such as the Nikon F-mount system. Performance of such high frame rate, back-illuminated CCD imagers is discussed by Lawer et al. (1994). In a digital optical acquisition system the image sensor is the central component. It mainly influences the quality of the obtained images. Possible error sources are dead pixels or bad columns whose number determine image quality and the price of the sensor. But future advances in image quality can be expected. Kodak offers a high-speed system with frame rates above 10 kHz.

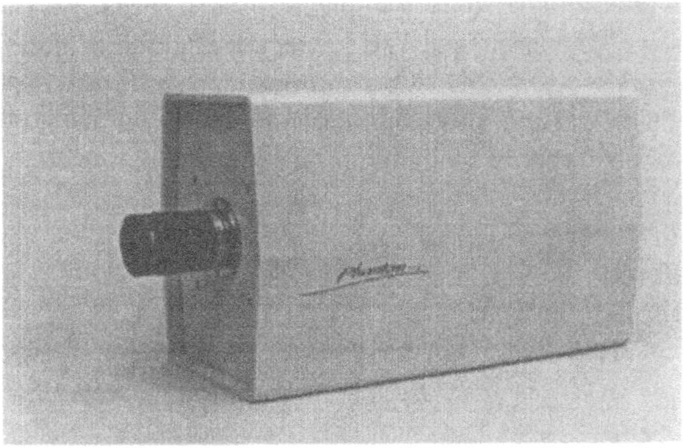

Figure 12. Phantom 2.0 high-speed motion analysis system

It should be kept in mind that the use of raw, unprocessed digital data is in the interest of data integrity. This is true especially for LIF images which must be post-corrected for attenuation losses and illumination non-uniformities, if necessary. Cameras that manipulate the images to mask defects in the CCD and analog chains before output may give raise to physically false, unreliable data. Another flaw of directly digitizing cameras may be that the array of parallel AD-converters is not tuned identically in gain and offset, which results in unwanted patterns in the image, e.g. stripes or fields. Maas (1993) examines this problem and discusses some smoothing algorithms. Smoothing images by interpolation may be tolerable for neat presentations, but it is a crude praxis to use such data for the precise determination of scalar gradients.

Low f-number lenses are desirable because they collect more photons through a wider aperture than normal lenses, but they have less depth of focus and must therefore be focused accurately.

The parameters describing a planar LIF technique, first specified by van Cruyningen et al. (1990), are strongly dependent on the performance of the imaging system: (i) spatial resolution, (ii) spatial dynamic range, (iii) minimum detectable signal, (iv) signal dynamic range, (v) temporal resolution. These parameters can equally be applied for 3D LIF imaging, however temporal resolution must be replaced by the acquisition time for an entire volume when calculating the sampling frequency.

An upper constraint for imaging speed in the linear fluorescence regime is the minimum scan time for one pixel t_i to be one order of magnitude above the natural fluorescence lifetime, i.e. $t_i \geq 5 \cdot 10^{-7}$s. Thus, for a 512x512 pixel imaging array, for instance, a scanned laser beam must cover all pixels for a sufficiently long time, i.e. $t_s/512 \geq 5 \cdot 10^{-7}$s, which constrains the scan time to $t_s \geq 2.56 \cdot 10^{-4}$s. A (future) digital camera with 512x512 pixels should have 4000 fps imaging speed to resolve this limiting scan speed. Assuming, say 50 planes are stacked to one tomographic volume, the total scan time per volume is 50/4,000 = 0.0125s, which corresponds to a 80 Hz sampling rate. This would be enough to measure turbulent flow at Reynolds numbers of several 10,000. However, such a camera will not be available in the near future.

5. References

Beiser L., 1991, Design equations for a polygon laser scanner, *SPIE*, **1454**, 60

Beiser, L., Johnson, R.B., 1995, Scanners, chap.19 of *Handbook of Optics*, Vol.**2**, McGraw Hill

Beiser, L., 1995, Fundamental architecture of optical scanning systems, *Applied Optics*, **34**, 7307

Berlman, I.D., 1971, *Handbook of fluorescence spectra of aromatic molecules,* Academic Press

Dahm, W.J.A., Southerland, K.B., Buch, K.A., 1991, Direct high resolution, four-dimensional measurements of the fine scale structure of Sc>>1 molecular mixing in turbulent flows, *Phys. Fluids* **A 3** (5), 1115

Dahm, W.J.A., Dimotakis, P.E., 1990, Mixing at large Schmidt number in the self-similar far field of turbulent jets, *J.Fluid Mech.,* **217**, 299

Gray.,C., Greated, C.A., McCluskey, D.R., Easson, W.J.; An analysis of the scanning beam PIV illumination system, *Meas.Sci.Technol.*, **2**, 717 (1991)

Handbook of Optics, 1995, Optical Society of America, McGraw Hill

Hecht, E., Zajak, A., 1974, *Optics*, Addison-Wesley

Hecht, J., *1992, The Laser Guidebook,* 2nd ed., Mc Graw Hill

Horikawa, H., Sugisaki, I., Tashiro, M., 1991, Relationship between fluctuations in mirror radius (within one polygon) and the jitter, *SPIE*, **1454**, 46

Lakowicz, J.R., 1991, *Topics in Fluorescence Spectroscopy*, Vol.**1-4**, Plenum Press

Laser Focus World, also *Laser Focus World Buyer's Guide*, PennWell Publ.

Lawler, W., Harrison, L., Levine, P.A., Sauer, D.J., Hseuh, F.L., Shallcross, F.V., Meary, G.M., Taylor, G.C., 1994, Performance of High Frame Rate, Back-Illuminated CCD Imagers, Proc. *SPIE* **2172**: *Charge-Coupled-Devices and Solid State Optical Sensors IV*, 90-99, San Jose

Maas, H.G., 1993, Destriping of digital images, *Int. Society for Photogrammetry and Remote Sensing (ISPRS), Com.I Workshop on Digital Sensors and Systems*, Trento, Italy, June 21-25

Marshall, G.F., 1985, Laser Beam Scanning, *Optical Engineering*, vol. **8**, M.Dekker

Marshall, G.F., Sherman, R.J., Beiser, L., 1991, Beam deflection and scanning technologies, *SPIE*, **1454**

Marshall, G.F., Montagu, J., 1995, Advances in Oscillatory Optical Scanners, *SPIE*, **2383**, 440

Montagu, J., 1991, Galvanometric and Resonant Low-Inertia Scanners, in *Optical Scanning*, ed. by Marshall, G.F., Marcel Dekker Inc. N.Y.

Preston, R.G., Colquham, A.B., Shepherd, J., 1991, Bearings for rotary scanners - an overview, *SPIE*, **1454**, 124

Rockwell, D., Magness, C., Towfighi, J., Akin, O., Corcoran, T., 1993, High image-density particle image velocimetry using laser scanning techniques, *Exp. Fluids*, **14**, 181

Seitz, P., Vietze, O., Spirig, T., 1995, From Pixels to Answers - Recent Developments and Trends in Electronic Imaging, From Pixels to Sequences, *Int. Society for Photogrammetry and Remote Sensing (ISPRS)*, **30**, part 5W1, ed. Baltsavias, E.P.

Seitzman, J.M., Hanson, R.K., 1993, Planar Fluorescence Imaging in Gases, in Instrumentation *for Flows with Combustion*, ed. Taylor, A.M.K.P, Academic Press

Slavik, J., 1994, *Fluorescent Probes in Cellular and Molecular Biology*, CRC Press

Tredwell, T.J., 1995, Visible Array Detectors, *Handbook of Optics*, Vol. **1**, Chap.22, McGraw Hill

van Cruyningen, I., Lozano, A., Hanson, R.K., 1990, Quantitative imaging of concentration by planar laser-induced fluorescence, *Exp. Fluids* **10**, 41

Wolfbeis, O.S., 1993, *Fluorescence Spectroscopy*, Springer

APPLICATION AND PERFORMANCE OF LIF-VELOCIMETRY

G.J. Merkel, Th. Dracos[1], P. Rys, F.S. Rys
Department of Industrial Chemistry & Chemical Engineering
[1] Prof. em. ETH Zürich
Swiss Federal Institute of Technology Zürich, Switzerland

1. Introduction

The videogrammetric Least Square Matching Technique (LSM) presented by A. Grün in Chapter three of this volume can be applied to tomographically obtained three-dimensional LIF images as shown by H.-G. Maas in this chapter. The method yields three shift components of a small cuboid and the nine components of the tensor describing its transformation, which is assumed to be affine. Provided a good estimate of it can be made for starting the iteration, no restriction is imposed on the magnitude of the shift by videogrammetry.

Some basic information on LIF and its tomographic application is given by Deusch et al. in this chapter. To prove the applicability of LIF-Velocimetry, LIF-Tomography will be applied to the flow field produced by a free, round, turbulent jet in water. A round jet was chosen because of its properties, which are favourable for an application of LIF-Velocimetry.

2. Determination of the Velocity and the Velocity Gradient Tensor

Tomographically obtained three-dimensional LIF images contain a number N_I of plane LIF images. The time needed to produce such a three-dimensional image is $\Delta t=N_I/f_c$, where f_c is the framing frequency of the camera used. This time is equal to the time separating consecutive 3-D images. LSM is applied to small cuboids cut our of these 3-D images and yields first the shift or translation components a_0, b_0, c_0 of such a cuboid in object space during time Δt. The determination of the translation velocity of the cuboid is straight forward and gives the flow velocity averaged over the volume of the cuboid in object space and over time Δt.

$$u=a_0/\Delta t, \quad v=b_0/\Delta t, \quad w=c_0/\Delta t \tag{1}$$

Th. Dracos (ed.), Three-Dimensional Velocity and Vorticity Measuring and Image Analysis Techniques, 299–310.

The elements a_1, b_2, c_3 in the diagonal of the transformation matrix give the edge lengths of the transformed cuboid normalised by the original lengths $a_1^{(0)}$, $b_2^{(0)}$, $c_3^{(0)}$. The assumption that the transformation of the cuboid is affine, is equivalent to the assumption that the transformation, and in particular the elongation of the edges of the cuboid, are small, i.e. $|a_1 - 1| << 1, |b_2 - 1| << 1, |c_3 - 1| << 1$. The off-diagonal elements of the transformation matrix contain rotation and angular deformation of the cuboid. Division of the deformation matrix by Δt yields the velocity-gradient tensor.

$$\nabla \boldsymbol{u} = \frac{1}{\Delta t}\begin{pmatrix} a_1 - 1 & a_2 & a_3 \\ b_1 & b_2 - 1 & b_3 \\ c_1 & c_2 & c_3 - 1 \end{pmatrix} \tag{2}$$

which can be decomposed to a symmetrical part $S_{ij} = 1/2\left[\nabla \boldsymbol{u} + \nabla \boldsymbol{u}^T\right]$ corresponding to the rate of strain tensor

$$S_{ij} = \frac{1}{\Delta t}\begin{pmatrix} a_1 - 1 & \frac{1}{2}(a_2 + b_1) & \frac{1}{2}(a_3 + c_1) \\ \frac{1}{2}(b_1 + a_2) & b_2 - 1 & \frac{1}{2}(b_3 + c_2) \\ \frac{1}{2}(c_1 + a_3) & \frac{1}{2}(c_2 + b_3) & c_3 - 1 \end{pmatrix} \tag{3}$$

and an anti-symmetrical part $\omega_{ij} = 1/2\left[\nabla \boldsymbol{u} - \nabla \boldsymbol{u}^T\right]$ corresponding to the rate of rotation tensor ω_{ij}, i.e. the vorticity vector ω_i.

$$\omega_{ij} = \frac{1}{\Delta t}\begin{pmatrix} 0 & \frac{1}{2}(a_2 - b_1) & \frac{1}{2}(a_3 - c_1) \\ \frac{1}{2}(b_1 - a_2) & 0 & \frac{1}{2}(b_3 - c_2) \\ \frac{1}{2}(c_1 - a_3) & \frac{1}{2}(c_2 - b_3) & 0 \end{pmatrix} \tag{4}$$

Notice that no differentiation of the velocity vector is needed to compute S_{ij} and ω_i, which are directly determined from the transformation matrix of the cuboid. This matrix is determined by maximising the correlation of grey-levels in the template cuboid at time t and its transformed conjugate at time t+Δt. First derivatives of grey-levels are involved only in the determination of the coefficient matrix A used for the computation of the alteration vector x_i of the parameters of the affine transformation. Hence, errors in grey-levels can influence the convergence of the iteration but not the quality of the result (see Maas, this chapter). This is a considerable advantage of this method. As a consequence, in most cases corrections of grey-levels can be reduced to such due to the variation of the intensity of the incident light (Deusch et al., this chapter).

The rate of strain S_{ij} and the vorticity vector ω_i determined by LSM from three-dimensional LIF tomographic images are averages over volumes in object space corresponding to the cuboids and over the time step Δt. The quantities u_i, S_{ij}, ω_i are assigned to the centre of gravity of the cuboid at time t.

It is obvious that the spatial resolution of the determined quantities increases with decreasing size of the cuboid. On the other hand, least square techniques demand redundancy in the information used and sufficient variation of the grey-levels in a cuboid. Some trials are therefore necessary to determine the optimum size of cuboids. Usually this size varies between $\Delta V = l_c^3 = 7^3$ voxels and 17^3 voxels.

Once the appropriate size of a cuboid, whose linear dimension should be an odd number of voxels, is chosen, a three-dimensional grid is constructed inside the tomographic volumes at the nodes of which one wishes to determine u_i, S_{ij}, ω_i. The mesh size l_m^3 can be chosen equal to or smaller than the size of a cuboid l_c^3. When $l_m < l_c$, cuboids overlap. All lengths are expressed in number of lengths of voxel edges l_v. As a rule, the incompressibility constraint should not be used in the LSM procedure (Maas, this chapter), so that it can be used as an à posteriori check of the results. The minimum distance of a node from the boundary of the LIF tomographic volume must be equal to the number of voxels along an edge of the cuboid (Deusch et al., this chapter). The number of nodes in a volume containing
$N_x \cdot N_y \cdot N_z$ voxels is:

$$N_n = \left(\frac{N_x - 2l_c}{l_m} + 1 \right) \cdot \left(\frac{N_y - 2l_c}{l_m} + 1 \right) \cdot \left(\frac{N_z - 2l_c}{l_m} + 1 \right) \tag{5}$$

provided the ratios in this expression are integer. Otherwise they must be rounded off. The number N_I of plain LIF images needed to determine u_i, S_{ij}, ω_i.in N_{nz} planes is:

$$N_I = N_{nz} \cdot l_m + 2l_c \tag{6}$$

N_{nz} is the number of nodes in z-direction and l_m and l_c must be integral multiples of a voxel length l_v. This relation allows to determine the number of plane images in a tomographically constructed volume. For $N_{nz} = 1$, $l_m = 0$, and the minimum number of planes is $2l_c$.

3. Experiments

3.1 THE ROUND JET

To check the LSM method described in this chapter a round turbulent jet in water was used. Jets, like all free shear flows, entrain ambient fluid, which subsequently is mixed with jet fluid. If either the jet fluid or the entrained fluid are marked by a conservative fluorescence tracer, the entrainment and mixing produce dye concentration variations.

This is a necessary condition for the application of all LIF-Velocimetry methods (Dracos, this chapter).Round jets have, in addition, some advantageous properties.

(i) In the self-similar far field of the jet, the Reynolds number, $Re=u_c\delta\nu^{-1}$, determined by using the centreline velocity u_c and the width δ of the jet (ν is the kinematic viscosity of the fluid) remains constant with increasing distance x from the virtual origin.

(ii) All relevant scales increase linearly with x.

(iii) The centreline velocity decreases inversely proportional to x

These properties allow to perform measurements at small local velocity and large scales if the observation is made at sufficiently large distance from the orifice.

One relevant length scale is the viscous length λ_v

$$\lambda_v=\nu/u_\tau \tag{7}$$

where u_τ is the shear velocity $u_\tau=(\tau/\rho)^{1/2}$, with τ the shear stress and ρ the density of the fluid. λ_v is related to Re by

$$\lambda_v=\delta Re^{-3/4} \tag{8}$$

To resolve this length the following condition must be fulfilled:

$$\lambda_v>l_v=l_pM \tag{9}$$

l_p is the edge length of a pixel in the sensor and M is the mapping scale. The size $L_x \cdot L_y$ of a LIF plane image is then fixed by the size of the sensor or $L_x=N_{px}\, l_v$, $L_y=N_{py}l_v$. N_{px} and N_{py} are the number of pixels in the sensor in x- and y-direction, respectively.

Another relevant length scale is the diffusion length of a scalar.

$$\lambda_D=\lambda_v Sc^{-1/2} \tag{10}$$

If the concentration in a voxel is not allowed to change during the exposure time t_e, $l_v>\lambda_D\Phi_t$, $\Phi=(t_e/t_D)^{1/2}$ and $t_D=\lambda_D{}^2/D=\lambda_v{}^2/\nu$. t_D is the diffusion time.

This condition depends on the exposure time t_e, which is equal to the integration time of the camera t_c when stationary illumination is used, e.g. a light sheet produced by a cylindrical lens. When sweep illumination is used, this time becomes much shorter. $t_e=t_c/N_p$, where N_p is the number of pixels in the sensor in sweep direction. This is another advantage of sweep illumination (Deusch et al., this chapter). To resolve λ_D the exposure time $t_e \ll t_D$, so that $\lambda_D\Phi_t<l_v<\lambda_D$.

Other important scales of the flow are the Kolmogorov scales

$$\lambda_K=(\nu^3/\varepsilon)^{1/4}, \qquad t_K=(\nu/\varepsilon)^{1/2} \tag{11}$$

ε is the rate of turbulence energy dissipation.

The grey-level patterns in consecutive tomographic observations should not differ considerably. This is equivalent to the assumption that the deformation of conjugate cuboids is small. It implies that a Kolmogorov time scale $t_k > \Delta t$.

The entrainment of the jet $Q_E(x)$ can be estimated from

$$Q_E(x) = C_E (m_0/\rho)^{1/2} x \qquad (12)$$

where $m_0 = \pi \rho d_0^2 u_0^2 / 4$ is the momentum flux at the nozzle (Schlichting 1979), u_0 is the velocity at the orifice and d_0 is the diameter of the orifice.

3.2. EXPERIMENTAL FACILITY

The experiments were conducted in a steel tank with acid/base resistant coating. The dimensions of the tank are L=3.0m, W=1.5m, D=1.0m. This tank serves as a recipient for the ambient fluid.

One of the long sides of the tank is equipped with a glass window of 2.0m x 1.0m, allowing illumination and observation of the flow. The jet is discharged at the centre of one of the narrow sides of the tank. Along the four corners of the long sides of the tank the estimated entrainment is supplied by four perforated pipes covered with filter material. The discharge in these pipes is controlled individually to guarantee an equal and stable supply. The jet and the entrainment supply are controlled by constant head vessels and separately measured. At the down-stream end of the tank the total amount of fluid introduced into the tank is withdrawn through an array of perforated pipes connected to a constant head vessel to keep the water level in the tank constant. The temperature of the entrainment and jet fluid is controlled and kept equal within 0.1 deg. C^0. The dye is injected by a volumetric pump in the duct supplying the jet fluid. Special care is taken to achieve complete mixing of the dye in the jet plenum. For more details see Merkel (1995). A schematic representation of the experimental facility is given in Fig.1 on the next page.

An Argon ion laser is used as a light source. Its total light power is 26W with 10W at 514.5nm and 8W at 488nm. The laser beam has a diameter of 1.9mm. The sensor of the camera used has a photodiod array of 256 x 256 square pixels. The size of a pixel is l_p=40µm. The size of the plane mapped on the array is 15mm x 15mm, giving a mapping scale of M=1.5. To get cubic voxels, the light sheet must be on the average 60µm thick. To achieve this the beam is first expanded by a factor of 16 to a diameter of approximately 30mm. The beam is then focused on the observation location in the jet by an adjustable focusing lens system. The focus beam is deflected upwards in vertical direction by a fixed mirror. At the desired elevation a second mirror deflects the beam in direction perpendicular to the axis of the jet. This mirror is mounted on a piezo scanner

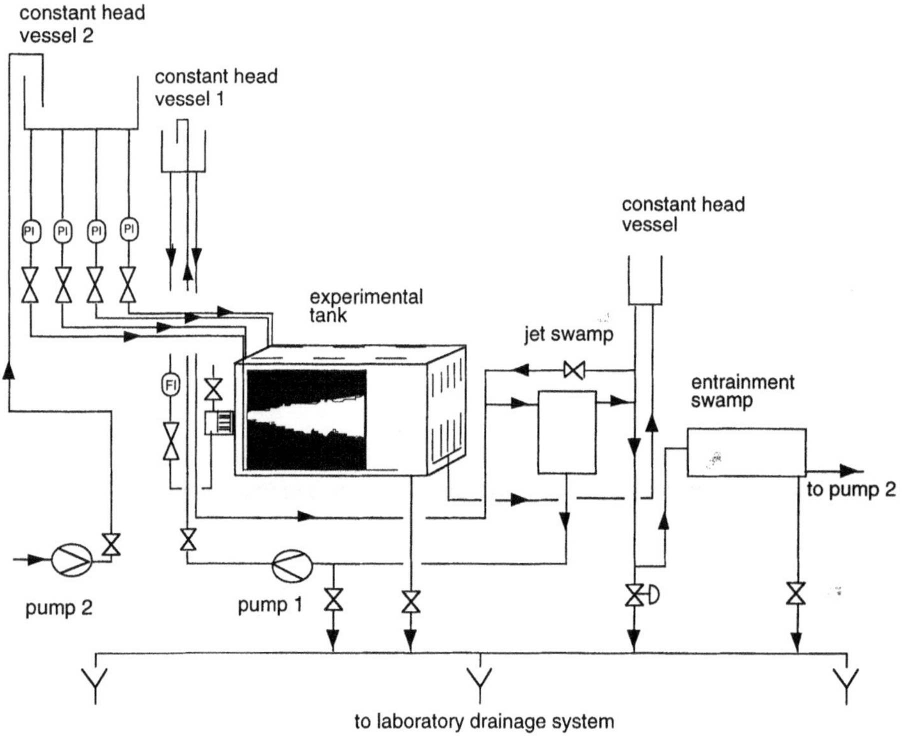

Figure 1 Sketch of the experimental facility

unit, which has a maximum angle of deflection α=2.23mrad. This angle can be subdivided into 4095 steps with a maximum stepping frequency of 5kHz. To produce tomographically a 3-D image the range of 2.2mrad was divided into 49 steps of about 45µrad. The mirror was moved with a frequency of 488 steps/s. After 49 steps the mirror is brought back to its iniitial position in 1.5ms and the procedure is repeated. The laser beam is expanded by a cylindrical lens to form a sheet of light of mean thickness 60µm. A tomographically produced image thus consists of 50 plane LIF images and the time necessary to produce them is Δt=0.102s. The storage capacity of the high-speed camera allows to perform 20 repetitions in 2.04s.

3.3. PARAMETERS OF THE EXPERIMENT

The experiment is conducted in water. The jet fluid discharges through an orifice of d_0=5mm diameter. The velocity at the orifice is u_0=0.85m/s. The coordinate system is defined as follows: x coincides with the jet axis and is positive in flow direction, y is horizontal and z is vertical, both normal to the jet axis and forming a right-handed coordinate system. The tomographic observation is performed in the x,y-plane at

$X=x/d_0=280$ and $\eta=y/b=1.18$. The half-width is $b=0.11x=0.154$m and corresponds to the distance from the axis at which $u/u_c=0.5$. The width of the jet is $\delta=0.22x=0.308$m. The centreline velocity at $X=280$ is $u_c=19.5$mm/s and the Reynolds number of the fully developed jet is Re=6000. The mean velocity u at the location of the observation is u=9.5mm/s. The experimental arrangement is shown in Fig.2

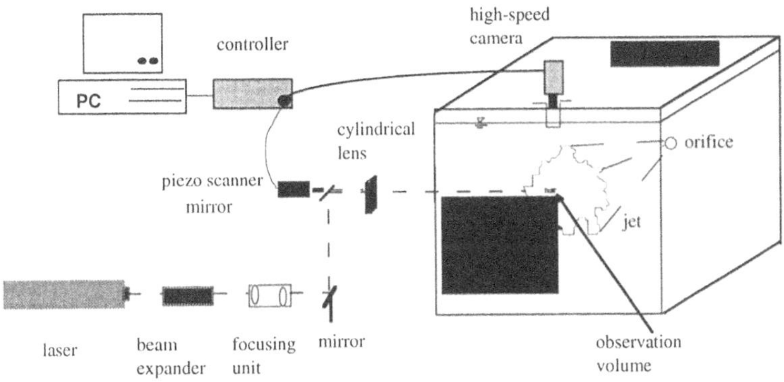

Figure 2 Experimental arrangement

Disodium fluorescein is used as a tracer. The Schmidt number is Sc=2075. The relevant time and length scales of the experiment are summarised in Table I.

TABLE I Time and length scales of the experiment

Δt [s]	t_{Dv} [s]	$t_{Dv}/\Delta t$	t_A [s]	$\Delta t/t_A$	t_k [s]	$t_k/\Delta t$	λ_v/l_v	l_k[mm]	λ_T[mm]
0.104	1.7	17	0.0063	16	0.9	9.0	7.5	0.99	13

The time t_{Dv} is the diffusion time related to the voxel size l_v and the advection time is $t_A=l_v/u$. The ratio $\Delta t/t_A$ gives the number of voxels. The first and last plane images in a tomographically produced 3-D image would be shifted if the camera was stationary. To avoid this shift the camera was moved with u=9.5mm/s.

The dye concentration c at the observation location is low. It varies in the range 10^{-8}mol/l $\leq c \leq 10^{-5}$mol/l. For such a small concentration a linearisation of the attenuation equation for the incident and the emitted light intensity is permitted (Deusch et al., this chapter). The relation between a grey-level GV registered on a pixel of the sensor in the camera and the concentration of dye c becomes in this case

$$GV-GV_{(noise)}=Kc \tag{13}$$

The constant K can be determined by calibration. For more details see Merkel (1995).

4. Application of LIF-Velocimetry by LSM

The size of the tomographically obtained LIF three-dimensional images of dye concentration is 15mm x 15mm x 3mm. Each contains 256 x 256 x 50=3276800 voxels. 20 volumes are recorded in each experiment. The velocity and the velocity gradient are determined in a grid with mesh size 0.3mm x 0.3mm x 0.3mm. 0.3mm correspond to $5l_v$. Eq.(5) yields for the total number of nodes 46 x 46 x 5=10170. The number of nodes per mm^3 is 37. The size of the cuboids is 15^3 voxels or l_c=0.9mm and $V_c=l_c^3$=0.729mm^3. This is the volume over which the measured velocities and related quantities are averaged. The edge length of this volume is equal to Kolmogorov's length scale $l_c=l_k=15l_v$.

Fig.3 shows a cuboid and its transformed conjugate determined by applying the Least Squares Matching (LSM) procedure. As the camera is moved with the local average velocity, the resulting translation of the cuboid is small and represents essentially the deviation of the velocity from the local mean, due to turbulent fluctuations of the velocity.The first estimate for this shift to be used in the iteration procedure in LSM is in this case zero.

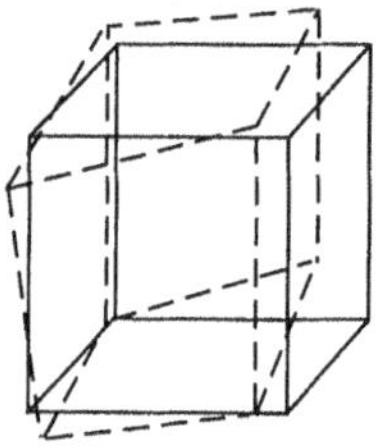

Figure 3 A cuboid at time t and its conjugate at time t+Δt

The number of nodes in a plane is 2116, the number of vectors determined in this plane is ≈1700 and the total number of vectors is ≈8400. Visual inspection of the velocity vectors determined showed that no spurious vectors exist (Fig.5, Maas, this chapter). A post-processing to eliminate spurious vectors is therefore not necessary. This shows the high reliability of the LSM technique. From the grey-level distribution superimposed to the vector field one easily sees that the areas in which the LSM method failed are the areas with very weak grey-level gradients corresponding to patches of ambient fluid engulfed in the jet . About 83% of the velocity vectors potentially determinable in the grid could be determined. This gives an average of 31 vectors per mm^3.

5. Performance tests

One disadvantage of LIF-Velocimetry is that the velocity and its derivatives cannot be determined over the whole field in the interior of a jet. As shown previously, the method fails when engulfed fluid is encountered. Therefore a comparison with conventional measuring techniques like Hotfilm Anemometry or Laser Doppler Anemometry is not possible. An indirect judgement of the accuracy of the method will be given in the following.

One stringent control of the accuracy with which velocity derivatives are determined is the continuity condition $\nabla \cdot \boldsymbol{u} = 0$. This condition can be optionally incorporated in the LSM procedure (Maas, this chapter). This was not done in the present case. The continuity condition can be expressed as

$$\left|\frac{\partial u}{\partial x}+\frac{\partial v}{\partial y}\right| \Big/ \left|\frac{\partial w}{\partial z}\right| = 1 \tag{14}$$

Values of $\left|\frac{\partial u}{\partial x}+\frac{\partial v}{\partial y}\right|$ versus $\left|\frac{\partial w}{\partial z}\right|$ are presented in Fig.4 on the next page. These values should, if correctly determined, lie on a straight line with slope 1 and go through the origin. A few of the 3500 points presented in this figure fall off this line and are obviously erroneous. The regression line through these data has a slope of 0.991 and goes through the origin. The correlation coefficient is 0.973. This result compares favourably with the results presented by Lemonis & Dracos in Chapter one of this volume as well as with the results of Tsinober et al. (1992) and Ong (1992). These

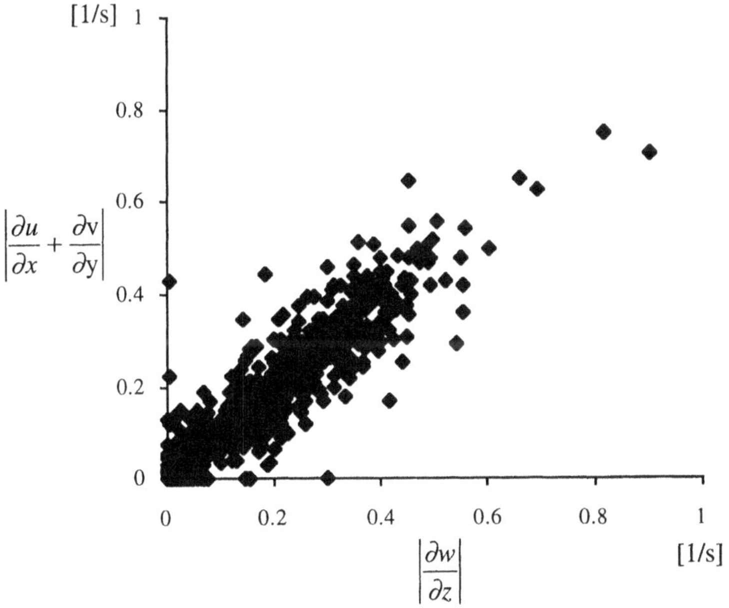

Figure 4 Correlation of $\left|\frac{\partial u}{\partial x}+\frac{\partial v}{\partial y}\right|$ with $\left|\frac{\partial w}{\partial z}\right|$.

authors measured this relation by using multi-hotwire anemometry, which at the moment is the only method which can measure the derivatives necessary to determine it. The correlation coefficient achieved by these authors is 0.7.

Batchelor (1952) and later Ashurst et al. (1987) showed that the vorticity vector ω_i is tangential to material surfaces. At large Schmidt numbers the material surfaces coincide with isoconcentration surfaces on which the gradient of concentration $\nabla \cdot c$ is normal. If this is the case, the angle between $\nabla \cdot c$ and ω should be $\pi/2$ and $cos(\nabla c, \omega) = 0$. Both quantities can be determined. Fig.5 on the next page shows the pdf of the $cos(\nabla c, \omega)$, respectively the angle θ formed by the vectors $\nabla \cdot c$ and ω. The average of this angle is θ =90.06 deg. and its standard deviation is $\sigma_\theta = \pm 7.2$ deg. This result is astonishingly good and proves not only the reliability of the method but also that the assumption that isoconcentration surfaces are material surfaces is valid at the high Schmidt number Sc=2075 of this experiment.

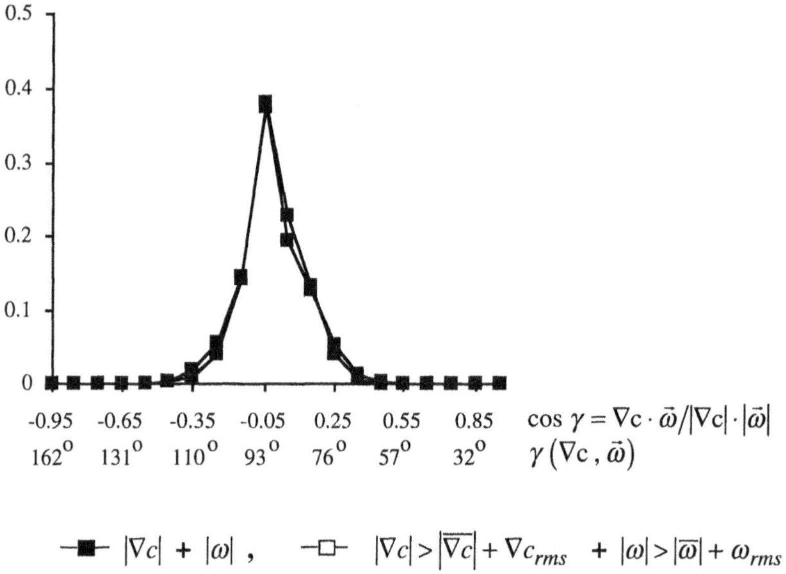

by

Figure 5 Pdf of the cosine of the angle between concentration gradient ∇c and vorticity vector $\vec{\omega}$.

One can also determine at each node at which LSM was successful the principal axes of the rate of strain tensor, i.e. their eigenvectors α, β, γ and the corresponding eigenvalues (see also Lemonis & Dracos, Chapter one). Using DNS results Ashurst et al. (1987)

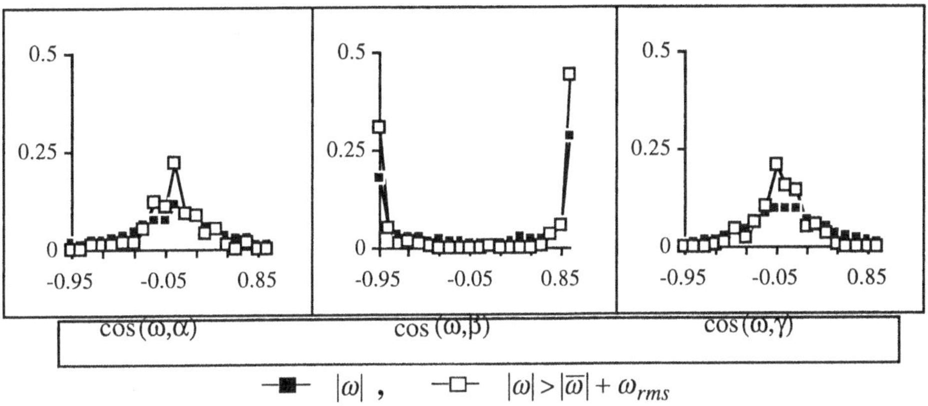

Figure 6: Pdf of the cosine of the angle between vorticity vector **w** and eigenvectors α (elongation), γ(compression) and β (intermediate).

showed that the vorticity vector preferably aligns with the intermediate eigenvector β. The present data confirm this result and show that this alignment becomes more pronounced for higher values of vorticity (Fig.6). They are in good agreement with the results of Lemonis & Dracos, Chapter one) and with the results of Tsinober et al. (1992).

6. Concluding remarks

LIF-Velocimetry proves to be a powerful method for the study of turbulence and turbulence mixing. The interesting feature of this method is that it allows to determine the velocity, vorticity and rate of strain fields in one single step. At the same time the concentration distribution of the scalar as well as its gradients and the dissipation of the scalar can be determined. This opens new possibilities for the study of mixing processes for conservative scalars.

To achieve reliable results the microscales of turbulence in the flow studied must be resolved. The optical resolution of the camera used must be of order $l_p \leq 0.1 l_k$. The time resolution must also be high. Only high speed cameras can be used in LIF-Velocimetry.

A disadvantage of the LSM technique is that it requires tomographically obtained 3-D LIF images consisting of a rather large number of plane images. This increases the time Δt and demands a large focus depth. The high spatial resolution required limits the size of the volume imaged. In the present application the size of the plane mapped on the sensor of the camera, which has 256 x 256 pixels, is of order λ_T^2, where λ_T is the Taylor-microscale of the flow. It is to be expected that the velocity in flow patterns of this size is well correlated. Statistical results can therefore be biased. It is therefore desirable to use cameras of higher resolution. Cameras with 512 x 512 pixels, capable of taking 500 frames/s are already available (Deusch et al., this chapter) and cameras with 1024 x 1024 pixels and the same speed will be available in the near future.

Cameras with a higher number of pixels can be run at higher speeds when the resolution is reduced. Thus, with the resolution used at present the speed can be increased and consequently the time Δt decreased. This will increase the accuracy of LSM or/and increase the size of the observation volume, which will lead to more reliable statistical analyses of the results.

Another problem inherent to LIF-Velocimetry is that it fails when grey-level gradients do not exist. As shown earlier, this is the case at each location where ambient non-mixed fluid is encountered. A solution for this inconvenience must be found. One possibility is to mark the jet and entrained fluid with fluorescent dyes emitting at different wavelengths. Use of a beam-splitter behind the objective and two cameras with appropriate filters allows to image on the one the part of the flow field consisting of mixed fluid and on the other the part consisting of ambient fluid. This will allow to determine the velocity field over the whole domain imaged, provided grey-level gradients exist everywhere. This is not certain for engulfed fluid.

References

Ashurst, W.T., Derstein, A.R., Kerr, R.M., Gibson, C.H. 1987. Alignment of vorticity and scalar gradient with strain rate in simulated Navier-Stokes turbulence, *Phys. Fluids,* **30**, 2343-2353

Batchelor, G.K. 1952. The effect of homogeneous turbulence on material lines and surfaces, *Proc. R. Soc. London,* **A 213**, 349 - 366

Merkel, G.J. 1995. Tomograpie in einem turbulenten Freistrahl mit Hilfe von PH-abhängiger Laser induzierter Fluoreszenz, *Ph.D. thesis, Nr. 11174,* ETH Zürich

Ong, L. 1992. Visualisation of turbulent flows with simultaneous velocity and vorticity measurements, *Ph.D. thesis, University of Maryland,* College Park, USA

Schlichting, H. 1979. Boundary layer theory, *McGraw Hill,* New York

Tsinober, A., Kit, E., Dracos, T. 1992. Experimental investigation of the field of velocity gradients in turbulent flows, *J. Fluid Mech.,***242**, 169- 192

ERCOFTAC SERIES

1. A. Gyr and F.-S. Rys (eds.): *Diffusion and Transport of Pollutants in Atmospheric Mesoscale Flow Fields.* 1995 ISBN 0-7923-3260-1
2. M. Hallbäck, D.S. Henningson, A.V. Johansson and P.H. Alfredsson (eds.): *Turbulence and Transition Modelling.* Lecture Notes from the ERCOFTAC/IUTAM Summerschool held in Stockholm. 1996 ISBN 0-7923-4060-4
3. P. Wesseling (ed.): *High Performance Computing in Fluid Dynamics.* Proceedings of the Summerschool held in Delft, The Netherlands. 1996 ISBN 0-7923-4063-9
4. Th. Dracos (ed.): *Three-Dimensional Velocity and Vorticity Measuring and Image Analysis Techniques.* Lecture Notes from the Short Course held in Zürich, Switzerland. 1996 ISBN 0-7923-4256-9

KLUWER ACADEMIC PUBLISHERS – DORDRECHT / BOSTON / LONDON

GPSR Compliance

The European Union's (EU) General Product Safety Regulation (GPSR) is a set of rules that requires consumer products to be safe and our obligations to ensure this.

If you have any concerns about our products, you can contact us on ProductSafety@springernature.com

In case Publisher is established outside the EU, the EU authorized representative is:

Springer Nature Customer Service Center GmbH
Europaplatz 3
69115 Heidelberg, Germany

Batch number: 10761599

Printed by Printforce, the Netherlands